CAMBRIDGE LIBRARY COLLECTION

Books of enduring scholarly value

Mathematics

From its pre-historic roots in simple counting to the algorithms powering
modern desktop computers, from the genius of Archimedes to the genius of
Einstein, advances in mathematical understanding and numerical techniques
have been directly responsible for creating the modern world as we know it.
This series will provide a library of the most influential publications and writers
on mathematics in its broadest sense. As such, it will show not only the deep
roots from which modern science and technology have grown, but also the
astonishing breadth of application of mathematical techniques in the
humanities and social sciences, and in everyday life.

Formal Logic

From the end of antiquity to the middle of the nineteenth century it was
generally believed that Aristotle had said all that there was to say concerning
the rules of logic and inference. One of the ablest British mathematicians
of his age, Augustus De Morgan (1806–71) played an important role in
overturning that assumption with the publication of this book in 1847.
He attempts to do several things with what we now see as varying degrees of
success. The first is to treat logic as a branch of mathematics, more specifically
as algebra. Here his contributions include his laws of complementation and
the notion of a universe set. De Morgan also tries to tie together formal
and probabilistic inference. Although he is never less than acute, the major
advances in probability and statistics at the beginning of the twentieth century
make this part of the book rather less prophetic.

Cambridge University Press has long been a pioneer in the reissuing of out-of-print titles from its own backlist, producing digital reprints of books that are still sought after by scholars and students but could not be reprinted economically using traditional technology. The Cambridge Library Collection extends this activity to a wider range of books which are still of importance to researchers and professionals, either for the source material they contain, or as landmarks in the history of their academic discipline.

Drawing from the world-renowned collections in the Cambridge University Library and other partner libraries, and guided by the advice of experts in each subject area, Cambridge University Press is using state-of-the-art scanning machines in its own Printing House to capture the content of each book selected for inclusion. The files are processed to give a consistently clear, crisp image, and the books finished to the high quality standard for which the Press is recognised around the world. The latest print-on-demand technology ensures that the books will remain available indefinitely, and that orders for single or multiple copies can quickly be supplied.

The Cambridge Library Collection brings back to life books of enduring scholarly value (including out-of-copyright works originally issued by other publishers) across a wide range of disciplines in the humanities and social sciences and in science and technology.

Formal Logic

Or, The Calculus of Inference,
Necessary and Probable

Augustus De Morgan

CAMBRIDGE
UNIVERSITY PRESS

University Printing House, Cambridge, CB2 8BS, United Kingdom

Cambridge University Press is part of the University of Cambridge.
It furthers the University's mission by disseminating knowledge in the pursuit of
education, learning and research at the highest international levels of excellence.

www.cambridge.org
Information on this title: www.cambridge.org/9781108070782

© in this compilation Cambridge University Press 2014

This edition first published 1847
This digitally printed version 2014

ISBN 978-1-108-07078-2 Paperback

This book reproduces the text of the original edition. The content and language reflect
the beliefs, practices and terminology of their time, and have not been updated.

Cambridge University Press wishes to make clear that the book, unless originally published
by Cambridge, is not being republished by, in association or collaboration with,
or with the endorsement or approval of, the original publisher or its successors in title.

FORMAL LOGIC:

OR,

The Calculus of Inference,

Neceſſary and Probable.

BY

AUGUSTUS DE MORGAN

Of Trinity College Cambridge,

Fellow of the Cambridge Philoſophical Society, Secretary of the Royal
Aſtronomical Society, Profeſſor of Mathematics in
Univerſity College London.

Καλὸς ὁ νόμος ἐάν τις αὐτῷ νομίμως χρῆται.

LONDON:

TAYLOR AND WALTON,

Bookſellers and Publiſhers to Univerſity College,

28, *Upper Gower Street.*

M DCCC XLVII

PREFACE.

THE fyftem given in this work extends beyond that commonly received, in feveral directions. A brief ftatement of what is now fubmitted for adoption into the theory of inference will be the matter of this preface.

In the form of the propofition, the copula is made as abftract as the terms: or is confidered as obeying only thofe conditions which are neceffary to inference.

Every name is treated in connection with its *contrary* or *contradictory* name; the diftinction between thefe words not being made, and others fupplied in confequence. Eight really feparable forms of predication are thus obtained, between any two names: the eight of the common fyftem amounting only to fix, when, as throughout my work, the two forms of a convertible propofition are confidered as identical.

The complex propofition is introduced, confifting in the coexiftence of two fimple ones. The theory of the fyllogifm of complex propofitions is made to precede that of the fimple or ordinary fyllogifm; which laft is deduced from it. I have only ufed the word *complex*, becaufe *fimple* was already appropriated (fee page 85).

By the introduction of contraries, the number of valid syllogistic forms is increased to thirty-two, connected together by many rules of relation, but all shewn to contain, each with reference to its own disposition of names and contraries, only one form of inference.

The distinction of figure is avoided from the beginning by introducing into every proposition an order of reference to its terms.

A simple notation, which includes the common one, gives the means of representing every syllogism by three letters, each accented above or below. By inspection of one of these symbols it is seen immediately, 1. What syllogism is represented, 2. Whether it be valid or invalid, 3. How it is at once to be written down, 4. What axiom the inference contains, or what is the act of the mind when it makes that inference (chapter XIV).

A subordinate notation is used (page 60) in abbreviation of the proposition at length.

Compound names are considered, both when the composition is conjunctive, and when it is disjunctive. Distinct notation and rules of transformation are given, and the compound syllogisms are treated as reducible to ordinary ones, by invention of compound names.

The theory of the numerical syllogism is investigated, in which, upon the hypothesis of numerical quantity in both terms of every proposition, a numerical inference is made.

But, when the numerical relations of the several terms are fully known, all that is unusual in the quantity of the predicate is shown to be either superfluous, or else, as I have called it, spurious.

The old doctrine of modals is made to give place to the numerical theory of probability. Many will object to this theory as extralogical. But I cannot fee on what definition, founded on real diftinction, the exclufion of it can be maintained. When I am told that logic confiders the validity of the inference, independently of the truth or falfehood of the matter, or fupplies the conditions under which the hypothetical truth of the matter of the premifes gives hypothetical truth to the matter of the conclufion, I fee a real definition, which propounds for confideration the forms and laws of inferential thought. But when it is further added that the only hypothetical truth fhall be abfolute truth, certain knowledge, I begin to fee arbitrary diftinction, wanting the reality of that which preceded. Without pretending that logic can take cognizance of the probability of any given matter, I cannot underftand why the ftudy of the effect which partial belief of the premifes produces with refpect to the conclufion, fhould be feparated from that of the confequences of fuppofing the former to be abfolutely true. Not however to difpute upon names, I mean that I fhould maintain, againft thofe who would exclude the theory of probability from logic, that, call it by what name they like, it fhould accompany logic as a ftudy.

I have, of courfe, been obliged to exprefs, in my own manner, my own convictions on points of mental philofophy. But any one will fee that, in all which I have propofed for adoption, it matters nothing whether my views of the phenomena of thought, or others, be made the bafis of the explanation. So far therefore, as I am confidered as propofing forms of fyllogifm, &c. to the logician, and not giving inftruction to the ftudent of the fcience, the reader has nothing to do with my choice of the terms in which mental operations are fpoken of.

In the appendix will be found fome remarks on the perfonal controverfy between Sir W. Hamilton of Edinburgh and myfelf, of which I fuppofe the celebrity of my opponent, and the appearance of part of it in a journal fo widely circulated as the *Athenæum*, has caufed many ftudents of logic to hear or read fomething.

At the end of the contents of fome chapters in the following table, are a few additions and corrections, to which I requeft the reader's attention.

A. De Morgan.

Univerfity College, London,
    October 14, 1847.

TABLE OF CONTENTS.

*** The articles entered in *Italic*, are thofe, the contents of which belong to the peculiar fyftem prefented in this work.

> *** This chapter may be omitted by thofe who have fome knowledge of the ordinary definitions and phrafeology of logic. It is ftrictly confined to the Ariftotelian forms and fyllogifms, and is the reprint of a tract publifhed in 1839, under the title of 'Firft Notions of Logic (preparatory to the ftudy of Geometry)': the only alterations are ;— the change of phrafeology, as altering 'fome X is Y' into 'fome Xs are Ys,' &c.; the correction of a faulty demonftration; and a few omiffions, particularly of fome infufficient remarks on the probability of arguments.

Additions and corre&ions. Page 56, *line* 7, *inſert* except only one which conſiſts of four ſimple propoſitions. *Page* 62, *line* 23; Say X and Y are not *complements* (inſtead of *contraries*) that is, do not together either fill, or more than fill, the univerſe. *Page* 72, *lines* 4 *and* 3, *from the bottom;* The oppoſitions are incorre&. It ought to be *cannot do without* and *cannot fail with :* muſt precede, and *muſt follow.* The reader may eaſily identify the eight forms of predication as having X for ſubje&, Y for predicate, with the copulæ, cannot be without, can be without, cannot be with, can be with, cannot fail without, can fail without, cannot fail with, can fail with.

CHAPTER V.—*On the Syllogiſm* (pages 76—106).

of the terms simple and complex, 85; *Denial of the simplicity of the simple proposition*, 85, 86; *Are not disjunctive and conjunctive the proper words?* 86; The denial of a conclusion, coupled with one of the premises, denies the other, 86; The simple syllogism, 86; Demonstration that a particular cannot lead to a universal, and that two particulars are inconclusive, *by help of the complex syllogism*, 86, 87; Opponent syllogisms, 87, 88; *Rules for the symbols of opponent syllogisms*, 87, 88; *Of fundamental syllogisms, there must be twice as many particular as universal*, 88; *Deduction of the fundamental simple syllogisms, eight universal, and sixteen particular, from the eight affirmatory complex syllogisms*, 88, 89; *Deduction of the eight strengthened syllogisms from the limiting forms of the affirmatory complex ones*, 90, 91; *Connexion of the two modes of strengthening a premise*, 90, 91; *The conclusion is never strengthened by strengthening the middle term, nor only weakened by weakening it*, 91; *Table of connexion of the strengthened syllogisms with the rest*, 91; *deduction of the strengthened syllogisms from the negatory complex ones, and dismissal of the latter as of no more logical effect than the former*, 92; *Direct rule of notation, applying to syllogisms which begin and conclude with like quantity*, 92; *Inverse rule of notation*, [N.B. the word *inverse* should have been *contrary*] *applying to syllogisms which begin and conclude with unlike quantity*, 93; *Rules for all the retained syllogisms*, 93; *Sub-rules for the particular syllogisms* [they would have done as remarks, but are needless as rules] 94; *Remarks, partly recapitulatory*, 94, 95, 96; *In all fundamental syllogisms*, the middle term is universal in one premise, *and particular in the other*, 95; *distinction thence arising*, 95; *rule for connecting the syllogisms which are formed by interchanging the concluding terms*, 96; *conversion of a particular into a universal*, 96; *distinction of the particular quantity in a conclusion into intrinsic and extrinsic*, 97; *the quantity of one term always intrinsic, and hence the syllogism can always be made universal*, 97; *Nominal mode of notation for, and representation of, a fundamental syllogism*, 98; *connexion of the nominal system with the former (or proponent) system*, 99; *mode of deriving concomitants and weakened forms*, 100; *more abstract mode of representation derived from the nominal*, 100; *nominal system of strengthened syllogisms*, 101; *mixed complex syllogism*, 101; *opponent forms*, 102; *verbal description of the simple syllogism*, 103; *new view of the syllogism, in which all is referred to the middle term*, 104; *rules thence derived*, 105; *compound names, and expulsion of quantity by reference of the proposition to possibility or impossibility of a compound name*, 105; *system of syllogism thence arising*, 106.

Additions and corrections.—*Page* 79, in the first diagram, for D₁D₁D, read D₁D₁D₁; *page* 88, *line* 23, *instead of* has the other two for its opponents, *read* has its opponents in the set; *page* 90, *line* 4, *from the bottom, for* premiss *read* premise: the first spelling has been common enough, but it seems strange that the cognate words promise, surmise, demise, &c. should not have dictated the second. *Page* 96;

The inverted forms of the ſtrengthened ſyllogiſms are omitted: of theſe, four are their own inverſions, namely, $A_{\iota}A'\mathrm{I}'$, $A'A\,I_{\iota}$, $E'E'\mathrm{I}_{\iota}$, and $E_{\iota}E_{\iota}\mathrm{I}'$: of the remainder, $A_{\iota}E'\mathrm{O}'$ and $E'A'\mathrm{O}_{\iota}$ are inverſions; and alſo $A'E_{\iota}\mathrm{O}_{\iota}$ and $E_{\iota}A_{\iota}\mathrm{O}'$. *Page* 100, *line* 12, *from the bottom*; for -011 *read* -011), the firſt time it occurs. *Page* 101: Read the ſymbols of the ſtrengthened ſyllogiſms ſo as to begin from the middle in both premiſes: thus, Xyzl is y)X$+$y)z$=$Xz. *Page* 101. I might have ſaid a word or two on the caſe in which a complex particular is combined with a univerſal; to form the reſults will be an eaſy exerciſe for the reader. *Page* 102, *line* 7, *from the bottom*, for $\mathrm{I}_{\iota}A'\mathrm{I}_{\iota}$ *read* $\mathrm{I}_{\iota}A_{\iota}\mathrm{I}_{\iota}$.

CHAPTER VI.—*On the Syllogiſm* (pages 107—126).

Remarks connected with the exiſtence of the terms, 107, 108, 109, 110, 111, 112, 113. *The concluſion not ſeparable from the premiſes except as to truth*, 107, 108; conditions, and conditional ſyllogiſm, 109; *incompleteneſs of reduction of conditional to categorical*, 109, 110; *univerſe of propoſitions*, 110; *exiſtence of the terms of a propoſition*, 111; *its aſſumption in ſyllogiſm, particularly as to the middle term*, 112, 113; *poſtulate more extenſive than the* dictum de omni et nullo, *involved as well in the formation of premiſes as in ſyllogiſm*, 114, 115; Invention of names, 115; *notation for conjunctive and disjunctive names*, 115, 116; *expreſſion of complex relations and their contraries*, 116; copulative and disjunctive ſyllogiſms and dilemma, 117; *Conjunctive poſtulate*, 117; *deduction of other evident propoſitions from it*, 118, 119; *The collective* and, *as conjunctive, oppoſed to the disjunctives* and *and or diſtributively uſed in univerſals*, and *or disjunctive (in the common ſenſe) in particulars*, 119; *Disjunctives may be rejected from univerſals, and conjunctives from particulars*, 119; *Tranſpoſition, introduction of, and rules for*, 120; *Table of the tranſpoſed forms of A and E with compound names*, 121; Examples of disjunctive ſyllogiſms, dilemmas, &c. *treated by the above method*, 122, 123, 124; Sorites, 124; *Extended rules for the formation of the various claſſes of Sorites*, 125, 126.

Additions and corrections. Page 121, *line* 8, *from the bottom.* For [x,y][p,q])u *read* [X,Y][p,q])u.

CHAPTER VII.—*On the Ariſtotelian Syllogiſm* (127—141).

Limitations impoſed either by Ariſtotle or his followers, 127; Dictum de omni et nullo, 127; *defect of this*, 128; excluſion of contraries, 128; Standard forms, 129; Major and minor terms, and diſtinction of figure, 129; Selection of the Ariſtotelian ſyllogiſms *from among thoſe of this work*, 130 131; Symbolic words, and meaning of their letters, 131; Reduction to the firſt figure, 131, 132; Old form of the fourth figure, 132, 133; *Suggeſtion as to two figures ſubdivided*, 133; Poſſible uſe of the diſtinction of figure, 133, 134; Collection of

Additions and corrections. Page 143, *line* 12 : Supply the propofitions X)M,P and Y)N,Q, as deducible from the numbers of inftances in the feveral names. *Page* 148, *line* 10, *from the bottom :
for* propofitions *read* prepofitions. *Page* 152, *line* 4: *for* m *read*
m. *Page* 153, *line* 22 : *for* will prefently fhow us, *read* have fhown
us in page 145. *Page* 154, *line* 2, *from the bottom, for* ys *read* zs.
Page 155, *line* 6 *from the bottom, for* mXY *read* mXY. *Page* 162,
line 2, *after the table: for* laft chapter *read* chapter V. *Page* 166,
line 17, *for* m¹xy *read* m¹xy. *Page* 167, *line* 24 : *for* 62 *read* 92.

CHAPTER XI.—*On Induction* (pages 211—226).

Explanation of induction, 211 ; Reduction of the procefs to a fyllogifm, 211 ; Induction by connexion, and inftance, 212 ; Ordinary
induction not a demonftrative procefs, 212, 213 ; Pure induction,
incomplete, probability of it, 213, 214; Ordinary miftakes on this
fubject, 215 ; Examination of Mr. T. B. Macaulay's enumeration of
inftances in which fcientific analyfis is ufelefs, 216, 217, 218, 219,
220, 221, 222, 223, 224; *probability of fyllogifms with particular
premifes,* 224, 225, 226; Circumftantial evidence, 226.

CHAPTER XII.—*On old logical Terms* (pages 227—237).

Dialectics, 227 ; fimple and complex terms, 227; apprehenfion,
judgment, difcourfe, 227 ; Univerfal and fingular, 228 ; Individuals,
228 ; categories, predicaments, 228 ; fubftance, 228 ; firft and fecond
fubftance, 229 ; quantity, continuous and difcrete, 229 ; Quality,
habit, difpofition, paffion, 229 ; Relation, 229 ; Action, paffion, immanent, tranfient, univocal, equivocal, 229, 230 ; Remaining categories,
230; predicables, genus, fpecies, 230 ; difference, property, accident, 231 ; caufe, material, formal, efficient, final, 231 ; form, motion, fubject, object, 231 ; Subjective, objective, adjunct, 232 ;
modals, fubftitution of the theory of probabilities for them, 232 ;
Their ufe in the old philofophy, 232, 233 ; Notions of old logicians
on quantity, 234 ; Intenfion or comprehenfion, and extenfion, objections to their oppofition as quantities, and references to places in
this work where the diftinction has occurred, 234, 235, 236 ; Inftance, 236 ; Enthymeme, Ariftotle's, and modern, 236, 237.

Additions and corrections. Page 230, *lines* 16 *and* 15, *from the
bottom; tranfpofe the words* former *and* latter. *Page* 234 *line* 2 *from
bottom, for* after *read* before. *Page* 237, *note ;* I find that etymolo-
gifts are decidedly of opinion that ῥῆσις, fpeech, and 'ῥέω, flow, have
different roots, and that the former is *fpeech* in its primitive meaning.
The reader muft make the alteration, which however does not affect
my fuggeftion.

ELEMENTS OF LOGIC.

CHAPTER I.

Firſt Notions.

THE firſt notion which a reader can form of Logic is by viewing it as the examination of that part of reaſoning which depends upon the manner in which inferences are formed, and the inveſtigation of general maxims and rules for conſtructing arguments, ſo that the concluſion may contain no inaccuracy which was not previouſly aſſerted in the premiſes. It has ſo far nothing to do with the truth of the facts, opinions, or preſumptions, from which an inference is derived; but ſimply takes care that the inference ſhall certainly be true, if the premiſes be true. Thus, when we ſay that all men will die, and that all men are rational beings, and thence infer that ſome rational beings will die, the *logical* truth of this ſentence is the ſame whether it be true or falſe that men are mortal and rational. This logical truth depends upon the *ſtructure of the ſentence,* and not upon the particular matters ſpoken of. Thus,

Inſtead of	Write,
All men will die.	Every Y is X.
All men are rational beings.	Every Y is Z.
Therefore ſome rational beings will die.	Therefore ſome Zs are Xs.

The ſecond of theſe is the ſame propoſition, logically conſidered, as the firſt; the conſequence in both is virtually contained in, and rightly inferred from, the premiſes. Whether the premiſes be true or falſe, is not a queſtion of logic, but of morals, philoſophy, hiſtory, or any other knowledge to which their ſubject-matter belongs: the queſtion of logic is, does the concluſion certainly follow if the premiſes be true?

Every act of reaſoning muſt mainly conſiſt in comparing together different things, and either finding out, or recalling from previous knowledge, the points in which they reſemble or differ from each other. That particular part of reaſoning which is called *inference*, conſiſts in the compariſon of ſeveral and different things with one and the ſame other thing; and aſcertaining the reſemblances, or differences, of the ſeveral things, by means of the points in which they reſemble, or differ from, the thing with which all are compared.

There muſt then be ſome propoſitions already obtained before any inference can be drawn. All propoſitions are either aſſertions or denials, and are thus divided into *affirmative* and *negative*. Thus, X is Y, and X is not Y, are the two forms to which all propoſitions may be reduced. Theſe are, for our preſent purpoſe, the moſt ſimple forms; though it will frequently happen that much circumlocution is needed to reduce propoſitions to them. Thus, ſuppoſe the following aſſertion, 'If he ſhould come to-morrow, he will probably ſtay till Monday;' how is this to be reduced to the form X is Y? There is evidently ſomething ſpoken of, ſomething ſaid of it, and an affirmative connection between them. Something, if it happen, that is, the happening of ſomething, makes the happening of another ſomething probable; or *is* one of the things which render the happening of the ſecond thing probable.

$$X \qquad\qquad \text{is} \qquad\qquad Y$$

| The happening of his arrival to-morrow | is | an event from which it may be inferred as probable that he will ſtay till Monday. |

The forms of language will allow the manner of aſſerting to be varied in a great number of ways; but the reduction to the preceding form is always poſſible. Thus, 'ſo he ſaid' is an affirmation, reducible as follows:

| What you have juſt ſaid (or whatever elſe 'ſo' refers to) | is | the thing which he ſaid. |

By changing ' is ' into ' is not,' we make a negative propofition ; but care muft always be taken to afcertain whether a
propofition which appears negative be really fo. The principal
danger is that of confounding a propofition which is negative
with another which is affirmative of fomething requiring a negative to defcribe it. Thus, ' he refembles the man who was not
in the room,' is affirmative, and muft not be confounded with
' he does not refemble the man who was in the room.' Again,
' if he fhould come to-morrow, it is probable he will not ftay till
Monday,' does not mean the fimple denial of the preceding propofition, but the affirmation of a directly oppofite propofition.
It is,

$$X \qquad\text{is}\qquad Y$$

$$\text{The happening of his arrival to-morrow,} \quad \text{is} \quad \begin{cases} \text{an event from which it may be} \\ \text{inferred to be } \textit{im}\text{probable that} \\ \text{he will ftay till Monday:} \end{cases}$$

whereas the following,

$$\text{The happening of his arrival to-morrow,} \quad \text{is } \textit{not} \quad \begin{cases} \text{an event from which it may be} \\ \text{inferred as probable that he} \\ \text{will ftay till Monday,} \end{cases}$$

would be expreffed thus : ' If he fhould come to-morrow, that is
no reafon why he fhould ftay till Monday.'

Moreover, the negative words not, no, &c.; have two kinds of
meaning which muft be carefully diftinguifhed. Sometimes they
deny, and nothing more : fometimes they are ufed to affirm the
direct contrary. In cafes which offer but two alternatives, one
of which is neceffary, thefe amount to the fame thing, fince the
denial of one, and the affirmation of the other, are obvioufly
equivalent propofitions. In many idioms of converfation, the
negative implies affirmation of the contrary in cafes which offer
not only alternatives, but degrees of alternatives. Thus, to the
queftion, ' Is he tall ? ' the fimple anfwer, ' No,' moft frequently
means that he is the contrary of tall, or confiderably under the
average. But it muft be remembered, that, in all logical reafoning, the negation is fimply negation, and nothing more, never
implying affirmation of the contrary.

The common propofition that two negatives make an affirmative, is true only upon the fuppofition that there are but two

possible things, one of which is denied. Grant that a man must be either able or unable to do a particular thing, and then *not unable* and able are the same things. But if we suppose various degrees of performance, and therefore degrees of ability, it is false, in the common sense of the words, that two negatives make an affirmative. Thus, it would be erroneous to say, ' John is able to translate Virgil, and Thomas is not unable; therefore, what John can do Thomas can do,' for it is evident that the premises mean that John is so near to the best sort of translation that an affirmation of his ability may be made, while Thomas is considerably lower than John, but not so near to absolute deficiency that his ability may be altogether denied. It will generally be found that two negatives imply an affirmative of a weaker degree than the positive affirmation.

Each of the propositions, ' X is Y,' and ' X is not Y,' may be subdivided into two species: the *universal*, in which every possible case is included ; and the *particular*, in which it is not meant to be asserted that the affirmation or negation is universal. The four species of proposition are then as follows, each being marked with the letter by which writers on logic have always distinguished it.

A	*Universal Affirmative*	Every X is	Y
E	*Universal Negative*	No X is	Y
I	*Particular Affirmative*	Some Xs are	Ys
O	*Particular Negative*	Some Xs are not Ys	

In common conversation the affirmation of a part is meant to imply the denial of the remainder. Thus, by ' some of the apples are ripe,' it is always intended to signify that some are not ripe. This is not the case in logical language, but every proposition is intended to make its amount of affirmation or denial, and no more. When we say, ' Some X is Y,' or, more grammatically, ' Some Xs are Ys,' we do not mean to imply that some are not : this may or may not be. Again, the word some means, ' one or more, possibly all.' The following table will shew the bearing of each proposition on the rest.

Every X is Y affirms *Some Xs are Ys* and denies $\begin{cases} No & X \text{ is } Y \\ Some & Xs \text{ are not } Ys \end{cases}$

No X is Y affirms *Some Xs are not Ys* and denies $\begin{cases} Every & X \text{ is } Y \\ Some & Xs \text{ are } Ys \end{cases}$

Some Xs are Ys does not contradict $\begin{cases} Every & X \text{ is } Y \\ Some & Xs \text{ are not } Ys \end{cases}$ but denies *No X is Y*

Some Xs are not Ys does not contradict $\begin{cases} No & X \text{ is } Y \\ Some & Xs \text{ are } Ys \end{cases}$ but denies *Every X is Y*

Contradictory propositions are those in which one denies *any thing* that the other affirms; *contrary* propositions are those in which one denies *every thing* which the other affirms, or affirms every thing which the other denies. The following pair are contraries,

Every X is Y and No X is Y

and the following are contradictories,

Every X is Y to Some Xs are not Ys
No X is Y to Some Xs are Ys

A contrary, therefore, is a complete and total contradictory; and a little confideration will make it appear that the decifive diftinction between contraries and contradictories lies in this, that contraries may both be falfe, but of contradictories, one muft be true and the other falfe. We may fay, ' Either P is true, or *fomething* in contradiction of it is true;' but we cannot fay, ' Either P is true, or *every thing* in contradiction of it is true.' It is a very common miftake to imagine that the *denial* of a propofition gives a right to *affirm* the contrary; whereas it fhould be, that the *affirmation* of a propofition gives a right to *deny* the contrary. Thus, if we deny that Every X is Y, we do not affirm that No X is Y, but only that Some Xs are not Ys; while, if we affirm that Every X is Y, we deny No X is Y, and alfo Some Xs are not Ys.

But, as to contradictories, affirmation of one is denial of the other, and denial of one is affirmation of the other. Thus, either Every X is Y, or Some Xs are not Ys: affirmation of either is denial of the other, and *vice verfâ*.

Let the ftudent now endeavour to fatisfy himfelf of the following. Taking the four preceding propofitions, A, E, I, O, let the fimple letter fignify the affirmation, the fame letter in parenthefes the denial, and the abfence of the letter, that there is neither affirmation nor denial.

From A follow (E), I, (O)	From (A) follow O
From E .... (A), (I), O	From (E) I
From I .... (E)	From (I) .... (A), E, O
From O .... (A)	From (O) ... A, (E), I

Theſe may be thus ſummed up : The affirmation of a univerſal propoſition, and the denial of a particular one, enable us to affirm or deny all the other three ; but the denial of a univerſal propoſition, and the affirmation of a particular one, leave us unable to affirm or deny two of the others.

In ſuch propoſitions as ' Every X is Y,' ' Some Xs are not Ys,' &c., X is called the *ſubjeƈt*, and Y the *predicate*, while the verb ' is ' or ' is not,' is called the *copula*. It is obvious that the words of the propoſition point out whether the ſubjeƈt is ſpoken of univerſally or partially, but not ſo of the predicate, which it is therefore important to examine. Logical writers generally give the name of *diſtributed* ſubjeƈts or predicates to thoſe which are ſpoken of univerſally ; but as this word is rather technical, I ſhall ſay that a ſubjeƈt or predicate enters wholly or partially, according as it is univerſally or particularly ſpoken of.

1. In A, or ' Every X is Y,' the ſubjeƈt enters wholly, but the predicate only partially. For it obviouſly ſays, ' Among the Ys are all the Xs,' ' Every X is part of the colleƈtion of Ys, ſo that all the Xs make a part of the Ys, the whole it *may* be.' Thus, ' Every horſe is an animal,' does not ſpeak of all animals, but ſtates that all the horſes make up a portion of the animals.

2. In E, or ' No X is Y,' both ſubjeƈt and predicate enter wholly. ' No X whatſoever is any one out of all the Ys ; ' ' ſearch the whole colleƈtion of Ys, and *every* Y ſhall be found to be ſomething which is not X.'

3. In I, or ' Some Xs are Ys,' both ſubjeƈt and predicate enter partially. ' Some of the Xs are found among the Ys, or make up a part (the whole poſſibly, but not known from the preceding) of the Ys.'

4. In O, or ' Some Xs are not Ys,' the ſubjeƈt enters partially, and the predicate wholly. ' Some Xs are none of them any whatſoever of the Ys ; every Y will be found to be no one out of a certain portion of the Xs.'

It appears then that,

In affirmatives, the predicate enters partially.

In negatives, the predicate enters wholly.

In contradictory propositions, both subject and predicate enter differently in the two.

The *converse* of a proposition is that which is made by interchanging the subject and predicate, as follows:

	The proposition.	Its converse.
A	Every X is Y	Every Y is X
E	No X is Y	No Y is X
I	Some Xs are Ys	Some Ys are Xs
O	Some Xs are not Ys	Some Ys are not Xs

Now, it is a fundamental and self-evident proposition, that no consequence must be allowed to assert more widely than its premises; so that, for instance, an assertion which is only of some Ys can never lead to a result which is true of all Ys. But if a proposition assert agreement or disagreement, any other proposition which asserts the same, to the same extent and no further, must be a legitimate consequence; or, if you please, must amount to the whole, or part, of the original assertion in another form. Thus, the converse of A is not true: for, in ' Every X is Y,' the predicate enters partially; while in ' Every Y is X,' the subject enters wholly. ' All the Xs make up a part of the Ys, then a part of the Ys are among the Xs, or some Ys are Xs.' Hence, the only *legitimate* converse of ' Every X is Y ' is, ' Some Ys are Xs.' But in ' No X is Y,' both subject and predicate enter wholly, and ' No Y is X ' is, in fact, the same proposition as ' No X is Y.' And ' Some Xs are Ys ' is also the same as its converse ' Some Ys are Xs :' here both terms enter partially. But ' Some Xs are not Ys ' admits of no converse whatever; it is perfectly consistent with all assertions upon Y and X in which Y is the subject. Thus neither of the four following lines is inconsistent with itself.

Some Xs are not Ys	and	Every Y is X
Some Xs are not Ys	and	No Y is X
Some Xs are not Ys	and	Some Ys are Xs
Some Xs are not Ys	and	Some Ys are not Xs.

Having thus discussed the principal points connected with the simple assertion, I pass to the manner of making two assertions give a third. Every inſtance of this is called a *ſyllogiſm*, the two aſſertions which form the baſis of the third are called *premiſes*, and the third itſelf the *concluſion*.

If two things both agree with a third in any particular, they agree with each other in the ſame; as, if X be of the ſame colour as Y, and Z of the ſame colour as Y, then X is of the ſame colour as Z. Again, if X differ from Y in any particular in which Z agrees with Y, then X and Z differ in that particular. If X be not of the ſame colour as Y, and Z be of the ſame colour as Y, then X is not of the colour of Z. But if X and Z both differ from Y in any particular, nothing can be inferred; they may either differ in the ſame way and to the ſame extent, or not. Thus, if X and Z be both of different colours from Y, it neither follows that they agree, nor differ, in their own colours.

The paragraph preceding contains the eſſential parts of all inference, which conſiſts in comparing two things with a third, and finding from their agreement or difference with that third, their agreement or difference with one another. Thus, Every X is Y, every Z is Y, allows us to infer that X and Z have all thoſe qualities in common which are neceſſary to Y. Again, from every X is Y, and ' No Z is Y,' we infer that X and Z differ from one another in all particulars which are eſſential to Y. The preceding forms, however, though they repreſent common reaſoning better than the ordinary ſyllogiſm, to which we are now coming, do not conſtitute the ultimate forms of inference. Simple *identity* or *non-identity* is the ultimate ſtate to which every aſſertion may be reduced; and we ſhall, therefore, firſt aſk, from what identities, &c., can other identities, &c., be produced ? Again, ſince we name objects in ſpecies, each ſpecies conſiſting of a number of individuals, and ſince our aſſertion may include all or only part of a ſpecies, it is further neceſſary to aſk, in every inſtance, to what extent the concluſion drawn is true, whether of all, or only of part ?

Let us take the ſimple aſſertion, ' Every living man reſpires ;' or every living man is one of the things (however varied they may be) which reſpire. If we were to encloſe all living men in a large triangle, and all reſpiring objects in a large circle, the preceding aſſertion, if true, would require that the whole of the triangle ſhould be contained in the circle. And in the ſame way we may reduce any affertion to the expreffion of a coincidence, total or partial, between two figures. Thus, a point in a circle may reprefent an individual of one fpecies, and a point in a triangle an individual of another fpecies: and we may exprefs that the whole of one fpecies is afferted to be contained or not contained in the other by fuch forms as, 'All the △ is in the ○'; 'None of the △ is in the ○'.

Any two affertions about X and Z, each expreffing agreement or difagreement, total or partial, with or from Y, and leading to a conclufion with refpeċt to X or Z, is called a fyllogifm, of which Y is called the *middle term.* The plaineft fyllogifm is the following :—

Every X is Y	All the △ is in the ○
Every Y is Z	All the ○ is in the □
Therefore Every X is Z	Therefore All the △ is in the □

In order to find all the poffible forms of fyllogifm, we muft make a table of all the elements of which they can confift ; namely—

X and Y		Z and Y	
Every X is Y	A	Every Z is Y	
No X is Y	E	No Z is Y	
Some Xs are Ys	I	Some Zs are Ys	
Some Xs are not Ys	O	Some Zs are not Ys	
Every Y is X	A	Every Y is Z	
Some Ys are not Xs	O	Some Ys are not Zs	

Or their fynonymes,

△ and ○		□ and ○
All the △ is in the ○	A	All the □ is in the ○
None of the △ is in the ○	E	None of the □ is in the ○
Some of the △ is in the ○	I	Some of the □ is in the ○
Some of the △ is not in the ○	O	Some of the □ is not in the ○
All the ○ is in the △	A	All the ○ is in the □
Some of the ○ is not in the △	O	Some of the ○ is not in the □

Now, taking any one of the fix relations between X and Y, and combining it with either of thofe between Z and Y, we have fix pairs of premifes, and the fame number repeated for every different relation of X to Y. We have then thirty-fix

forms to conſider : but, thirty of theſe (namely, all but (A, A)
(E, E), &c.,) are half of them repetitions of the other half. Thus,
'Every X is Y, no Z is Y,' and 'Every Z is Y, no X is Y,'
are of the ſame form, and only differ by changing X into Z and
Z into X. There are then only 15+6, or 21 diſtinct forms,
ſome of which give a neceſſary concluſion, while others do not.
We ſhall ſelect the former of theſe, claſſifying them by their
concluſions ; that is, according as the inference is of the form
A, E, I, or O.

I. In what manner can a univerſal affirmative concluſion be
drawn ; namely, that one figure is entirely contained in the other ?
This we can only aſſert when we know that one figure is entirely
contained in the circle, which itſelf is entirely contained in the
other figure. Thus,

Every X is Y	All the △ is in the ○	A
Every Y is Z	All the ○ is in the □	A
Every X is Z	All the △ is in the □	A

is the only way in which a univerſal affirmative concluſion can
be drawn.

II. In what manner can a univerſal negative concluſion be
drawn ; namely, that one figure is entirely exterior to the other ?
Only when we are able to aſſert that one figure is entirely within,
and the other entirely without, the circle. Thus,

Every X is Y	All the △ is in the ○	A
No Z is Y	None of the □ is in the ○	E
No X is Z	None of the △ is in the □	E

is the only way in which a univerſal negative concluſion can be
drawn.

III. In what manner can a particular affirmative concluſion be
drawn ; namely, that part or all of one figure is contained in the
other ?   Only when we are able to aſſert that the whole circle is
part of one of the figures, and that the whole, or part of the circle, is part of the other figure. We have then two forms.

Every Y is X	All the ○ is in the △	A
Every Y is Z	All the ○ is in the □	A
Some Xs are Zs	Some of the △ is in the □	I

Every Y is X	All the ○ is in the △	A
Some Ys are Zs	Some of the ○ is in the □	I
Some Xs are Zs	Some of the △ is in the □	I

The second of these contains all that is strictly necessary to the conclusion, and the first may be omitted. That which follows when an assertion can be made as to some, must follow when the same assertion can be made of all.

IV. How can a particular negative proposition be inferred; namely, that part, or all of one figure, is not contained in the other? It would seem at first sight, whenever we are able to assert that part or all of one figure is in the circle, and that part or all of the other figure is not. The weakest syllogism from which such an inference can be drawn would then seem to be as follows.

Some Xs are Ys	Some of the △ is in the ○
Some Zs are not Ys	Some of the □ is not in the ○
∴Some Zs are not Xs	∴Some of the △ is not in the □

But here it will appear, on a little consideration, that the conclusion is only thus far true; that those Xs which are Ys cannot be *those* Zs which are not Ys; but they may be *other* Zs, about which nothing is asserted when we say that *some* Zs are not Ys. And further consideration will make it evident, that a conclusion of this form can only be arrived at when one of the figures is entirely within the circle, and the whole, or part of the other without; or else when the whole of one of the figures is without the circle, and the whole or part of the other within; or lastly, when the circle lies entirely within one of the figures, and not entirely within the other. That is, the following are the distinct forms which allow of a particular negative conclusion, in which it should be remembered that a particular proposition in the premises may always be changed into a universal one, without affecting the conclusion. For that which necessarily follows from " some," follows from " all."

Every X is Y	All the △ is in the ○	A
Some Zs are not Ys	Some of the □ is not in the ○	O
∴ Some Zs are not Xs	Some of the □ is not in the △	O

No X is Y	None of the △ is in the ○	E
Some Zs are Ys	Some of the □ is in the ○	I
∴ Some Zs are not Xs	Some of the □ is not in the △	O

Every Y is X	All the ○ is in the △	A
Some Ys are not Zs	Some of the ○ is not in the □	O
∴ Some Xs are not Zs	Some of the △ is not in the □	O

It appears, then, that there are but fix diftinct fyllogifms. All others are made from them by ftrengthening one of the premifes, or converting one or both of the premifes, where fuch converfion is allowable; or elfe by firft making the converfion, and then ftrengthening one of the premifes. And the following arrangement will fhow that two of them are univerfal, three of the others being derived from them by weakening one of the premifes in a manner which does not deftroy, but only weakens, the conclufion.

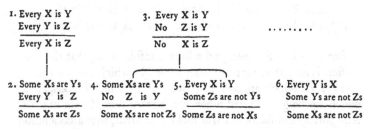

1. Every X is Y
   Every Y is Z
   ————
   Every X is Z

3. Every X is Y
   No Z is Y
   ————
   No X is Z

. . . .

2. Some Xs are Ys
   Every Y is Z
   ————
   Some Xs are Zs

4. Some Xs are Ys
   No Z is Y
   ————
   Some Xs are not Zs

5. Every X is Y
   Some Zs are not Ys
   ————
   Some Zs are not Xs

6. Every Y is X
   Some Ys are not Zs
   ————
   Some Xs are not Zs

We may fee how it arifes that one of the partial fyllogifms is not immediately derived, like the others, from a univerfal one. In the preceding, A E E may be confidered as derived from A A A, by changing the term in which Y enters univerfally into a univerfal negative. If this be done with the other term inftead, we have

No X is Y ⎱ from which univerfal premifes we cannot deduce a
Every Y is Z ⎰ univerfal conclufion, but only fome Zs are not Xs.

If we weaken one and the other of thefe premifes, as they ftand, we obtain

Some Xs are not Ys
Every Y is Z
————
No conclufion

and

No X is Y
Some Ys are Zs
————
Some Zs are not Xs

equivalent to the fourth of the preceding: but if we convert the firſt premiſe, and proceed in the ſame manner,

From No Y is X	we obtain	Some Ys are not Xs
Every Y is Z		Every Y is Z
Some Zs are not Xs		Some Zs are not Xs

which is legitimate, and is the ſame as the laſt of the preceding liſt, with X and Z interchanged.

Before proceeding to ſhow that all the uſual forms are contained in the preceding, let the reader remark the following rules, which may be proved either by collecting them from the preceding caſes, or by independent reaſoning.

1. The middle term muſt enter univerſally into one or the other premiſe. If it were not ſo, then one premiſe might ſpeak of one part of the middle term, and the other of another; ſo that there would, in fact, be no middle term. Thus, 'Every X is Y, Every Z is Y,' gives no concluſion: it may be thus ſtated;

All the Xs make up *a part* of the Ys
All the Zs make up *a part* of the Ys

And, before we can know that there is any common term of compariſon at all, we muſt have ſome means of ſhowing that the two parts are to ſome extent the ſame; or the preceding premiſes by themſelves are inconcluſive.

2. No term muſt enter the concluſion more generally than it is found in the premiſes; thus, if X be ſpoken of partially in the premiſes, it muſt enter partially into the concluſion. This is obvious, ſince the concluſion muſt aſſert no more than the premiſes imply.

3. From premiſes both negative no concluſion can be drawn. For it is obvious, that the mere aſſertion of diſagreement between each of two things and a third, can be no reaſon for inferring either agreement or diſagreement between theſe two things. It will not be difficult to reduce any caſe which falls under this rule to a breach of the firſt rule: thus, No X is Y, No Z is Y, gives

Every X is (ſomething which is not Y)
Every Z is (ſomething which is not Y)

in which the middle term is not fpoken of univerfally in either. Again, ' No Y is X, fome Ys are not Zs,' may be converted into

> Every X is (a thing which is not Y)
> Some (things which are not Zs) are Ys

in which there is no middle term.

4. From premifes both particular no conclufion can be drawn. This is fufficiently obvious when the firft or fecond rule is broken, as in 'Some Xs are Ys, Some Zs are Ys.' But it is not immediately obvious when the middle term enters one of the premifes univerfally. The following reafoning will ferve for exercife in the preceding refults. Since both premifes are particular in form, the middle term can only enter one of them univerfally by being the predicate of a negative propofition; confequently (Rule 3) the other premife muft be affirmative, and, being particular, neither of its terms is univerfal. Confequently both the terms as to which the conclufion is to be drawn enter partially, and the conclufion (Rule 2) can only be a particular *affirmative* propofition. But if one of the premifes be. negative, the conclufion muft be *negative* (as we fhall immediately fee). This contradiction fhows that the fuppofition of particular premifes producing a legitimate refult is inadmiffible.

5. If one premife be negative, the conclufion, if any, muft be negative. If one term agree with a fecond and difagree with a third, no agreement can be inferred between the fecond and third.

6. If one premife be particular, the conclufion muft be particular. This may be fhown as follows. If two propofitions P and Q, together prove a third, R, it is plain that P and the denial of R, prove the denial of Q. For P and Q cannot be true together without R. Now if poffible, let P (a particular) and Q (a univerfal) prove R (a univerfal). Then P (particular) and the denial of R (particular) prove the denial of Q But two particulars can prove nothing.

In the preceding fet of fyllogifms we obferve one form only which produces A, or E, or I, but three which produce O.

Let an affertion be faid to be weakened when it is reduced from univerfal to particular, and ftrengthened in the contrary cafe. Thus, 'Every X is Z' is called ftronger than 'Some Xs are Zs.'

Every ufual form of fyllogifm which can give a legitimate refult is either one of the preceding fix, or another formed from
one of the fix, either by changing one of the affertions into its
converfe, if that be allowable, or by ftrengthening one of the
premifes, without altering the conclufion, or both.   Thus,

Some Xs are Ys ⎱
Every Y is Z   ⎰ may be written ⎱ Some Ys are Xs
                                   ⎰ Every Y is Z

What follows will ftill follow from ⎱ *Every* Y is X
                                     ⎰ Every Y is Z

for all which is true when 'Some Ys are Xs,' is not lefs true when
' Every Y is X.'

It would be poffible alfo to form a legitimate fyllogifm by
weakening the conclufion, when it is univerfal, fince that which
is true of all is true of fome.   Thus, ' Every X is Y, Every Y
is Z,' which yields ' Every X is Z,' alfo yields ' Some Xs are Zs.'
But writers on logic have always confidered thefe fyllogifms as
ufelefs, conceiving it better to draw from any premifes their
ftrongeft conclufion.   In this they were undoubtedly right; and
the only queftion is, whether it would not have been advifable
to make the premifes as weak as poffible, and not to admit any
fyllogifms in which more appeared than was abfolutely neceffary
to the conclufion.   If fuch had been the practice, then

Every Y is X, Every Y is Z, therefore Some Xs are Zs

would have been confidered as formed by a fpurious and unneceffary excefs of affertion.   The minimum of affertion would be
contained in either of the following,

Every Y is X, Some Ys are Zs, therefore Some Xs are Zs
Some Ys are Xs, Every Y is Z, therefore Some Xs are Zs

In this chapter, fyllogifms have been divided into two claffes :
firft, thofe which prove a univerfal conclufion ; fecondly, thofe
which prove a partial conclufion, and which are (all but one)
derived from the firft by weakening one of the premifes, in fuch
manner as to produce a legitimate but weakened conclufion.
Thofe of the firft clafs are placed in the firft column, and of the
other in the fecond.

Universal.		Particular.	
A	Every X is Y	Some Xs are Ys	I
A	Every Y is Z ——	Every Y is Z	A
A	Every X is Z	Some Xs are Zs	I
		Some Xs are Ys	I
		No Y is Z	E
A	Every X is Y	Some Xs are not Zs	O
E	No Y is Z ——	Every X is Y	A
E	No X is Z	Some Zs are not Ys	O
		Some Zs are not Xs	O
		Every Y is X	A
....		Some Ys are not Zs	O
		Some Xs are not Zs	O

In all works on logic, it is cuſtomary to write that premiſe firſt which contains the predicate of the concluſion. Thus,

Every Y is Z		Every X is Y
Every X is Y	would be written, and not	Every Y is Z
Every X is Z		Every X is Z

The premiſes thus arranged are called major and minor; the predicate of the concluſion being called the major term, and its ſubject the minor. Again, in the preceding caſe we ſee the various ſubjects coming in the order Y, Z; X, Y; X, Z: and the number of different orders which can appear is four, namely—

Y Z	Z Y	Y Z	Z Y
X Y	X Y	Y X	Y X
X Z	X Z	X Z	X Z

which are called the four *figures*, and every kind of ſyllogiſm in each figure is called a *mood*. I now put down the various moods of each figure, the letters of which will be a guide to find out thoſe of the preceding liſt from which they are derived. Co means that a premiſe of the preceding liſt has been converted; + that it has been ſtrengthened; Co+, that both changes have taken place. Thus,

A	Every Y is Z		A	Every Y is Z
I	Some Xs are Ys	becomes	A	Every Y is X : (Co +)

| I | Some Xs are Zs | | I | Some Xs are Zs |

And Co + points out the following: If ſome Xs be Ys, then ſome Ys are Xs (Co); and all that is true when Some Ys are Xs, is true when Every Y is X (+); therefore the ſecond ſyllogiſm is legitimate, if the firſt be ſo.

Firſt Figure.

A	Every Y is Z		A	Every Y is Z
A	Every X is Y		I	Some Xs are Ys

| A | Every X is Z | | I | Some Xs are Zs |

E	No Y is Z		E	No Y is Z
A	Every X is Y		I	Some Xs are Ys

| E | No X is Z | | O | Some Xs are not Zs |

Second Figure.

E	No Z is Y (Co)		E	No Z is Y (Co)
A	Every X is Y		I	Some Xs are Ys

| E | No X is Z | | O | Some Xs are not Zs |

A	Every Z is Y		A	Every Z is Y
E	No X is Y (Co)		O	Some Xs are not Ys

| E | No X is Z | | O | Some Xs are not Zs |

Third Figure.

A	Every Y is Z		E	No Y is Z
A	Every Y is X (Co +)		A	Every Y is X (Co +)

| I | Some Xs are Zs | | O | Some Xs are not Zs |

I	Some Ys are Zs (Co)		O	Some Ys are not Zs
A	Every Y is X		A	Every Y is X

| I | Some Xs are Zs | | O | Some Xs are not Zs |

A	Every Y is Z		E	No Y is Z
I	Some Ys are Xs (Co)		I	Some Ys are Xs (Co)

| I | Some Xs are Zs | | O | Some Xs are not Zs |

Fourth Figure.

A Every Z is Y (+)   I Some Zs are Ys
A Every Y is X   A Every Y is X
———————————   ———————————
I Some Xs are Zs   I Some Zs are Xs

A Every Z is Y   E No   Z is Y (Co)
E No Y is X   A Every Y is X (Co +)
———————————   ———————————
E No X is Z   O Some Xs are not Zs

E No   Z is Y (Co)
I Some Ys are Xs (Co)
———————————
O Some Xs are not Zs

The above is the ancient method of dividing syllogisms; but, for the present purpose, it will be sufficient to consider the six from which the rest can be obtained. And since some of the six have X in the predicate of the conclusion, and not Z, I shall join to them the six other syllogisms which are found by transposing Z and X. The complete list, therefore, of syllogisms with the weakest premises and the strongest conclusions, in which a comparison of X and Z is obtained by comparison of both with Y, is as follows:

Every X is Y	Every Z is Y	Some Xs are Ys	Some Zs are Ys
Every Y is Z	Every Y is X	No Y is Z	No Y is X
Every X is Z	Every Z is X	Some Xs are not Zs	Some Zs are not Xs
Every X is Y	Every Z is Y	Every X is Y	Every Z is Y
No Y is Z	No Y is X	Some Zs are not Ys	Some Xs are not Ys
No X is Z	No Z is X	Some Zs are not Xs	Some Xs are not Zs
Some Xs are Ys	Some Zs are Ys	Every Y is X	Every Y is Z
Every Y is Z	Every Y is X	Some Ys are not Zs	Some Ys are not Xs
Some Xs are Zs	Some Zs are Xs	Some Xs are not Zs	Some Zs are not Xs

In the list of page 12, there was nothing but recapitulation of forms, each form admitting a variation by interchanging X and Z. This interchange having been made, and the results collected as above, if we take every case in which Z is the predicate, or can be made the predicate by allowable conversion, we

have a collection of all possible *weakeſt* forms in which the reſult
is one of the four 'Every X is Z,' 'No X is Z,' 'Some Xs are Zs,'
' Some Xs are not Zs ;' as follows. The premiſes are written
in what appeared the moſt natural order, without diſtinction of
major or minor.

Every X is Y
Every Y is Z
———————
Every X is Z

Some Xs are Ys	Some Zs are Ys
Every Y is Z	Every Y is X
———————	———————
Some Xs are Zs	Some Xs are Zs
Every X is Y	Every Z is Y
No　Z is Y	No　X is Y
———————	———————
No　X is Z	No　X is Z

Some Xs are Ys	Every Z is　Y	Every Y is X
No　Z is Y	Some Xs are not Ys	Some Ys are not Zs
———————	———————	———————
Some Xs are not Zs	Some Xs are not Zs	Some Xs are not Zs

Every aſſertion which can be made upon two things by compariſon with any third, that is, every ſimple inference, can be
reduced to one of the preceding forms. Generally ſpeaking, one
of the premiſes is omitted, as obvious from the concluſion ; that
is, one premiſe being named and the concluſion, that premiſe is
implied which is neceſſary to make the concluſion good. Thus,
if I ſay, " That race muſt have poſſeſſed ſome of the arts of life,
for they came from Aſia," it is obviouſly meant to be aſſerted,
that all races coming from Aſia muſt have poſſeſſed ſome of the
arts of life. The preceding is then a ſyllogiſm, as follows :

　　That race is ' a race of Aſiatic origin :'
　　Every ' race of Aſiatic origin' is ' a race which muſt
　　　have poſſeſſed ſome of the arts of life :'
　Therefore, That race *is* a race which muſt have poſſeſſed
　　　ſome of the arts of life.

A perſon who makes the preceding aſſertion either means to
imply, antecedently to the concluſion, that all Aſiatic races muſt
have poſſeſſed arts, or he talks nonſenſe if he aſſert the conclu-

ſion poſitively. 'X muſt be Z, for it is Y,' can only be an inference when 'Every Y is Z.' This latter propóſition may be called the ſuppreſſed premiſe; and it is in ſuch ſuppreſſed propoſitions that the greateſt danger of error lies. It is alſo in ſuch propoſitions that men convey opinions which they would not willingly expreſs. Thus, the honeſt witneſs who ſaid, 'I always thought him a reſpectable man—he kept his gig,' would probably not have admitted in direct terms, 'Every man who keeps a gig muſt be reſpectable.'

I ſhall now give a few detached illuſtrations of what precedes.

"His imbecility of character might have been inferred from his proneneſs to favourites; for all weak princes have this failing." The preceding would ſtand very well in a hiſtory, and many would paſs it over as containing very good inference. Written, however, in the form of a ſyllogiſm, it is,

All weak princes are prone to favourites
He was prone to favourites
_____
Therefore He was a weak prince

which is palpably wrong. (Rule 1.) The writer of ſuch a ſentence as the preceding might have meant to ſay, 'for all who have this failing are weak princes;' in which caſe he would have inferred rightly. Every one ſhould be aware that there is much falſe form of inference ariſing out of badneſs of ſtyle, which is juſt as injurious to the habits of the untrained reader as if the errors were miſtakes of logic in the mind of the writer.

'X is leſs than Y; Y is leſs than Z: therefore X is leſs than Z.' This, at firſt ſight, appears to be a ſyllogiſm; but, on reducing it to the uſual form, we find it to be,

X is (a magnitude leſs than Y)
Y is (a magnitude leſs than Z)
Therefore X is (a magnitude leſs than Z)

which is not a ſyllogiſm, ſince there is no middle term. Evident as the preceding is, the following additional propoſition muſt be formed before it can be made explicitly logical. 'If Y be a magnitude leſs than Z, then every magnitude leſs than Y is alſo leſs than Z.' There is, then, before the preceding can be reduced to a ſyllogiſtic form, the neceſſity of a deduction from the ſecond premiſe, and the ſubſtitution of the reſult inſtead of that premiſe. Thus,

$$X \text{ is leſs than } Y$$
Leſs than Y is leſs than Z : following from Y is leſs than Z.

Therefore X is leſs than Z

But, if the additional argument be examined—namely, if Y be leſs than Z, then that which is leſs than Y is leſs than Z—it will be found to require preciſely the ſame conſiderations repeated ; for the original inference was nothing more. In faɛt, it may eaſily be ſeen as follows, that the propoſition before us involves more than any ſimple ſyllogiſm can expreſs. When we ſay that X is leſs than Y, we ſay that if X were applied to Y, every part of X would match a part of Y, and there would be parts of Y remaining over. But when we ſay, ' Every X is Y,' meaning the premiſe of a common ſyllogiſm, we ſay that every inſtance of X is an inſtance of Y, without ſaying any thing as to whether there are or are not inſtances of Y ſtill left, after thoſe which are alſo X are taken away. If, then, we wiſh to write an ordinary ſyllogiſm in a manner which ſhall correſpond with ' X is leſs than Y, Y is leſs than Z, therefore X is leſs than Z,' we muſt introduce a more definite amount of aſſertion than was made in the preceding forms. Thus,

Every X is Y, and there are Ys which are not Xs
Every Y is Z, and there are Zs which are not Ys

Therefore Every X is Z, and there are Zs which are not Xs
Or thus :

The Ys contain all the Xs, and more
The Zs contain all the Ys, and more

The Zs contain all the Xs, and more

The moſt technical form, however, is,

From Every X is Y ; [Some Ys are not Xs]
Every Y is Z ; [Some Zs are not Ys]
Follows Every X is Z ; [Some Zs are not Xs]

This ſort of argument is called *à fortiori* argument, becauſe the premiſes are more than ſufficient to prove the concluſion, and the extent of the concluſion is thereby greater than its mere form would indicate. Thus, ' X is leſs than Y, Y is leſs than Z,

therefore, *à fortiori*, X is lefs than Z,' means that the extent to
which X is lefs than Z muft be greater than that to which X is
lefs than Y, or Y than Z.   In the fyllogifm laft written, either
of the bracketted premifes might be ftruck out without deftroying
the conclufion ; which laft would, however, be weakened.   As
it ftands, then, the part of the conclufion, ' Some Zs are not
Xs,' follows *à fortiori*.

The argument *à fortiori* may then be defined as a univerfally
affirmative fyllogifm, in which both of the premifes are fhewn to
be lefs than the whole truth, or greater.   Thus, in ' Every X is
Y, Every Y is Z, therefore Every X is Z,' we do not certainly
imply that there are more Ys than Xs, or more Zs than Ys, fo
that we do not know that there are more Zs than Xs.   But if
we be at liberty to ftate the fyllogifm as follows,

All the Xs make up part (and part only) of the Ys
Every Y is Z ;

then we are certain that

All the Xs make up part (and part only) of the Zs.

But if we be at liberty further to fay that

All the Xs make up part (and part only) of the Ys
All the Ys make up part (and part only) of the Z

then we conclude that

All the Xs make up *part of part* (only) of the Zs

and the words in Italics mark that quality of the conclufion from
which the argument is called *à fortiori*.

Moft fyllogifms which give an affirmative conclufion are generally meant to imply *à fortiori* arguments, except only in mathematics.   It is feldom, except in the exact fciences, that we meet
with a propofition, ' Every X is Z,' which we cannot immediately
couple with ' fome Zs are not Xs.'

When an argument is completely eftablifhed, with the exception of one affertion only, fo that the inference may be drawn as
foon as that one affertion is eftablifhed, the refult is ftated in a
form which bears the name of an *hypothetical* fyllogifm.   The
word hypothefis means nothing but fuppofition ; and the fpecies
of fyllogifm juft mentioned firft lays down the affertion that a
confequence will be true if a certain condition be fulfilled, and

then either afferts the fulfilment of the condition, and thence the confequence, or elfe denies the confequence, and thence denies the fulfilment of the condition. Thus, if we know that

When X is Z, it follows that P is Q ;

then, as foon as we can afcertain that X is Z, we can conclude that P is Q ; or, if we can fhew that P is not Q, we know that X is not Z. But if we find that X is not Z, we can infer nothing ; for the preceding does not affert that P is Q *only* when X is Z. And if we find out that P is Q we can infer nothing. This conditional fyllogifm may be converted into an ordinary fyllogifm, as follows. Let K be any 'cafe in which X is Z,' and V, a 'cafe in which P is Q ;' then the preceding affertion amounts to ' Every K is V ' Let L be a particular inftance, the X of which may or may not be Z. If X be Z in the inftance under difcuffion, or if X be not Z, we have, in the one cafe and the other,

	Every K is V	Every K is V
	L is a K	L is not a K
Therefore	L is a V	No conclufion

Similarly, according as a particular cafe (M) is or is not V, we have

Every K is V	Every K is V
M is a V	M is not a V
No conclufion	M is not a K

That is to fay : the affertion of an hypothefis is the affertion of its neceffary confequence, and the denial of the neceffary confequence is the denial of the hypothefis : but the affertion of the neceffary confequence gives no right to affert the hypothefis, nor does the denial of the hypothefis give any right to deny the truth of that which would (were the hypothefis true) be its neceffary confequence.

Demonftration is of two kinds : which arifes from this, that every propofition has a contradictory ; and of thefe two, one muft be true and the other muft be falfe. We may then either prove a propofition to be true, or its contradictory to be falfe. ' It is true that every X is Z,' and ' it is falfe that there are fome Xs which are not Zs,' are the fame propofition ; and the proof of either is called the indirect proof of the other.

But how is any propofition to be proved falfe, except by prov-ing a contradiction to be true? By proving a neceffary confe-quence of the propofition to be falfe. But this is not a complete anfwer, fince it involves the neceffity of doing the fame thing; or, fo far as this anfwer goes, one propofition cannot be proved falfe unlefs by proving another to be falfe. But it may happen, that a neceffary confequence can be obtained which is obvioufly and felf-evidently falfe, in which cafe no further proof of the falfehood of the hypothefis is neceffary. Thus the proof which Euclid gives that all equiangular triangles are equilateral is of the following ftructure, logically confidered.

(1.) If there be an equiangular triangle not equilateral, it fol-lows that a whole can be found which is not greater than its part.*

(2.) It is falfe that there can be any whole which is not greater than its part (felf evident).

(3.) Therefore it is falfe that there is any equiangular triangle which is not equilateral; or all equiangular triangles are equila-teral.

When a propofition is eftablifhed by proving the truth of the matters it contains, the demonftration is called *direct*; when by proving the falfehood of every contradictory propofition, it is called *indirect*. The latter fpecies of demonftration is as logical as the former, but not of fo fimple a kind; whence it is defira-ble to ufe the former whenever it can be obtained.

The ufe of indirect demonftration in the Elements of Euclid is almoft entirely confined to thofe propofitions in which the con-verfes of fimple propofitions are proved. It frequently happens that an eftablifhed affertion of the form

$$\text{Every X is Z} \dots\dots\dots\dots (1)$$

may be eafily made the means of deducing,

$$\text{Every (thing not X) is not Z} \dots (2)$$

which laft gives

$$\text{Every Z is X} \dots\dots\dots\dots (3)$$

---

* This is the propofition in proof of which nearly the whole of the de-monftration of Euclid is fpent.

The converfion of the fecond propofition into the third is
ufually made by an indirect demonftration, in the following manner:
If poffible, let there be one Z, which is not X, (2) being true.
Then there is one thing which is not X and is Z; but every
thing not X is not Z; therefore there is one thing which is Z
and is not Z : which is abfurd. It is then abfurd that there
fhould be one fingle Z which is not X ; or, Every Z is X.

The following propofition contains a method which is of frequent ufe.

HYPOTHESIS.—Let there be any number of propofitions or
affertions,—three for inftance, X, Y, and Z,—of which it is the
property that one or the other muft be true, *and one only.* Let
there be three other propofitions, P, Q, and R, of which it is
alfo the property that one, and one only, muft be true. Let it
alfo be a connexion of thofe affertions, that

When X is true, P is true
When Y is true, Q is true
When Z is true, R is true

CONSEQUENCE : then it follows that

When P is true, X is true
When Q is true, Y is true
When R is true, Z is true

For, when P is true, then Q and R muft be falfe; confequently,
neither Y nor Z can be true, for then Q or R would be true.
But either X, Y, or Z muft be true, therefore X muft be true ;
or, when P is true, X is true. In a fimilar way the remaining
affertions may be proved.

Cafe 1.   If   When P is Q,   X is Z
              When P is not Q, X is not Z
It follows that   When X is Z,   P is Q
              When X is not Z, P is not Q

Cafe 2.   If   { When X is greater than Z, P is greater than Q
             { When X is equal to   Z, P is equal to   Q
             { When X is lefs than   Z, P is lefs than   Q

It follows that   { When P is greater than Q, X is greater than Z
                 { When P is equal to   Q, X is equal to   Z
                 { When P is lefs than   Q, X is lefs than   Z

CHAPTER II.

On Objects, Ideas, and Names.

LOGIC is derived from a Greek word (λόγος) which fignifies communication of thought, ufually by fpeech. It is the name which is generally given to the branch of inquiry (be it called fcience or art), in which the act of the mind in reafoning is confidered, particularly with reference to the connection of thought and language. But no definition yet given in few words has been found fatisfactory to any confiderable number of thinking perfons.

All exifting things upon this earth, which have knowledge of their own exiftence, poffefs, fome in one degree and fome in another, the power of *thought*, accompanied by *perception*, which is the awakening of thought by the effect of external objects upon the fenfes. By thought I here mean, all mental action, not only that comparatively high ftate of it which is peculiar to man, but alfo that lower degree of the fame thing which appears to be poffeffed by brutes.

With refpect to the mind, confidered as a complicated apparatus which is to be ftudied, we are not even fo well off as thofe would be who had to examine and decide upon the mechanifm of a watch, merely by obfervation of the functions of the hands, without being allowed to fee the infide. A mechanician, to whom a watch was prefented for the firft time, would be able to give a good guefs as to its ftructure, from his knowledge of other pieces of contrivance. As foon as he had examined the law of the motion of the hands, he might conceivably invent an inftrument with fimilar properties, in fifty different ways. But in the cafe of the mind, we have manifeftations only, without the fmalleft power of reference to other fimilar things, or the leaft knowledge of ftructure or procefs, other than what may be derived from thofe manifeftations. It is the problem of the watch to thofe who have never feen any mechanifm at all.

We have nothing more to do with the science of mind, usually called *metaphysics*,* than to draw a very few necessary distinctions, which, whatever names we use to denote them, are matters of fact connected with our subject. Some modes of expressing them favor one system of metaphysics, and some another; but still they are matters of observed fact. Our words must be very imperfect symbols, drawn from comparison of the manifestations of thought with those of things in corporeal existence. For instance, I just now spoke of the mind as an apparatus, or piece of mechanism. It is a structure of some sort, which has the means of fulfilling various purposes; and so far it resembles the hand, which by the disposition of bone and muscle, can be made to perform an immense variety of different motions and grasps. Where the resemblance begins to be imperfect, and why, is what we cannot know. In all probability we should need new modes of perception, other senses besides sight, hearing, and touch, in order to know thought as we know colour, size, or motion. But the purpose of the present treatise is only the examination of some of the manifestations of thinking power in their relation to the language in which they are expressed. Knowledge of thought and knowledge of the results of thought,

* All systems make an assumption of the uniformity of process in all minds, carried to an extent the propriety of which ought to be a matter of special discussion. There are no writers who give us so much *must* with so little *why*, as the metaphysicians. If persons who had only seen the outside of the timepiece, were to invent machines to answer its purpose, they might arrive at their object in very different ways. One might use the pendulum and weight, another the springs and the balance: one might discover the combination of toothed wheels, another a more complicated action of lever upon lever. Are we *sure* that there are not differences in our minds, such as the preceding instance may suggest by analogy; if so, *how* are we sure? Again, if our minds be as tables with many legs, do we know that a weight put upon different tables will be supported in the same manner in all. May not the same leg support much or all of a certain weight in one mind, and little or nothing in another? I have seen striking instances of something like this, among those who have examined for themselves the grounds of the mathematical sciences.

I would not dissuade a student from metaphysical inquiry; on the contrary, I would rather endeavour to promote the desire of entering upon such subjects: but I would warn him, when he tries to look down his own throat with a candle in his hand, to take care that he does not set his head on fire.

are very different things. The watch abovementioned might have the functions of its hands difcovered, might be ufed in finding longitude (and even latitude) all over the world, without the parties ufing it having the fmalleft idea of its interior ftructure.

That our minds, fouls, or thinking powers (ufe what name we may) exift, is the thing of all others of which we are moft certain, each for himfelf. Next to this, nothing can be more certain to us, each for himfelf, than that other things alfo exift; other minds, our own bodies, the whole world of matter. But between the character of thefe two certainties there is a vaft difference. Any one who fhould deny his own exiftence would, if ferious, be held beneath argument: he does not know the meaning of his words, or he is falfe or mad. But if the fame man fhould deny that any thing exifts except himfelf, that is, if he fhould affirm the whole creation to be a dream of his own mind, he would be abfolutely unanfwerable. If I (who *know* he is wrong, for *I* am certain of *my own* exiftence) argue with him, and reduce him to filence, it is no more than might* happen in his dream. A celebrated metaphyfician, Berkeley, maintained that with regard to *matter*, the above is the ftate of the cafe: that our impreffions of matter are only impreffions, communicated by the Creator without any intervening caufe of communication.

Our moft convincing communicable proof of the exiftence of other things, is, not the appearance of objects, but the neceffity of admitting that there are *other minds* befides our own. The external inanimate objects might be creations of our own thought, or thinking and perceptive function: they are fo fometimes, as in the cafe of infanity, in which the mind has frequently the appearance of making the whole or part of its own external world. But when we fee other beings, performing fimilar functions to thofe which we ourfelves perform, we come fo irrefiftibly to the conclufion that there muft be other fentients like ourfelves, that we fhould rather compare a perfon who doubted it to one who denied his own exiftence, than to one who fimply denied the real external exiftence of the *material* world.

---

* It is not impoffible that in a real dream of fleep, fome one may have created an antagonift who beat him in an argument to prove that he was awake.

When once we have admitted different and independent minds, the reality of external objeⱥs (external to all thofe minds) follows as of courfe. For different minds receive impreſſions at the fame time, which their power of communication enables them to know are fimilar, fo far as any impreſſions, one in each of two different minds, can be known to be fimilar. There muſt be a *fomewhat* independent of thofe minds, which thus aⱥs upon them all at once, and without any choice of their own. This *fomewhat* is what we call an external objeⱥ: and whether it arife in Berkeley's mode, or in any other, matters nothing to us here.

We ſhall then, take it for granted that external *objeⱥs* aⱥually exiſt, independently of the mind which perceives them. And this brings us to an important diſtinⱥion, which we muſt carry with us throughout the whole of this work. Befides the aⱥual external objeⱥ, there is alfo the mind which perceives it, and what (for want of better words or rather for want of knowing whether they be good words or not) we muſt call the *image of that objeⱥ in the mind,* or the *idea* which it communicates. The term *fubjeⱥ* is applied by metaphyficians to the perceiving mind: and thus it is faid that a thing may be confidered *fubjeⱥively* (with reference to what it is in the mind) or *objeⱥively* (with reference to what it is independently of any particular mind). But logicians ufe the word fubjeⱥ in another fenfe. In a propofition fuch as ' bread is wholefome', the thing fpoken of, ' bread', is called the fubjeⱥ of the propofition: and in faⱥ the word *fubjeⱥ* is in common language fo frequently confounded with *objeⱥ*, that it is àlmoſt hopelefs to fpeak clearly to beginners about themfelves as *fubjeⱥs*. I ſhall therefore adopt the words *ideal* and *objeⱥive*, *idea* and *objeⱥ*, as being, under explanations, as good as any others: and better than *fubjeⱥ* and *objeⱥ* for a work on logic.

The word *idea*, as here ufed, does not enter in that vague fenfe in which it is generally ufed, as if it were an opinion that might be right or wrong. It is that which the objeⱥ gives to the mind, or the ſtate of the mind produced by the objeⱥ. Thus the idea of a horfe is *the horfe in the mind:* and we know no other horfe. We admit that there is an external *objeⱥ*, a horfe, which may give a *horfe in the mind* to twenty different perfons: but no one of thefe twenty knows the objeⱥ; each one only knows his *idea*.

There is an object, becauſe each of the twenty perſons receives an idea without communicating with the others : ſo that there is ſomething external to give it them. But when they talk about it, under the name of a horſe, they talk about their ideas. They all refer to the object, as being the thing they are talking about, until the moment they begin to differ: and then they begin to ſpeak, not of external horſes, but of impreſſions on their minds ; at leaſt this is the caſe with thoſe who know what knowledge is ; the poſitive and the unthinking part of them ſtill talk of the *horſe*. And the latter have a great advantage* over the former with thoſe who are like themſelves.

Why then do we introduce the term *object* at all, ſince all our knowledge lies in ideas ? For the ſame reaſon as we introduce the term *matter* into natural philoſophy, when all we know is form, ſize, colour, weight, &c., no one of which is matter, nor even all together. It is convenient to have a word for that external ſource from which *ſenſible* ideas are produced : and it is juſt as convenient to have a word for the external ſource, material or not, from which *any* idea is produced. Again, why do we ſpeak of our power of conſidering things either ideally or objectively, when as we can know nothing but ideas, we can have no right to ſpeak of any thing elſe ? The anſwer is that, juſt as in other things, when we ſpeak of an object, we ſpeak of the *idea of an object*. We learn to ſpeak of the external world, becauſe there are others like ourſelves who evidently draw ideas from the ſame ſources as ourſelves : hence we come to have the idea of thoſe ſources, the idea of external objects, as we call them. But we do not know thoſe ſources ; we know only our ideas of them.

We can even uſe the terms ideal and objective in what may appear a metaphorical ſenſe. When we ſpeak of ourſelves in the manner of this chapter, we put ourſelves, as it were, in the poſition of ſpectators of our own minds : we ſpeak and think of our

---

* One man aſſerts *a fact* on his own knowledge, another aſſerts his full *conviction* of the contrary fact. Both uſe the evidence of their ſenſes: but the ſecond knows that *full conviction* is all that man can have. The firſt will carry it hollow in a court of juſtice, in which perſons are conſtantly compelled to ſwear, not only that they have an impreſſion, but that the impreſſion is correct ; that is to ſay, is the impreſſion which mankind in general would have, and muſt have, and ought to have.

own minds objectively. And it muſt be remembered that by the word object, we do not mean *material* object only. The mind of another, any one of its thoughts or feelings, any relation of minds to one another, a treaty of peace, a battle, a diſcuſſion upon a controverted queſtion, the right of conveying a freehold, —are all objects, independently of the perſons or things engaged in them. They are things external to our minds, of which we have ideas.

An object communicates an idea: but it does not follow that every idea is communicated by an object. The mind can create ideas in various ways ; or at leaſt can derive, by combinations which are not found in external exiſtence, new collections of ideas. We have a perfectly diſtinct idea of a unicorn, or a flying dragon : when we ſay there are no ſuch things, we ſpeak objectively only : ideally, they have as much exiſtence as a horſe or a ſheep; to a herald, more. Add to this, that the mind can ſeparate ideas into parts, in ſuch manner that the parts alone are not ideas of any exiſting ſeparate material objects, any more than the letters of a word are conſtituent parts of the meaning of the whole. Hence we get what are called *qualities* and *relations*. A ball· may be hard and round, or may have hardneſs and roundneſs : but we can not ſay that hardneſs and roundneſs are ſeparate external material objects, though they are objects the ideas of which neceſſarily accompany our perception of certain objects. Theſe ideas are called *abſtract* as being removed or abſtracted from the complex idea which gives them : the abſtraction is made by compariſon or obſervation of reſemblances. If a perſon had never ſeen any thing round except an apple, he would perhaps never think of roundneſs as a diſtinct object of thought. When he ſaw another round body, which was evidently not an apple, he would immediately, by perception of the reſemblance, acquire a ſeparate idea of the thing in which they reſemble one another.

Abſtraction is not performed upon the ideas of material objects only. For inſtance, from conduct of one kind, running through a number of actions, performed by a number of perſons, we get the ideas of goodneſs, wickedneſs, talent, courage. But we muſt not imagine that we can make ideally external repreſentation of theſe words. They are *objects*, that is to ſay, the mind conſiders them as external to itſelf : but they are not material objects.

Some people deny their exiftence, and look upon them as only abftract words, or words under which we fpeak of minds or bodies without fpecifying any more than one of the ideas produced by thefe minds or bodies. For inftance, they affert that when we fay ' knowledge gives power ' it ·is really that perfons with knowledge are therefore able, or have power, to produce, or to do, what perfons without it cannot. This is a queftion which it does not concern me here to difcufs.

Seeing that the mind poffeffes a power of originating new combinations of ideas, and alfo of abftracting from complex ideas the more fimple ones of which, it feems natural to fay, they are compofed, it has long been a queftion among metaphyficians whether the mind has any ideas of its own which it poffeffes independently of all fuggeftion from external objects. It is not neceffary that I fhould attempt to lead the ftudent to any conclufion* on this fubject: for our purpofe, the diftinction between ideas and objects, though it were falfe, is of more importance than that between innate and acquired ideas, though it be true. But one of thefe two things muft be true : either we have ideas which we do *not* acquire from or by means of communication with the external world (experience, trial of our fenfes) or there is a power in the mind of acquiring a certainty and a generality which experience alone could not properly give. For inftance, we are fatisfied as of our own exiftence that feven and three collected are the fame as five and five, *whatever the objects may be*

* It has always appeared to me much fuch a queftion as the following. There are hooks which certainly catch fifh if put into the water; and moft certainly they have been put into the water. There are then fifh upon them. But thefe fifh might have been on fome of them when they were put into the water. It is to no purpofe to inquire whether it was fo or not, unlefs there be fome diftinction between the fifhes which may make it a queftion whether fome of them could have been bred in the river into which the hooks were put. The mind has certainly a power of acquiring and retaining ideas, which power, when put into communication with the external world, it muft exercife. There is no mind to experiment upon, except thofe which have had fuch communication. Are there found any ideas which we have reafon to think *could* not have been acquired by this communication? any fifhes which *could* not have come out of the river? Metaphyficians feem to admit that if any ideas be innate, they are thofe of fpace, time, and of caufe and effect: they feem alfo to admit, that if there be any ideas, which, not being innate, are fure to be acquired, are thefe very ones.

which are counted: the thing is true of fingers, pebbles, counters, fheep, trees, &c. &c. &c. We cannot have affured ourfelves of this by experience : for example, we *know* it to be true of peb- bles at the North Pole, though we have never been there ; we are as fure of it as of our own exiftence. I do not mean that we have a rational conviction only, fit to act upon, that it is fo at the North Pole, becaufe it is fo in every place in which it has been tried : if we had nothing elfe, we fhould have this ; but we feel that this leffer conviction is fwallowed up by a greater. We have the leffer conviction that the pebbles at the pole fall to the ground when they are let go : we are very fure of this, without afferting that it cannot be otherwife : we fee no *impoffibility* in thofe peb- bles being fuch as always to remain in air wherever they are placed.* But that feven and three are no other than five and five is a matter which we are prepared to affirm as pofitively of the pebbles at the North Pole as of our own fingers, both that it is fo, and that it *muft* be fo. Whence arifes this actual difference in point of fact, between our mode of viewing and knowing

---

* Metaphyficians, in their fyftems, have often taken this diftinction to be one of fyftem only, treating it as a thing to be accepted or rejected with the fyftem, inftead of an actual and indifputable phenomenon which re- quires explanation under any fyftem. Dr. Whewell, of all Englifh writers on natural fcience I know, is the one who has made the fact, as a fact, per- vade his writings, fometimes attached to a fyftem, fometimes not. The following remarks on the general fubject are worth confideration : " It is indeed, extremely difficult to find, in fpeaking of this fubject, expreffions which are fatisfactory. The reality of the objects which we perceive is a profound, apparently an infoluble problem. We cannot but fuppofe that exiftence is fomething different from our knowledge of exiftence :—that what exifts, does not exift merely in our knowing that it does :—truth is truth whether we know it or not. Yet how can we conceive truth, otherwife than as fomething known ? How can we conceive things as exifting, without conceiving them as objects of perception ? Ideas and Things are conftantly oppofed, yet neceffarily coexiftent. How they are thus oppofite and yet iden- tical, is the ultimate problem of all philofophy. The fucceffive phafes of philofophy have confifted in feparating and again uniting thefe two oppofite elements ; in dwelling fometimes upon the one and fometimes upon the other, as the principal or original or only element ; and then in difcovering that fuch an account of the ftate of the cafe was infufficient. Knowledge requires ideas. Reality requires things. Ideas and things coexift. Truth *is*, and is known. But the complete explanation of thefe points appears to be beyond our reach."

different species of affertions? the truth of the laft named affertion is not born with us, for children are without it, and learn it
by experience, as we know. The *muft be fo* cannot be acquired
from experience in the common way, for that fame experience
on which we rely tells us that however often a thing may have
been found true, whatever rule may have been eftablished by repeated inftances, an exception may at laft occur. There feems
then to be in the mind a power of developing, from the ideas
which experience gives, a real and true diftinction of neceffary
and not neceffary, poffible and impoffible. The things which
are without us always confirm our *neceffary* propofitions: but
how we derive that complete affurance that they *will* do fo as
faithfully as hitherto they *have* done fo, is not within our power
to fay.

Connected with ideas are the *names* we give them; the fpoken
or written founds by which we think of them, and communicate
with others about them. To have an idea, and to make it the
fubject of thought as an idea, are two perfectly diftinct things:
the *idea of an idea* is not the idea itfelf. I doubt whether we
could have made thought itfelf the fubject of thought without
language. As it is, we give names to our ideas, meaning by a
name not merely a fingle word, but any collection of words which
conveys to one mind the idea in another. Thus a-man-in-a-
black-coat-riding-along-the high-road-on-a-bay-horfe is as much
the name of an idea as man, black, or horfe. We can coin
words at pleafure; and, were it worth while, might invent a
fingle word to ftand for the preceding phrafe.

Names are ufed indifferently, both for the objects which produce ideas, and for the ideas produced by them. This is a dif-
advantage, and it will frequently be neceffary to fpecify whether
we fpeak ideally or objectively. In common converfation we
fpeak ideally and think we fpeak objectively: we take for granted
that our own ideas are fit to pafs to others, and will convey to
them the fame ideas as the objects themfelves would have done.
That this may be the cafe, it is neceffary firft, that the object
fhould really give us the fame ideas as to others; fecondly, that
our words fhould carry from us to our correfpondents the fame
ideas as thofe which we intended to exprefs by them. How,
and in what cafes, the firft or the fecond condition is not ful-

filled, it is impoffible to know or to enumerate. But we have nothing to do here except to obferve that we are only incidentally concerned with this queftion in a work of logic. We prefume fixed and, if objective, objectively true ideas, with certain names attached : fo that it is never in doubt whether a name be or be not properly attached to any idea. This method muft be followed in all works of fcience : a conceivably attainable end is firft prefumed to be attained, and the confequences of its attainment are ftudied. Then, afterwards, comes the queftion whether this end is always attained, and if not, why. The way to mend bad roads muft come at the end, not at the beginning, of a treatife on the art of making good ones.

Every name has a reference to every idea, either affirmative or negative. The term *horfe* applies to every thing, either pofitively or negatively. This (no matter what I am fpeaking of) either *is* or *is not* a horfe. If there be any doubt about it, either the idea is not precife, or the term *horfe* is ill underftood. A name ought to be like a boundary, which clearly and undeniably either fhuts in, or fhuts out, every idea that can be fuggefted. It is the imperfection of our minds, our language, and our knowledge of external things, that this clear and undeniable inclufion or exclufion is feldom attainable, except as to ideas which are *well within the boundary :* at and near the boundary itfelf all is vague. There are decided greens and decided blues : but between the two colours there are fhades of which it muft be unfettled by univerfal agreement to which of the two colours they belong. To the eye, green paffes into blue by imperceptible gradations : our fenfes will fuggeft no place on which all agree, at which one is to end and the other to begin.

But the advance of knowledge has a tendency to fupply means of precife definition. Thus, in the inftance above cited, Wollafton and Fraunhofer have difcovered the black lines which always exift in the fpectrum of folar colours given by a glafs prifm, in the fame relative places. There are definite places in the fpectrum, by the help of which the place of any fhade of colour therein exifting may* be afcertained, and means of definition given.

When a name is complex, it frequently admits of definition,

---

* It is quite within the poffibilities of the application of fcience to the

nominal or real. A name may be faid to be *defined nominally* when we can of right fubftitute for it other terms. In fuch a cafe, a perfon may be made to know the meaning of the word without accefs to the objeĉt of which it is to give the idea. Thus, an *ifland* is completely defined in ' land furrounded by water.' In definition, we do not mean that we are neceffarily to have very precife terms in which to explain the name defined : but, as the terms of the definition fo is the name which is defined ; according as the firft are precife or vague, clear or obfcure, fo is the fecond. Thus there may be a queftion as to the meaning of *land :* is a marfh fticking up out of the water an ifland? Some will fay that, as oppofed to water, a marfh is land, others may confider marfh as intermediate between what is commonly called [dry] land and water. If there be any vaguenefs, the term ifland muft partake of it : for ifland is but fhort for ' land furrounded by water,' whether this phrafe be vague or precife. This fort of definition is *nominal*, being the fubftitution of names for names. It is complete, for it gives all that the name is to mean. An ifland, as fuch, can have nothing neceffarily belonging to it except what neceffarily belongs to ' land furrounded by water.' By *real* definition, I mean fuch an explanation of the word, be it the whole of the meaning or only part, as will be fufficient to feparate the things contained under that word from all others. Thus the following, I believe, is a complete definition of an *elephant ;* 'an animal which naturally drinks by drawing the water into its nofe, and then fpirting it into its mouth.' As it happens, the animal which does this is the elephant only, of all which are known upon the earth : fo long as this is the cafe, fo long the above definition anfwers every purpofe ; but it is far from involving all the ideas which arife from the word. Neither fagacity, nor utility, nor the produĉtion of ivory, are neceffarily conneĉted with drinking by help of the nofe. And this definition is purely objeĉtive ; we do not mean that every idea we could form of an animal fo drinking is to be called an elephant. If a new animal were to be difcovered, having the fame mode of drinking, it would be a matter of pure choice whether it fhould be called elephant or not. It

---

arts that the time fhould come when the fpeĉtrum, and the lines in it, will be ufed for matching colours in every linen-draper's fhop.

muſt then be ſettled whether it ſhall be called an elephant, and that race of animals ſhall be divided into two ſpecies, with diſtinctive definitions; or whether it ſhall have another name, and the definition above given ſhall be incomplete, as not ſerving to draw an entire diſtinction between the elephant and all other things.

It will be obſerved that the nominal definition includes the real, as ſoon as the terms of ſubſtitution are really defined : while the real definition may fall ſhort of the nominal.

When a name is clearly underſtood, by which we mean when of every object of thought we can diſtinctly ſay, this name does or does not, contain that object—we have ſaid that the name applies to everything, in one way or the other. The word man has an application both to Alexander and Bucephalus : the firſt *was* a man, the ſecond *was not.* In the formation of language, a great many names are, as to their original ſignification, of a purely negative character : thus, parallels are only lines which do *not* meet, aliens are men who are *not* Britons (that is, in our country). If language were as perfect and as copious as we could imagine it to be, we ſhould have, for every name which has a poſitive ſignification, another which merely implies all other things : thus, as we have a name for a tree, we ſhould have another to ſignify every thing that is not a tree. As it is, we have ſometimes a name for the poſitive, and none for the negative, as in *tree :* ſometimes for the negative and none for the poſitive, as in *parallels :* ſometimes for both, as in a frequent uſe of *perſon* and *thing.* In logic, it is deſirable to conſider names of incluſion with the correſponding names of excluſion : and this I intend to do to a much greater extent than is uſual : inventing names of excluſion by the prefix not, as in tree and not-tree, man and not-man. Let theſe be called *contrary,** or *contradictory,* names.

Let us take a pair of contrary names, as man and not-man. It is plain that between them they repreſent everything imaginable or real, in the univerſe. But the contraries of common language uſually embrace, not the whole univerſe, but ſome one general idea. Thus, of men, Briton and alien are contraries : every man muſt be one of the two, no man can be both. Not-Briton and alien are identical names, and ſo are not-alien and Briton.

* I intend to draw no diſtinction between theſe words.

The same may be said of integer and fraction among numbers, peer and commoner among subjects of the realm, male and female among animals, and so on. In order to express this, let us say that the whole idea under consideration is *the universe* (meaning merely the whole of which we are considering parts) and let names which have nothing in common, but which between them contain the whole idea under consideration, be called contraries *in, or with respect to, that universe*. Thus, the universe being mankind, Briton and alien are contraries, as are soldier and civilian, male and female, &c. : the universe being animal, man and brute are contraries, &c.

Names·may be represented by the letters of the alphabet: thus A, B, &c., may stand for any names we are considering, simple or complex. The contraries may be represented by not-A, not-B, &c., but I shall usually prefer to denote them by the small letters *a*, *b*, &c. Thus, everything in the universe (whatever that universe may embrace) is either A or not-A, either A or *a*, either B or *b*, &c. Nothing can be both B and *b* ; every not-B is *b*, and every not-*b* is B : and so on.

No language, as may well be supposed, has been constructed beforehand with any intention of providing for the wants of any metaphysical system. In most, it is seen that the necessity of providing for the formation of contrary terms has been obeyed. Our own language has borrowed from the Latin as well as from its parent : thus we have *imperfect*, *disagreeable*, as well as *unformed* and *witless*. There is a choice of contraries without very well settled modes of appropriation : standing for different degrees of contrariety. Thus we have *not perfect* which is not so strong a term as *imperfect*; and *not imperfect*, the contrary of a contrary, which is not so strong as *perfect*. The wants of common conversation have sometimes retained a term and allowed the contrary to sink into disuse ; sometimes retained the contrary and neglected the original term ; sometimes have even introduced the contrary without introducing any term for the original notion, and allowed no means of expressing the original notion except as the contrary of a contrary. If we could imagine a perfect language, we should suppose it would contain a mode of signifying the contrary of every name : this indeed our own language may be said to have, though sometimes in an awkward and unidiomatic manner. One inflexion, or one additional word, may ferve to fignify a contrary of any kind: thus *not man* is effective to denote all that is other than man. But there is a wider want, which can only be partially fupplied, for its complete fatisfaction would require words almoft beyond the power of arithmetic to count: and all that has been done to make it lefs confifts, in our language and in every other, moftly in the formation of compound terms, be they fubftantive and adjective, double fubftantives, or any others. A clafs of objects has a fub-clafs contained within it, the individuals of which are diftinguifhed from all others of the clafs by fomething common to them and them only. If the diftinguifhing characteriftic have been feparated, and a word formed to fignify the abftract idea, that word, or an adjective formed from it (if it be not an adjective) is joined with the general name of the clafs. Thus we have ftrong men, white horfes, &c. Or it may happen that the individuals of the fub-clafs take, in right of the diftinguifhing characteriftic, a perfectly new name, and by the moft varied rules. A corn-grinding man is called from the implement he ufes, a *miller;* a meat-killing man from the organ which he fupplies, a *butcher*, (if the firft idea of the etymology of this word be correct). Other men ufe mills and other trades feed the mouth: ftill cuftom has fettled thefe terms, though the firft is only connected with its origin by the fpelling, and the fecond by a derivation which muft be fought in another language. But again, it will more often happen that a diftinctive characteriftic, belonging to fome only, gives no diftinctive name to thofe *fome*, which ftill remain an unnamed *fome* out of the whole, to be feparated by the defcription of their characteriftic when wanted, inftead of being the *all* of a name invented to exprefs them, and them alone of their clafs. In fuch a predicament, for inftance, are men who have never feen the fea, as diftinguifhed from thofe who have feen it. Hence it appears that particular propofitions are not fo diftinct from univerfal ones in real character as they are generally made to be. If I fay ' fome As are Bs' the reader may well fuppofe that it is not often neceffary to advert to this fact: had it been fo, a name would have been invented fpecially to fignify ' As which are Bs.' If this name had been C, the propofition would have been ' *every* C is B.'

The fame convenience which dictates the formation of a name for one fub-clafs and not for another, rules in the formation of contrary terms, as already noted. And thefe caprices of language —for logically confidered they are nothing elfe, though their formation is far from lawlefs—make it defirable to include in a formal treatife the moft complete confideration of all propofitions, with reference not only to their terms, but alfo to the contraries of thofe terms. Every negative propofition is affirmative, and every affirmative is negative. Whatever completely does one of the two, include or exclude, alfo does the other. If I fay that ' no A is B,' then, *b* being the name of every thing not B in the univerfe of the propofition, I fay that ' every A is *b*:' and if I fay that ' every A is B,' I fay that ' no A is *b*.' Whether a language will happen to poffefs the name B, or *b*, or both, depends on circumftances of which logical preference is never one, except in treatifes of fcience. The Englifh may poffefs a term for B, the French only for *b* : fo that the fame idea muft be prefented in an affirmative form to an Englifhman, as in ' every A is B,' and in a negative one to a Frenchman, as ' no A is *b*.' From all this it follows that it is an accident of language whether a propofition is univerfal or particular, pofitive or negative. We, having the names A and B, may be able to fay ' every A is B :' another language, which only names the contrary of B, muft fay ' no A is *b*.' A third language, in which As have not a feparate name, but are only individuals of the clafs C, muft fay ' fome Cs are Bs ; ' while a fourth, which is in the further predicament of naming only *b*, muft have it ' fome Cs are not *b*s.' When we come to confider the fyllogifm, we fhall have full confirmation of the correctnefs and completenefs of this view.

It may be objected that the introduction of terms which are merely negations of the pofitive ideas contained in other terms is a fpecies of fiction. I anfwer, that, firft, the fiction, if it be a fiction, exifts in language, and produces its effects : nor will it eafily be proved more fictitious than the invention of founds to ftand for things. But, fecondly, there is a much more effective anfwer, which will require a little development.

When writers on logic, up to the prefent time, ufe fuch contraries as man and not-man, they mean by the alternative, man and everything elfe. There can be little effective meaning, and no use, in a classification which, because they are not men, includes in one word, *not-man*, a planet and a pin, a rock and a featherbed, bodies and ideas, wishes and things wished for. But if we remember that in many, perhaps most, propositions, the range of thought is much less extensive than the whole universe, commonly so called, we begin to find that the whole extent of a subject of discussion is, for the purpose of discussion, what I have called a *universe*, that is to say, a range of ideas which is either expressed or understood as containing the whole matter under consideration. In such universes, contraries are very common : that is, terms each of which excludes every case of the other, while both together contain the whole. And, it must be observed that the contraries of a limited universe, though it be a sufficient real definition of either that it is not the other, are frequently both of them the objects from which positive ideas are obtained. Thus, in the universe of property, personal and real are contraries, and a definition of either is a definition of the other. But though each be a negative term as compared with the other, no one will say that the idea conveyed by either is that of a mere negation. Money is *not* land, but it *is* something. And even when the contrary term is originally invented merely as a negation, it may and does acquire positive properties. Thus alien is strictly not-Briton : but suppose a man taken in arms against the crown on some spot within its dominions, and claiming to be a prisoner of war. The answer that he *is* a British subject is a negation : to establish his positive claim he first must prove himself an alien, and moreover that he is in another positive predicament, namely, that he is the subject of a power at war with Great Britain. Accordingly, of two contraries, neither must be considered as *only* the negation of the other : except when the universe in question is so wide, and the positive term so limited, that the things contained under the contrary name have nothing but the negative quality in common.

Perception of agreements and disagreements is the foundation of all *assertion :* the acquirement of such perception with respect to any two ideas by the comparison of both with a third, is the process of all *inference.* To infer, by comparison of abstract ideas, is the peculiar privilege of man ; to need inference is his imperfection. To what point man would carry inference if he wanted language, how much further the lower animals could carry what they have of it if they had language, are queftions on which it is vain to fpeculate. The words *is* and *is not*, which imply the agreement or difagreement of two ideas, muft exift, explicitly or implicitly, in every affertion. And what we call agreement or difagreement, may be reduced to identity or non-identity. When we fay John is a man, we have the firft and moft objective form of affertion. Looked at in the moft objective point of view it is only this, *John* is one of the individual objects who are called *man*. Looked at ideally, the propofition is more general. The idea of man, gathered from inftances, prefents itfelf as a collective mafs of ideas, of which we can figure to ourfelves an inftance without neceffarily calling up the idea of any man that ever ex-ifted. In the ideal conception of man, Achilles is a man as much as the Duke of Wellington, whether the former ever ex-ifted objectively or no : of all the ideas of man which the mind can imagine, the former is one as well as the latter.

The feparation of ideas, or formation of abftract ideas, and affertion by means of them, prefents nothing, for our purpofe, which differs from the former cafe. If we fay ' this picture is beautiful,' the mere phrafe is incomplete, for ' beautiful' is only an attribute, a purely ideal reference to a claffification which the mind makes, dictated by its own judgment. The picture being a material object, cannot *be* anything but an object, cannot be-long to any clafs of notions, unlefs that clafs contain objects. What the propofition may mean is to a certain extent dependent upon the implied fubftantive to which beautiful belongs : that is, to the clafs of objects which the propofition implies the mind to have feparated into beautiful and not beautiful. It may be that the picture is a beautiful picture : or a beautiful work of art, tak-ing its place in that divifion by which not only pictures, but fta-tues, buildings, reliefs, &c. are feparated into beautiful and other-wife : or a beautiful creation of human thought, placed among works of art, imagination, or fcience, &c. in the fubdivifion beau-tiful : or finally, it may be a beautiful *thing*, placed with all ob-jects of perception in a fimilar fubdivifion.

In all affertions, however, it is to be noted, once for all, that *formal logic*, the object of this treatife, deals with *names* and not with either the *ideas* or *things* to which thefe names belong. We

are concerned with the properties of 'A is B' and 'A is not B'
so far as they present an idea independently of any specification
of what A and B mean : with such ideas upon propositions as
are presented by their *forms*, and are common to all forms of the
same kind. The reality of logic is the examination of the use of
is and *is not :* the tracing of the consequences of the application
of these words. The argument ' when the sun shines it is day :
but it is not day, therefore the sun does not shine,' contains a
theory and 'two facts, the latter of which is made to follow from
the former by the theory. That inference is made is seen in the
word *therefore :* and the sentence is capable of being put upon its
trial for truth or falsehood by logical examination. But this examination rejects the meaning of sun and day, the truth of the
theory and of the facts ; and only inquires into the right which
the sentence, of its own structure, gives us to introduce the word
therefore. It merely enters upon ' when A *is*, B *is ;* but B *is
not ;* therefore A *is not :*' and decides that this is a correct junction of precedents and consequent, an exhibition of necessary connexion between what goes before and after *therefore*, and a development, in the latter, of what is virtually, though not actually,
expressed in the former. What A and B may mean is of no
consequence to the *inference*, or right to *bring in* ' A is *not.*'

Thus A and B, divested of all specific meaning, are really
names as names, independently of things : or at least may be so
considered. For the truth of the proposition, under all meanings, gives us a right to suppose, if we like, that *names are the
meanings*—that is to say, that we may put it thus, ' When the
name A is, the name B is: but the name B is not; therefore the
name A is not.

It is not therefore the object of logic to determine whether
conclusions be true or false ; but whether what are asserted to
be conclusions are *conclusions*. By a *conclusion* is meant that which
is and must be *shut in with* certain other preceding things put in
first : it is that which must have been put into a sentence because
certain other things were put in. To *infer a conclusion* is to *bring
in*, as it were, the direct statement of that which has been virtually stated already—has been *shut in*. When we say ' A is B,
B is C ; we *conclude* A is C' ; it would be more correct to say
' A is B, B is C ; we *have concluded* A is C'. We should never

think of faying ' we have put into a box a man's upper drefs of
the colour of the trees ; therefore we *muft put in* a green coat ' ;
—we fhould fay ' *we have put in.*' To infer the conclufion then
is to bring in a ftatement that we have *concluded.*

Inference does not give us more than there was before : but it
may make us *fee* more than we *faw* before : ideally fpeaking,
then, it does give us (in the mind) more than there was before.
But the homely truth that no more can come out than was in,
though accepted as to all material objects even by metaphyficians
—who are generally well pleafed to find the key of a box which
contains what they want, though fure that it will put in no more
than was there already—has been applied to logic, and even to
mathematics, in depreciation of their rank as branches of knowledge. Thofe who have made this ftrangeft of human errors
muft have affumed an ideal omnifcience, and looked at human
imperfection objectively. Omnifcience need neither compare
ideas, nor draw inferences : the conclufion which we deduce
from premifes, is always prefent *with* them ; truths are *concomitants,* not *confequences.* When we fay that one affertion *follows*
from another, we fpeak purely ideally, and defcribe an imperfection of our own minds : it is not that the confequence follows
from the premifes, but that our *perception* of the confequence follows our *perception* of the premifes : the confequence, objectively
fpeaking, is in, and with, and of, the premifes. We fpeak wrongly
if we fpeak ideally, when we fay that ' A is C,' *is* in ' A is B and
B is C' : in fact, it is only by giving an objective view to the
argument, that we can even affert that it *will be* feen. To uncultivated minds, this fimple conclufion is never concomitant
with the premifes, and only with fome difficulty a confequence.

From the certainty that a confequence may be made to come
out, which is an allegorical ufe of the word *out,* we affume a right
to declare, by the fame fort of allegory,* that it was *in.* The
premifes therefore contain the conclufion : and hence fome have
fpoken as if in ftudying how to draw the conclufion, we were
ftudying to know what we knew before. All the propofitions
of pure geometry, which multiply fo faft that it is only a fmall

* I am of opinion that it is more confiftent with analogy to fay that the
hypothefis *is contained in* its neceffary confequence, than to fay that the former contains the latter. My reafon will appear in the courfe of the work.

and isolated class even among mathematicians who know all that has been done in that science, are certainly contained in, that is necessarily deducible from, a very few simple notions. But *to be known from* these premises is very different from being *known with* them.

Another form of the assertion is that consequences are *virtually* contained in the premises, or (I suppose) *as good as* contained in the premises. Persons not spoiled by sophistry will smile when they are told that knowing two straight lines cannot enclose a space, the whole is greater than its part, &c.—they as good as knew that the three intersections of opposite sides of a hexagon inscribed in a circle must be in the same straight line. Many of my readers will learn this now for the first time : it will comfort them much to be assured, on many high authorities, that they virtually knew it ever since their childhood. They can now ponder upon the distinction, as to the state of their own minds, between virtual knowledge and absolute ignorance.

There must always be some contention as to the relative value of their knowledge between the students of the things which we can see must have been, and of the things which, for what we can see, might have been otherwise. How much of the distinction is due to our ignorance, no one can tell. In the mean time, it is of more use to point out the advantage, as things are, of studying both kinds of knowledge, than to attempt to institute a rivalry between them. Those who have undervalued the study of necessary consequences, have allowed themselves, in illustrating their argument, phrases* which taken literally, mean more perhaps than they intended.

* We might sometimes take them to mean that the study of necessary connexion in logic, mathematics, &c., is at least useless, if not pernicious. Now we should suppose, if this be what they mean, that close connexion, short of absolute necessity, must partake somewhat of the same character. If the absolute mathematical necessity that three angles of a triangle are equal to two right angles is therefore to be avoided, the study of physics, in which there are the necessities which we express by the term *laws of nature,* must do some harm. History, in which we may so often count upon the actions which motives will produce, cannot be quite faultless : and there are laws of formation in language which might as well be kept out of sight, for they act almost with the uniformity of laws of nature. True knowledge must consist in the study of the actions of madmen : that a certain man imagined

The study of logic, then, considered relatively to human knowledge, stands in as low a place as that of the humble rules of
arithmetic, with reference to the vast extent of mathematics and
their physical applications. Neither is the less important for its
lowliness : but it is not every one who can see that. Writers on
the subject frequently take a scope which entitles them to claim
for logic one of the highest places : they do not confine themselves to the connexion of premises and conclusion, but enter
upon the *periculum et commodum* of the formation of the premises
themselves. In the hands of Mr. Mill, for example (and to some
extent in those of Dr. Whateley) logic is the science of distinguishing truth from falsehood, so as both to judge the premises
and draw the conclusion, to compare name with name, not only
as to identity or difference, but in all the varied associations of
thought which arise out of this comparison.

CHAPTER III.

On the Abstract Form of the Proposition.

IN the preceding chapter, I have endeavoured to put together
such notions on the actual sources of our knowledge as may
give the reader the means of thinking upon points which any
system of logic, however restricted, must necessarily suggest.
We cannot attempt to connect our use of words with our notions
of things, without the occurrence of a great many difficulties, a
great many sources of adverse theories, and of never-ending dis-
putes. We cannot even represent phænomena, as phænomena,
except in the language of some system, and it may be of a wrong
one. The confidence which the favourers of these several theories place in their correctness is a sufficient reason for keeping
the account of the process of the understanding, so far as it can
be made an exact science, as distinct as possible from all of them :
for they differ widely, and if they agree in anything which can

himself to be Cæsar, when he might just as well have been Newton or Ne-
buchadnezzar, must be a real bit of knowledge, not virtually contained in
anything else, wholly or partially.

be diſtinctly apprehended, it is only in having names of great authority enrolled among the partiſans of every one.

In order to examine the laws of inference, of the way of diſtinctly perceiving the right to ſay 'therefore,' 'ſo that,' 'whence it muſt be,' &c., &c., in a manner which may be admitted, ſo far as we go, by all, we muſt make this ſeparation very complete. All admit propoſitions, as 'man is animal,' 'no man is faultleſs;' all are, after a little thought, agreed upon the modes of inference : but upon the import of a ſimple propoſition, there is every kind of difference. How much we mean, when we ſay 'man is animal,' and how we arrive at our meaning, is matter for volumes on different ſides of unſettled queſtions.

In order properly to examine the laws of inference, or of any thing elſe, we muſt firſt endeavour to arrive at a diſtinct abſtraction of ſo much of the idea we are concerned with, as is itſelf the precedent reaſon, if it be right ſo to ſpeak, of the law in queſtion. This is an eaſy proceſs upon familiar things. We do not give the carriers of goods much credit for profundity, in ſeeing that, on a given road, there is only the difference of weight by which they are concerned to know how one parcel differs from another; and further that, as long as they have to carry a pound, it matters nothing whether it be of ſugar or iron. It is this proceſs which we want to perform to the utmoſt, upon the ſimple propoſition. Writers on logic, from Ariſtotle downwards, have made a large and important ſtep in ſubſtituting for ſpecific names, with all their ſuggeſtions about them, the mere letters of the alphabet, A, B, C, &c. Theſe letters are *ſymbols*, and *general* ſymbols : each of them ſtands for any one we pleaſe of its claſs. But what are they ſymbols of, names, ideas,* or the objects which give thoſe ideas ? The anſwer is, that this is preciſely one of thoſe conſiderations which we may leave behind, in abſtracting what is neceſſary to an examination of the laws of inference. The only condition is, that we are to confine ourſelves to one or the other. When we ſay man is animal, it may be that the name man is contained in the name animal, that the idea of man is contained in that of animal, or that the object man is in the object animal. Or if there were twenty more different appropriations of the

* Meaning of courſe (page 30) ideas of ideas, and ideas of objects.

symbols, the same thing might be said of each. This is, I believe, the first use of the general symbol in order of time; the algebraical use of letter or other symbol, to designate number, being both subsequent and derived.

When therefore we say 'Every X is Y', we understand that X is a symbol which represents an instance of a name, idea, object, &c., as the case may be. There may be more or fewer of such instances; they may be numerable or innumerable. And the same of Y. The language of logicians has generally been unfavourable to the distinct perception of their terms being distributively applicable to classes of instances. They have rather been *quantitative* than *quantuplicitative:* expressing themselves as if, in saying that animal is a larger or wider term than man, they would rather draw their language from the idea of two areas, one of which is larger than the other, than from two collections of indivisible units, one of which is in number more than the other. They have even carried this so far as to make it doubtful, except from context, whether their distinction between universal and particular is that of *all* and *some,* or of *the whole* and *part.* If their instances had been *white squares,* their ' all A is B' and ' some A is B' might have applied as well to ' All the square is white' and ' Some of the square is white' as to ' All the squares are white' and ' Some of the squares are white.' I shall take particular care to use numerical language, as distinguished from magnitudinal, throughout this work, introducing of course, the plurals Xs, Ys, Zs, &c.

I may mention here another mode of speaking, which will, I think, appear objectionable to all who are much used to consideration of quantity. When a compound idea contains two or more simpler ones, some logicians have spoken as if the combination were legitimately represented by *arithmetical addition.* Thus the combination of the ideas of *animal* and *rational* must give the idea of *man :* for the two notions co-exist in nothing else that we know of. Accordingly, some write *animal + rational = man.* If this be intended as an abstraction of the notation of arithmetic, for the purpose of fitting to it entirely different meaning, there is of course no objection which I need consider here: but it seems to me that more is meant, and that those who have used this notation imagine a great resemblance between *combining* ideas, and *cumulating* them. What the difference is, I cannot pretend to fay, any more than I can pretend to fay what the difference is between chemically combining volumes of oxygen and hydrogen, fo as to produce water, and fimple cumulation of them in the fame veffel, fo as to produce a mixed gas : every beginner knows that the electric fpark, or fome other inexplicable agency, is neceffary to turn the mixed gas into a new chemical combination. But that the difference exifts in the former cafe alfo, feems to me as clear as any thing I can imagine. Even in chemiftry the cumulative notation, which was once thought an all-fufficient mode of expreffing the refults of the atomic theory, has failed with the progrefs of knowledge. To a confiderable extent, the introduction of modes of cumulation as yet anfwers the purpofe : but there ftill remain *ifomeric* compounds, differing in properties, but of the fame compofition. For example, the tartaric and racemic acids: of which Profeffor Graham fays (*Elements of Chemiftry* p. 158), "A nearer approach to identity could fcarcely be conceived than is exhibited by thefe bodies, which are, indeed, the fame both in form and compofition. . . . . . But by no treatment can the one acid be tranfmuted into the other." If the above mode of confounding cumulation and combination be admiffible, I fuppofe we might eafily give ourfelves a right to fay that

$$2 + 2 + \text{addition} = 4$$

an equation at which the mathematician would ftare.

So much for the characteriftics of the *terms* of a propofition, as wanted for the abftract forms of inference. It remains to confider thofe of the connecting copulæ *is* and *is not*.

The complete attempt to deal with the term *is* would go to the form and matter of every thing in *exiftence*, at leaft, if not to the poffible form and matter of all that does not exift, but might. As far as it could be done, it would give the grand Cyclopædia, and its yearly fupplement would be the hiftory of the human race for the time. That logic exifts as a treated fcience, arifes from the characteriftics of the word, requifite to be abftracted in ftudying inference, being few and eafily apprehended. It may be ufed in many fenfes, all having a common property. Names, ideas, and objects, require it in three different fenfes. Speak of *names*, and fay ' man *is* animal': the *is* is here an *is* of applicability ; to

E

whatsoever (idea, object, &c.) man is a name to be applied, to that same (idea, object, &c.) animal is a name to be applied. As to ideas, the *is* is an *is* of possession of all essential characteristics; *man* is an idea which possesses, contains, presents, all that is constitutive of the idea *animal*. As to absolute external objects, the *is* is an *is* of identity, the most common and positive use of the word. Every man *is* one of the animals ; touch him, you touch an animal, destroy him, you destroy an animal.

These senses are not all interchangeable. Take the *is* of identity, and the name *man* is not, as a name, the name *animal :* the idea man is not, as an idea, the idea animal. Now we must ask, what common property is possessed by each of these three notions of *is*, on which the common laws of inference depend. Common laws of inference there certainly are. If the applicability of the name A be always accompanied by that of B, and that of B by that of C, then that of A is always accompanied by that of C. If the idea A contain all that is essential to the idea B, and B all that is essential to C, then A contains all that is essential to C. If the object A be actually the object B, and if B be actually C, then A is actually C.

The following are the characteristics of the word *is* which, existing in any proposed meaning of it, make that meaning satisfy the requirements of logicians when they lay down the proposition ' A is B. To make the statement distinct, let the proposition be doubly singular, or refer to one instance of each, one A and one B : let it be ' this one A is this one B.'

First, the double singular proposition above mentioned, and every such double-singular, must be indifferent to conversion : the ' A is B,' and the ' B is A' must have the same meaning, and be both true or both false.

Secondly, the connexion *is*, existing between one term and each of two others, must therefore exist between those two others ; so ' A is B' and ' A is C' must give ' B is C.'

Thirdly, the essential distinction of the term *is not* is merely that *is* and *is not* are *contradictory alternatives*, one must, both cannot, be true.

Every connexion which can be invented and signified by the terms *is* and *is not*, so as to satisfy these three conditions, makes all the rules of logic true. No doubt absolute identity was the sug-

gesting connexion from which all the others arose : just as arithmetic was the medium in which the forms and laws of algebra
were suggested. But, as now we *invent algebras* by abstract-
ing the forms and laws of operation, and fitting new meanings to
them, so we have power to invent new meanings for all the
forms of inference, in every way in which we have power to
make meanings of *is* and *is not* which satisfy the above conditions. For instance, let X, Y, Z, each be the symbol attached
to every instance of a class of *material* objects, let *is* placed between two, as in ' X *is* Y ' mean that the two are tied together,
say by a cord, and let X be considered as tied to Z when it is
tied to Y which is tied to Z, &c. There is no syllogism but
what remains true under these meanings. Thus

The syllogism	Is true in the sense
Every X is Y	Every X is tied to a Y
Some Zs are not Ys	Some Zs are not tied to Ys
∴ Some Zs are not Xs	∴ Some Zs are not tied to Xs

This last instance might be considered as a material representation of attachment together of ideas in the mind.

We must distinctly observe that it is not every case of inference which demands all the characteristics to be satisfied. Thus
in the most common case of all, ' Every A is B, every B is C,
therefore every A is C,' of all the three conditions only the second is wanted to secure the validity of this case. Though it be seldom
thought worth while to make this observation, yet it is universal practice to act upon it, and so as to introduce into formal
logic apparent contradictions of its own rules. For example, the
following are allowed to pass for syllogisms, in the ordinary definition of that word.

' Every A is greater than some one B ; every B is greater than
some one C, therefore every A is greater than some one C.'
And the same when instead of *greater than* is read *equal to* or
less than. The form which most commonly appears is the
pair of doubly singular propositions, ' A (one thing) is greater
than B ; B is greater than C ; therefore A is greater than C.'
Here ' greater than greater' is ' greater,' the second rule is satisfied,
and no other is wanted. But this meaning for *is* (or this substitute for it, if the reader like it better) will not satisfy all the con-

ditions, and therefore will not apply to all the forms of inference.

But *is* in the sense ʿis equal toʾ *does* satisfy all the conditions. This sense of *is*, namely agreement in magnitude, is the copula of the mathematician's syllogism, when he is reasoning on quantity only.

It will probably be affirmed that the generalization thus made, or shown to be possible, in the conception of the word *is* for purposes of inference, amounts only to a very frequent, if not most usual, use of the word, namely, as signifying a certain mode, not of identity, but of agreement in quality. As when we say ʿthese two things are the same—in colourʾ or ʿthe one thing *is* the other—in colour :ʾ that the name man *is* the name animal, in a certain respect, namely, in what the latter can be applied to : that the idea man is animal, in both possessing certain characteristics : that every object man is an object animal, in actual substance : that A is B in magnitude,—when we say A equals B ; and so on. But I admit only the converse, namely, that all these uses satisfy the conditions. It would hardly be for any one to say, that every possible use of *is* which satisfies three such simple requirements, has been or can be exhausted. Even the material example which was just now given, cannot be identified with any common use, or easily imaginable one, of the common verb. But if no invented meaning, proper to satisfy the conditions, can be found, other than already exists in more or less of use, still, these conditions are the laws to which the word must submit in its logical acceptation.

There are common uses of the word which are not admitted in logic : and among them, one of the most common, connection of an object with its quality, and of an idea with one of its constituent or associated ideas. As when we say, the rose *is* red, prudence *is* desirable. Here the logical conditions are not satisfied. For example, ʿ red is the rose,ʾ though a poetical inversion of the first assertion, is not logically true. It is usual to consider such propositions, in logic, as elliptical ; thus ʿthe rose is redʾ is considered as ʿthe rose is a red object, or an object of red colour ;ʾ in which the *is* now takes one of the senses which allows of conversion. Similarly, in all other cases, the subject and predicate are made to take the same character ; both names, both ideas, or both objects. This reduction renders unnecessary both the study of the varieties of meaning of the word is (meaning varieties out of the pale of the conditions above enumerated), and alfo that of the tranfitions of meaning within the circle of which the inference remains good.

The moft common ufes of the verb are ;—firft abfolute identity, as in 'the thing he fold you *is* the one I fold him :' fecondly, agreement in a certain particular or particulars underftood, as in 'He is a negro' faid of a European in reference to his colour : thirdly, poffeffion of a quality, as in 'the rofe is red :' fourthly, reference of a fpecies to its genus, as in 'man is an animal.' All thefe ufes are independent of the ufe of the verb alone, denoting exiftence, as in 'man is [i. e. exifts].' In all thefe fenfes, and in all which might be added confiftently with the conditions in page 50, fome propofitions fometimes admit of having the fenfe of *is* fhifted, and fome do not. Thus, in negative propofitions, the *is* of agreement in particulars may be lawfully converted into that of identity : if 'No A is B in colour,' then abfolutely 'No A is B.' But 'Every A is B' in colour, does not prove 'Every A is B. But the firft pair might be connected by a fyllogifm.

The *is* of agreement in particulars may always be reduced to the *is* of identity, by alteration of the predicate ; thus 'Every A is B in colour' is 'Every A is a thing having the colour of one of the Bs.' When a fyllogifm has a negative conclufion, and the middle term is, or can be made, the predicate of both premifes, then the whole fyllogifm can be transformed from one in which there is only the *is* of agreement to one in which there is no *is* but that of identity. For example, fuppofe the premifes to be 'No X is Y (in colour) ; every Z is Y (in colour),' not meaning neceffarily that all the Ys are of one colour, but reading it as 'No X is of the colour of any one of the Ys ; every Z is of the colour of one of the Ys.' The conclufion is that 'no Z is X (in colour),' or 'no Z is of the colour of any one of the Xs.' But from this it follows that no Z is X, for if any one Z were abfolutely X, it would have * the colour of that X. This

* The reader muft not paint any of the letters during the procefs. The fenfe in which we fay a door *is* the fame door as before, after it has been painted of a different colour, is not the fenfe of logical identity : it is the fame in all but colour and colouring matter ; and the *is* is one of agreement. Except as a joke in fufficient anfwer to a captious objection or a trap, no

laſt concluſion can be brought directly from altered premiſes: thus, *is* being that of identity, we have ' No X is [a thing having the colour of one of the Ys] ; every Z is [a. thing having the colour of one of the Ys]; therefore no Z is X.' But ſuppoſe we take the following premiſes, ' Some Ys are not Xs (in colour) ; every Y is Z (in colour).' From this it follows that ſome Zs are not Xs (in colour), and thence that ſome Zs *are not* Xs. But we cannot now alter the premiſes, ſo as to produce the laſt concluſion from X, Z, and a middle term.

CHAPTER IV.

On Propositions.

A NAME is a ſymbol which is attached to one or more objects of thought, on account of ſome reſemblance, or community of properties. Or elſe it is a ſymbol attached to ſome one or more objects of thought, to diſtinguiſh them from others having the ſame properties. Objects of the ſame name are, ſo far as that name is concerned, undiſtinguiſhable. And one object may have many names, as being one in each of many claſſes of objects of thought.

Names, as explained in chapter II, are excluſively the objects of formal logic. The identity and difference of things is deſcribed by aſſerting the right to aſſert, or the right to deny, the application of names. And names, whether ſimple or complex, will be repreſented by letters of the alphabet, as X, Y, Z.

A propoſition is the aſſertion of agreement, more or leſs, or diſagreement, more or leſs, between two names. It expreſſes that of the objects of thought called Xs, there are ſome which are, or are not, found among the objects of thought called Ys:

change whatever muſt take place in the terms of concluſion, during inference. The American calculating boy, Zerah Colburn, was aſked how many black beans it would take to make ten white ones; to which he very properly anſwered ' Ten, if you ſkin 'em : but the ten ſkinned beans would not be the *ſame beans* as before: except, indeed, to thoſe to whom black is white.

that there are objects which have both names, or which have one but not the other, or which have neither.

For the moft part, the objects of thought which enter into a propofition are fuppofed to be taken, not from the whole univerfe of poffible objects, but from fome more definite collection of them. Thus when we fay " All animals require air," or that the name *requiring air* belongs to every thing to which the name *animal* belongs, we fhould underftand that we are fpeaking of things on this earth : the planets, &c., of which we know nothing, not being included. By the *univerfe* of a propofition, I mean the whole range of names in which it is expreffed or underftood that the names in the propofition are found. If there be no fuch expreffion nor underftanding, then the univerfe of the propofition is the whole range of poffible names. If, the univerfe being the name U, we have a right to fay ' every X is Y,' then we can only extend the univerfe fo as to make it include all poffible names, by faying ' Every X which is U is one of the Ys which are Us,' or fomething equivalent.

Contrary names, with reference to any one univerfe, are thofe which cannot both apply at once, but one or other of which always applies. Thus, the univerfe being *man*, *Briton* and *alien* are contraries ; the univerfe being *property*, *real* and *perfonal* are contraries. Names which are contraries in one univerfe, are not neceffarily fo in a larger one. Thus in geometry, when the univerfe is *one plane*, pairs of ftraight lines are either parallels or interfectors, and never both : parallels and interfectors are then contraries. But when the ftudent comes to folid geometry, in which *all fpace* is the univerfe, there are lines which are neither parallels nor interfectors ; and thefe words are then not contraries. But names which are contraries in the larger and containing univerfe, are neceffarily contraries in the fmaller and contained, unlefs the fmaller univerfe abfolutely exclude one name, and then the other name is *the* univerfe.

In future, I always underftand fome one univerfe as being that in which all names ufed are wholly contained : and alfo (which it is very important to bear in mind) that no one name mentioned in a propofition fills this univerfe, or applies to everything in it. Nothing is more eafy than to treat the fuppofition of a name being the univerfe as an extreme cafe. And I fhall denote contraries by large and small letters: thus, X being a name, x is the contrary name. And everything (in the universe understood) is either X or x: and nothing is both.

A proposition may be either *simple* and *incomplete*, or *complex* and *complete.* The simple proposition only asserts that Xs are Ys, or are not Ys: the complex proposition, which always consists of two simple ones, disposes in one manner or the other of every X and every Y. Thus ' Every X is Y' is a simple proposition: but it forms a part of two complex propositions. It may belong either to ' every X is Y and every Y is X,' or to ' Every X is Y and some Ys are not Xs.'

The propositions advanced in common life are usually complex, with one simple proposition expressed and one understood: but books of logic have hitherto considered only the simple proposition. And this last should be considered before the complex form.

The simple proposition must be considered with respect to *sign*, *relative quantity*, and *order*.

Simple propositions are of two *signs: affirmative* and *negative.* It is either ' Xs are Ys,' or ' Xs are not Ys.' The phrases *are* and *are not*, or *is* and *is not*, which mark the distinction, are called *copulæ.*

The *relative quantity* of a proposition has reference to the numbers of instances of the different names which enter it. The distinctions of quantity usually recognized are *all* and *some*:* leading to the distinction of *universal* and *particular.* Thus ' Every X is Y' and ' Every X is not Y' are the universal affirmative and negative propositions: the latter is usually stated as ' No X is Y.' And ' some Xs are Ys' and ' some Xs are not Ys' are the particular affirmative and negative propositions. And when the propositions are reduced strictly to these four forms,

---

* *Some*, in logic, means *one or more, it may be all.* He who says that *some are* , is not to be held to mean that *the rest are not.* ' Some men breathe,' ' some horses are distinguishable by shape from their riders' would be held false in common language. The reason is, as above noted, that common language usually adopts the complex particular proposition, and implies that some are not in saying that some are. The student cannot be too careful to remember this distinction. A particular proposition is only a may be particular.'

the firſt named, X, is called the *ſubjeƈt*, and the ſecond named, Y, the *predicate*.

It has been propoſed to conſider the univerſal propoſitions as *definite* with reſpeƈt to *quantity*. but this is not quite correƈt. The phraſe ʻ all Xs are Ys' does not tell us how many Xs there are, but that, be the unknown number of Xs in *exiſtence* what it may, the unknown number mentioned in the propoſition is the ſame. That which is definite is the *ratio* of the number of Xs of the propoſition to the Xs of the univerſe. So under-ſtood, however, the ʻ definite quantity,' as an abbreviation, may be ſaid to belong to univerſals. And the indefiniteneſs of the parti-cular propoſition is only hypothetical. It is in our power to ſup-poſe the *ſome* to be one half of the whole, or two-thirds, or any other fraƈtion.

The quantity of the ſubjeƈt is expreſſed; that of the predicate, though not expreſſed, is neceſſarily implied by the meaning of language. The predicate of an *affirmative* is *particular :* the predicate of a *negative* is *univerſal.* If I ſay ʻ Xs are Ys,' even though I ſpeak of all the Xs, I only really ſpeak of ſo many Ys as are compared with Xs and found to agree: and theſe need not be all the Ys. ʻ Every horſe is an animal,' declares that ſo many horſes as there are to ſpeak of, ſo many animals are ſpoken of: and leaves it wholly unſettled whether there be or be not more animals left. But if I ſhould ſay ʻ Xs are not Ys,' though it ſhould be only one X, as in ʻ this X is not a Y,' yet I ſpeak of every Y which exiſts. The aſſertion is ʻ this X is not any one whatſoever of all the Ys in exiſtence.' A perſon who ſhould wiſh to verify by aƈtual inſpeƈtion, ʻ theſe 20 Xs are Ys' might perchance, be enabled to affirm the reſult upon the examination of only 20 Ys, if he came firſt upon the right ones. But he could not verify ʻ this one X is not a Y' until he had examined every Y in exiſtence. This is the common doƈtrine, but though admitting of courſe that the affirmative propoſition only enables us to infer of ſome inſtances of the predicate, yet I think it more correƈt to ſay that the predicate itſelf is ſpoken of univerſally, but *indiviſibly*, and that in the negative propoſition the predicate is ſpoken of univerſally *and diviſibly.* ʻ Some Xs are Ys' tells us that each X mentioned is *either* the firſt Y, *or* the ſecond Y, *or* the third Y, &c., no Y being excluded from compariſon. But

'Some Xs are not Ys' tells us that each X mentioned is abfolutely *not* the firft Y, *nor* the fecond, *nor* the third, &c; is not,
in fact, any one of all the Ys. Still, however, the predicate of
an affirmative yields no more than it would do if the Ys finally
accepted as Xs were fpecially feparated, and confidered as the
only Ys fpoken of.

The relation of the univerfal quantity to the whole quantity of
inftances in exiftence is *definite*, being that whole quantity itfelf.
But the particular quantity is wholly *indefinite:* 'Some Xs are
Ys' gives no clue to the fraction of all the Xs fpoken of, nor to
the fraction which they make of all the Ys. Common language
makes a certain conventional approach to definitenefs, which has
been thrown away in works of logic. 'Some,' ufually means a
rather fmall fraction of the whole; a larger fraction would be
expreffed by 'a good many'; and fomewhat more than half by
'moft'; while a ftill larger proportion would be 'a great majority' or 'nearly all'. A perfectly *definite particular*, as to quantity, would exprefs how many Xs are in exiftence, how many
Ys, and how many of the Xs are or are not Ys: as in '70 out
of the 100 Xs are among the 200 Ys'. In this chapter I fhall
treat only the *indefinite particular*, leaving the *definite particular*
for future confideration.

The *order* of a propofition has relation to the choice of fubject and predicate. Thus 'Every X is Y' and 'every Y is X'
though both eftablifh a univerfal affirmative relation between X
and Y, yet are in fact two different propofitions. They are called
converfe forms. When the fubject and predicate are of the fame
fort of quantity, both univerfal or both particular, the converfe
forms give the fame propofition. Thus 'No X is Y' and 'No
Y is X' are the fame; neither has any meaning, except perhaps
of emphafis, which the other has not. And 'Some Xs are Ys'
is the fame as 'Some Ys are Xs'. The univerfal negative, then,
in which both terms are univerfal, and the particular affirmative,
in which both are particular—are *neceffarily convertible propofitions.* But the univerfal affirmative, in which the fubject is univerfal and the predicate particular, and the particular negative, in
which the fubject is particular, and the predicate univerfal—are
not neceffarily convertible, and are generally called *inconvertible.*
They *may* be convertible, in one cafe, and inconvertible in an-

other. But the term *inconvertible* is not incorrect, for the following reason.

The agreements and disagreements which are treated in logic are of this character; there can only be agreement with one, but there may be disagreement with all. If 'this X be a Y' it is one Y only: it is 'this X is *either* the first Y, *or* the second Y, *or* the third Y, &c. If there be 100 Ys, there is, to those who can know it, 99 times as much negation as affirmation in the proposition: and yet most-assuredly it is properly called *affirmative.* But if it be 'this X is not a Y,' we have 'this X is not the first Y, *and* it is not the second Y, *and* it is not the third Y, &c.' The affirmation is what is commonly called disjunctive, the negation conjunctive. A disjunctive negation would be no proposition at all, except that one and the same thing cannot be two different things: any X is either not the first Y *or* not the second Y. And in like manner a conjunctive affirmation would be an impossibility: it would state that one thing is two or more different things.

We must be prepared, then, to consider cases of opposition in which on the one side there is fixed necessity, and on the other side possibility of alternatives: and we must be prepared to denote these by opposite terms, which, looking to etymology only, denote fixed necessities of opposite characters. This happens in the case above: *convertible* means absolutely and necessarily convertible, inconvertible means *convertible or inconvertible as the case may be.* Taking the four forms of one order, we find that each of the universals cannot exist with either proposition of opposite form. Thus 'Every X is Y' cannot be true if either 'No X is Y' or 'Some Xs are not Ys:' while 'No X is Y' cannot be true if either 'Every X is Y' or 'Some Xs are Ys.' But each of the particulars is necessarily inconsistent with nothing but the universal of opposite form. That 'Some Xs are Ys' cannot be true if 'No X is Y' but it may be true if 'Some Xs are not Ys.' And 'Some Xs are not Ys' cannot be true if 'Every X is Y,' but it may be true though 'Some Xs are Ys.'

The pair 'Every X is Y' and 'some Xs are not Ys' are called *contradictory:* and so are the pair 'No X is Y' and 'Some Xs are Ys.' Of each pair of contradictories, one must be true and

one muſt be falſe : ſo that the affirmation of either is the denial
of the other, and the denial of either is the affirmation of the
other.   The pair ' Every X is Y ' and ' No X is Y ' are uſually
called *contraries;* contrariety implying the utmoſt extreme of
contradiction.   Contraries may both be falſe, but cannot both be
true.   The pair ' Some Xs are Ys,' and ' Some Xs are not Ys,'
which may both be true, but cannot both be falſe, are uſually
called *ſubcontraries.*   But, for reaſons hereafter to be given, I
intend to abandon the diſtinction between the words *contrary*
and *contradictory*, and to treat them as ſynonymous.   And the
propoſitions uſually called *contraries*, ' Every X is Y ' and ' No
X is Y ' I ſhall call *ſubcontraries :* while thoſe uſually called *ſubcontraries* ' Some Xs are Ys ' and ' Some Xs are not Ys ' I ſhall
call *ſupercontraries.*

I ſhall now proceed to an enlarged view of the propoſition,
and to the ſtructure of a notation proper to repreſent its different
caſes.

As uſual, let the univerſal affirmative be denoted by A, the particular affirmative by I, the univerſal negative by E, and the particular negative by O.   This is the extent of the common ſym-
bolic expreſſion of propoſitions : I propoſe to make the following
additions for this work.   Let one particular choice of order, as
to ſubject and predicate, be ſuppoſed eſtabliſhed as a ſtandard of
reference.   As to the letters X, Y, Z, let the order be always
that of the alphabet, XY, YZ, XZ.   Let x, y, z, be the contrary names of X, Y, Z ; and let the ſame order be adopted
in the ſtandard of reference.   Let the four forms, when choice
is made out of X, Y, Z, be denoted by A_1, E_1, I_1, O_1 ; but when
the choice is made from the contraries, let them be denoted by
A', E', I', O'.   Thus, with reference to Y and Z, ' Every Y is
Z ' is the A_1 of that pair and order : while ' Every y is z ' is the
A'.   I ſhould recommend A_1 and A to be called the *ſub*-A and
the *ſuper*-A of the pair and order in queſtion : the helps which
this will give the memory will preſently be very apparent.   And
the ſame of I_1 and I', &c.

Let the following abbreviations be employed ;—

X)Y means ' Every X is Y '   X.Y means ' No X is Y '
X:Y   —   ' Some Xs are not Ys ' | XY   —   ' Some Xs are Ys '

There are eight diftin&t modes, independent of contraries, in which a fimple propofition may be made by means of X and Y. Thefe eight modes are X)Y and Y)X, X:Y and Y:X, X.Y and Y.X, and XY and YX. But the eight are equivalent only to fix: for X.Y and Y.X are the fame, and fo are XY and YX. Again, there are fix fimple propofitions between x and y, fix between X and y, fix between x and Y. Taking in contraries, there are then twenty-four apparent modes of forming a fimple propofition from X and Y: but thefe are not all diftinct. Eight of them contain all the reft: thefe eight being the A_i, E_i, I_i, O_i, A', E', I', O', above defcribed. This is feen in the following table, the ftudy of which fhould be carefully made,

A_i X)Y = X.y = y)x			A' x)y = x.Y = Y)X		
O_i X:Y = Xy = y:x			O' x:y = xY = Y:X		
E_i X.Y = X)y = Y)x			E' x.y = x)Y = y)X		
I_i XY = X:y = Y:x			I' xy = x:Y = y:X		

I fuppofe moft readers will readily fee the truth of the identities here affirmed: if not, the following mode of illuftration (which will be very ufeful when I come to treat of the fyllogifm) may be tried. Let U be the name which is the univerfe of the propo- fition: and write down in a line as many Us as there are diftinct objects to which this name applies. A dozen will do as well for illuftration as a million. Under every U which is an X write down X: and x, of courfe, under all the reft. Follow the fame plan with Y. The occurrence of letters in the fame column fhows that they are names of the fame object. The following are fpecimens of the eight ftandard varieties of affertion, to which all the reft may be referred.

A_i UUUUUUUUUUUU
XXXXX x x x x x x x
YYYYYYYY y y y y

A' UUUUUUUUUUUU
XXXXXXXX x x x x
YYYYY y y y y y y y

O_i ⎫ UUUUUUUUUUUU
⎬ XXXXXXX x x x x x
I_i ⎭ y y y y YYYYYY y y

O' ⎫ UUUUUUUUUUUU
⎬ XXXXX x x x x x x x
I' ⎭ YY y y y y y y YYYY

E_i UUUUUUUUUUUU
XXXX x x x x x x x x
y y y y y y y YYYYY

E' UUUUUUUUUUUU
XXXXXXXX x x x x
y y y y y YYYYYYYY

In the firſt ſcheme, A_1, there exiſt twelve Us, the firſt five of which are both Xs and Ys, the next three Ys but not Xs, the laſt four neither Xs nor Ys. This caſe, ſo conſtructed that X̣)Y is true, ſhòws X.y and y)x.

The propoſitions A_1 and A', X)Y and x)y, may be called *contranominal*, as having each names contrary of thoſe in the other. It appears, then, that as to inconvertibles, contranominal and converſe are terms of the ſame meaning, for X)Y and y)x are the ſame, and x:y and Y:X. And ſince it is more natural to ſpeak of direct names than of their contraries, it will be beſt to attach to A' and O' the ideas of Y)X and Y:X ; but not ſo as to forget their derivation from x)y and x:y. Obſerve alſo that each univerſal propoſition has converted contranominals for its *affirmative* forms. Thus X)Y $=$ y)x : and though X.Y is not y.x, yet if we make X.Y take the affirmative form X)y, it is equivalent to Y)x. In particular propoſitions, the negative forms have the ſame property. The contranominals of the convertible propoſitions E_1 and I are of totally different meaning. They have never till now been introduced into logic, and a few words of explanation are wanted.

Firſt as to I' or xy. We here expreſs that ſome not-Xs are not-Ys, or that there are things in the univerſe which are neither Xs nor Ys. That is, X and Y are not contraries. Next as to E' or x.y. We here expreſs that no not-X is not-Y, or that everything in the univerſe is either X or Y, *or both*. Theſe laſt words are important : by omitting them, we ſhould imagine that x.y ſignifies that X and Y are contraries ; which is not neceſſarily true.

Accordingly, the eight ſtandard forms of expreſſion, with reference to the order $\overline{XY}$, and exhibited in the form in which it will be moſt convenient to think and ſpeak of them, are as follows,

A_1 or X)Y Every X is Y	A' or Y)X Every Y is X	
O_1 or X:Y Some Xs are not Ys	O' or Y:X Some Ys are not Xs	
E_1 or X.Y No X is Y	E' or x.y Everything is either X or Y	
I_1 or XY Some Xs are Ys	I' or xy Some things are neither Xs nor Ys.	

Returning to the table, we now ſee the following general laws.

1. Each triad of equivalents contains two inconvertibles and one convertible. 2. Of the four, X, Y, x, y, each of the eight forms

ſpeaks univerſally of two, and particularly of two. 3. A propoſition ſpeaks in different ways of each name and its contrary; univerſally of one and particularly of the other. 4. The propoſitions called *contradictory*, from the common meaning of this word, may be ſo called in another ſenſe: for they ſpeak in the ſame manner of contraries. Thus X)Y ſpeaks univerſally of X, and particularly of Y: its denial, X:Y or y:x, ſpeaks univerſally of x, and particularly of y.

Any two of the eight forms being taken, it is clear either that they cannot exiſt together, or that one muſt exiſt when the other exiſts, or that one may exiſt either with or without the other. The alternatives of each caſe are preſented in the following table.

	Denies	Con-tains	Is indif-ferent to			Denies	Is con-tained in	Is indif-ferent to
A_1	O_1E_1E'	I_1I'	$A'O'$		O_1	A_1	E_1E'	$A'O'I_1I'$
A'	$O'E'E_1$	$I'I_1$	A_1O_1		O'	A'	$E'E_1$	$A_1O_1I'I_1$
E_1	I_1A_1A'	O_1O'	$E'I'$		I_1	E_1	A_1A'	$E'I'O_1O'$
E'	$I'A'A_1$	$O'O_1$	E_1I_1		I'	E'	$A'A_1$	$E_1I_1O'O_1$

Let the *concomitants* of a propoſition be thoſe to which it is wholly indifferent. Then it appears that each univerſal has for concomitants its contranominal and the contradictory of the laſt: but each particular has all for concomitants except only its own contradictory. Each univerſal denies, beſides its own contradictory, the two univerſals of oppoſite name; and contains the two particulars of the ſame name. The two concomitants of a univerſal may be deſcribed as its univerſal and its particular concomitant.

There is a certain ſort of repetition in our choice of the four forms, combined with the four ſelections XY, Xy, xy, xY. If *any one* of the four forms A_1 E_1 A E' be applied to all the above, it will give the four forms derived from XY. Thus the A_1 of XY, Xy, xy, xY, are ſeverally the A_1, E_1, A' and E' of XY; and the E' of XY, Xy, xy, and xY are ſeverally the E', A', E_1, and A_1 of XY: and ſo on. It will ſerve for exerciſe to verify the above, and ſtill more the caſes contained in the following.

There are four things in a propoſition, each of which may be changed into its contrary: ſubject, predicate, order, and copula. Let S be the direction to change the ſubject into its contrary: P

the fame for the predicate : let T be the direction to transform
the order : and F the direction to change the form, from affirmative to negative, or from negative to affirmative. When T enters,
let it be done laft, to avoid confufion. Thus SPT performed
upon X)Y gives x)Y from S, x)y, from P, and y)x from T;
which is X)Y, fo that in this cafe alteration of fubject, predicate,
and order, is no alteration at all. Let L be the reprefentation of
no alteration at all. To inveftigate equivalent alterations, obferve, firft, that F and P, fingly, are identical : thus F performed
on X.Y gives X)Y, and P on X.Y gives X.y. And X)Y =
X.y. This perfect identity of F and P in effect, remains in
all combinations into which T does not enter. But when T
enters, it is S and F which are identical. Thus ST performed
on Y)X gives X)y or X.Y : and FT performed on Y)X gives
X.Y. The reafon is, that T interchanges fubject and predicate ;
fo that F, after T, makes a change which is counterbalanced by
a change in what was the fubject. Accordingly, remembering
that each operation performed twice is no operation at all (thus
PP is L, and TT is L), we have in all cafes

$$P=F, \ SP=SF, \ PF=L, \ SPF=S$$
$$ST=FT, \ SPT=FPT, \ SFT=T, \ SPFT=PT$$

all which fhould be tried for exercife. Again, in a *convertible*
propofition, transformation is no alteration or T=L : in an *inconvertible* one, transformation changes it into its contranominal ;
or T=SP. Now fet out as follows ;—L, in *convertible* propofitions is T ; which in *inconvertibles*, is SP ; which, in *convertibles*
again, is SPT ; which, in *inconvertibles* again, is TT, or L.
Put thefe down as follows, writing under them the operations
which are always equivalent to them, as fhewn above,

$$\begin{array}{c|c|c|c|c} L & T & SP & SPT & L \\ PF & SFT & SF & PFT & PF \end{array}$$

The combinations written under one another are always the
fame in effect : thofe feparated by double lines have the fame
effect on convertibles : thofe feparated by fingle lines, have the
fame effect on inconvertibles. Again P, for convertibles, is the
fame as PT ; which, for inconvertibles is the fame as PSP, or
S ; which, for convertibles again, is the fame as ST ; which, for
inconvertibles, is SSP or P. Thefe treated as before, give the
table

$$\text{P} \parallel \text{PT} \mid \text{S} \parallel \text{ST} \mid \text{P}$$
$$\text{F} \parallel \text{SPFT} \mid \text{SPF} \parallel \text{FT} \mid \text{F}$$

In thefe two cycles there are L and all the fifteen felections which can be made out of S, P, F, T. And every poffible cafe of equivalent changes is contained in thefe two tables. Thus PT is in all cafes equivalent to SPFT; in convertible cafes, to P and to F; in inconvertible ones, to S and to SPF. And no other combination is in any cafe equivalent to PT. In verification of thefe tables, obferve that the operation F always occurs in the lower line, and never in the upper; and that this operation changes convertibles into inconvertibles, and *vice verfâ*. We ought then to expect, that the equivalences which, containing F, apply to inconvertibles, will be thofe which when F is ftruck out, apply to convertibles; and *vice verfâ*. And fo we fhall find it: for inftance, SPFT and SPF are equivalent when performed on inconvertibles; ftrike out F and we have SPT and SP, which are equivalent when performed on convertibles.

It appears, then, that any change which can be made on a propofition, amounts in effect to L, P, S, or PS. This is another verification of the preceding table: for all our forms may be derived from applying thofe which relate to XY in the cafes of Xy, xY, and xy.

We have seen that A_1 and A' both contain I_1 and I'; and that E_1 and E' both contain O_1 and O'. Hence each of the univerfals may be faid to be the *ftrengthened form* of either of its particulars of the fame fign: and each of the particulars the *weakened form* of its univerfals of the fame fign. The only diftinction which appears between the two forms of the convertible particulars, XY and YX, xy and yx, is that the ftrengthened forms derived from extending the fubjects are different. Thus xy gives x)y or Y)X; but yx gives y)x or X)Y.

A *complex propofition* is one which involves within itfelf the affertion or denial of each and all of the eight fimple propofitions. If thefe eight propofitions were all concomitants, or if any number of them might be true, and the reft falfe, there would be 256 poffible cafes of the complex propofition. As it is, owing to the connexion eftablifhed in the table of page 63, there are but *feven*.

F

First, let the names X and Y be so related that neither of the four universals are true. Then all the four particulars are true: and this is the first case. Let it be called a *complex particular*, and denoted by P. Then, denoting coexistence of simple propositions by writing + between their several letters, we have

$$P = O' + O_1 + I' + I_1$$

This case is of the least frequent mention in the theory of the syllogism.

Next, let one of the universal propositions be true. Then five of the other propositions are settled, either by affirmation or denial. There remain the two concomitants, which are contradictory; so that only one is true. Accordingly, with the exception of the complex particular just described, every complex proposition must consist of the coexistence of a universal and one of its concomitants. But there are not therefore eight more such propositions: for $A' + A_1$ and $A_1 + A'$ are the same, and so are $E_1 + E'$ and $E' + E_1$. The remaining number is then reduced to six, which are

$$A_1 + O', \qquad A_1 + A', \qquad A' + O_1$$
$$E_1 + I', \qquad E_1 + E', \qquad E' + I_1,$$

These must be separately examined.

First, take $A_1 + A'$ (the order XY always understood). We have then X)Y and Y)X. That is, there is no object whatsoever which has one of these names, but what also has the other. The names X and Y are then *identical*, not as names, but as subjects of application. Where either can be applied, there can the other also. Thus, in geometry (the universe being plane rectilinear figure) equilateral and equiangular are identical names. Not that they agree in etymology nor in meaning: more than this, a few words would explain the first to many who could not comprehend the second without difficulty. But they agree in that what figure soever has a right to either name, it has the same right to the other. It will tend to uniformity of language, if we call X, in this case, *an identical* of Y, and Y *an identical* of X. Let the symbol of an identical be D: then we have

$$D = A_1 + A'$$

Next, take $A_i + O'$. We have then X)Y and Y:X. Every X is Y, and fo far there is a character of identity. But fome Ys are not Xs; there are more Ys than Xs, and X ftops fhort of a complete claim of identity with Y   Let X be called a *fubidentical* of Y (thus *man* is a fubidentical of *animal*), and let D_i denote this cafe.   Then

$$D_i = A_i + O'$$

Let.$A' + O_i$ exift. We have then Y)X and X:Y. Every Y is X, and fo far there is identity. But fome Xs are not Ys, there are more Xs than Ys, or X goes beyond a claim of identity with Y. Let X be now called a *fuperidentical* of Y, and let it be denoted by D'. Then

$$D' = A' + O_i$$

The terms fuperidentical and fubidentical are obviously correlative. If X be either of Y, Y is the other of X. Now let us confider $E_i + E'$. We have then X.Y and x.y. There is nothing which is both X and Y, there is nothing which is neither. Confequently X and Y are *contraries*, or juft fill up the univerfe. Let C be the mark of this relation.   Then

$$C = E_i + E'$$

Next, take $E_i + I'$. We have then X.Y and xy. Nothing is both X and Y, but there are things which are neither. X and Y are clear of one another, but do not amount to contraries, for they do not fill up the univerfe. Let them be called *fubcontraries*, (thus in the univerfe *metal*, *gold* and *filver* are fubcontraries, and let C_i denote the relation.   Then

$$C_i = E_i + I'$$

Laftly, take $E' + I_i$. We have x.y and XY. The names fill the univerfe; for there is nothing but what is either X or Y. But they *overfill* it; for fome things are both Xs and Ys. There is then all the completenefs of a contrary and more. Let X and Y be called *fupercontraries*,* and let C' denote the relation. Then we have

$$C' = E' + I_i$$

* The *fupercontrary* relation, though effential to a complete fyftem of fyllogifm, is not frequently met with. The other extreme of the fupercon-

To complete our language, let A, or X)Y, with reference to the order XY, be called *sub-affirmative;* and A' or Y)X, *superaffirmative.* Let E, or X.Y be called *subnegative;* and E or x.y, *supernegative.* Let the particulars I,, I', and O,, O', have also these several names. This extension of our language will require a little explanation.

When I say that X is a subidentical of Y, I mean that the etymological suggestions are actually satisfied. The whole name X, and more, is contained in Y. But when I say that X is a universal *subaffirmative* of Y, or X)Y, I mean no more than that we have the proposition whose form is not superaffirmative, according to the etymology of that word. An algebraist would well understand the distinction at a glance. He has often to distinguish the case in which *a* is less than *b* from that in which *a* is less than or equal to *b :* the case in which the extreme limit of the assertion is not included from that in which it is included.

Again, the word negative had better be viewed as not so much presenting *exclusion* for its first idea, as *inclusion in the contrary.* Thus a subnegative, when universal, is to suggest complete inclusion in the contrary, meaning the extreme case, possibly; namely, that the subnegative names may be contraries. Again, supernegative is to suggest the idea of supercontrary, with the lowest extreme, the relation of contrary, possibly included.

For exercise in this language, and in the ideas which it is meant to present, I now state the following results.

Universal affirmation, though as a general term, it is to include super and sub affirmation, yet looked at as one of the three, and distinguished from the rest, it means identity. The same of negation and contrariety. Subidentity requires universal subaffirmation and particular supernegation. Identity is universal sub and super affirmation, both. Superidentity requires universal superaffirmation and particular subnegation. Subcontrariety requires universal subnegation and particular superaffirmation. Con-

trary, or the subidentical, is so much the easiest of all our complex relations, that the latter rarely allows the former to appear. The first instance that suggested itself to me was *man* and *irrational* (as descriptive of the quality of the individual and not of the species) in the universe *animal.* These more than fill that universe, *idiot* being common to both. But it is more natural to say that *rational* (in this sense) is subidentical of *man.*

trariety is univerfal fub and fuper negation, both. Supercontrariety requires univerfal fupernegation and particular fubaffirmation. Again, univerfal fubaffirmation is either fubidentity or identity : particular fubaffirmation is a denial of contrariety and fubcontrariety. Univerfal fuperaffirmation is either fuperidentity or identity : particular fuperaffirmation denies contrariety and fupercontrariety. Univerfal fubnegation is either fubcontrariety or contrariety : particular fubnegation denies fubidentity and identity. Univerfal fupernegation is either fupercontrariety or contrariety : particular fupernegation. denies fuperidentity and identity. All this is expreffed in the following table,

D_i affirms	A_i *and* O'		A_i affirms	D_i *or* D
D —	A_i and A'		A —	D_i or D or D'
D' —	A' and O_i		A' —	D' or D
C_i —	E_i and I'		E_i —	C_i or C
C —	E_i and E'		E —	C_i or C or C'
C' —	E' and I_i		E' —	C' or C
Denial of D_i —	A' *or* O_i		O_i denies	D_i *and* D
— D —	O' or O_i		O —	D_i and D, or D' and D
— D' —	A_i or O^i		O' —	D' and D
— C_i —	E' or I_i		I_i —	C_i and C
— C —	I' or I_i		I —	C_i and C or C' and C
— C' —	E_i or I'		I' —	C' and C

Every fubidentical of a name is the fubcontrary of its contrary ; every fubcontrary is the fubidentical of the contrary. Treat the word contrary as negative, the word identical as pofitive ; and the two as of different figns. Then the algebraical rule ' like figns give a pofitive, unlike figns a negative,' holds in every cafe : including the variety of it fo well known as ' two negatives make an affirmative.' When the modifying prepofition comes firft it muft be retained ; when it comes fecond, it muft be changed. Thus the fubcontrary of a contrary is a fubidentical : but the contrary of a fubcontrary is a fuperidentical. In putting two relations together, however, we have got into fyllogifm, as we fhall prefently fee.

The following tables will fhow a connexion between the expreffions, for different orders and felections, which it may be ufeful to verify.

XY	YX	xY	Yx	Xy	yX	xy	yx
$A_1O'D_1$	$A'O_1D'$	$E'I_1C'$	$E'I_1C'$	$E_1I'C_1$	$E_1I'C_1$	$A'O_1D'$	$A_1O'D_1$
$A'O_1D'$	$A_1O'D_1$	$E_1I'C_1$	$E_1I'C_1$	$E'I_1C'$	$E'I_1C'$	$A_1O'D_1$	$A'O_1D'$
$E_1I'C_1$	$E_1I'C_1$	$A'O_1D'$	$A_1O'D_1$	$A_1O'D_1$	$A'O_1D'$	$E'I_1C'$	$E'I_1C'$
$E'I_1C'$	$E'I_1C'$	$A_1O'D_1$	$A'O_1D'$	$A'O_1D'$	$A_1O'D_1$	$E_1I'C_1$	$E_1I'C_1$

This table only contains some of the rules already laid down in pp. 64, 65. It expreſſes that, for inſtance, the A_1, O', and D_1 of XY, are ſeverally the ſame as the E_1, I', and C_1 of yX. This table may be exhibited thus, the identicals counting as inconvertibles, the contraries as convertibles.

Change of	In Convertibles, changes	In Inconvertibles, changes
Subject	Sign and Prepoſition	Sign and Prepoſition
Predicate	Sign	Sign
Subject and Predicate	Prepoſition	Prepoſition
Order	Neither	Prepoſition
Subject and Order	Sign	Sign and Prepoſition
Predicate and Order	Sign and Prepoſition	Sign
Subject, Predicate, and Order	Prepoſition	Neither

In all caſes, change of ſubject is change both of ſign and prepoſition; change of predicate is change of ſign; change of ſubject and predicate is change of prepoſition. Theſe three caſes are of great importance in the ſyllogiſm: and the reader would do well to connect in his mind

Subject	with	*Sign and prepoſition*
Subject and Predicate	—	*Prepoſition*
Predicate	—	*Sign*

It is deſirable to conſider the ſeveral complex relations as to the continuous tranſition from one into another: *the growth of names* concerns not only the etymologiſt, but the logician alſo.

With the analogies and affinities by which the dominion of one name is extended to inſtance after inſtance, and claſs after claſs—and ſometimes, in ſcientific language at leaſt, deprived of a part of what it has held—I have here nothing to do. It is enough that the phenomena exiſt which may be deſcribed as the gradual transformation of one relation into another. The words *butt* and *bottle*, for example, are now ſubcontraries in the univerſe *receptacle*: but the etymology of the ſecond word ſhows

that it *was* a fubidentical of the firft, being a diminutive. And
if we were to take the whole clafs butt, bufs, boot, bufhel, box,
boat, bottle, pottle, &c, which are all of one origin, the number
of tranfitions would be found to be very large.

I affume that all the inftances of a name are counted and
arranged in its univerfe : a conceivable, though not attainable,
fuppofition. Alfo, that the inftances of the name are arranged
contiguoufly, as in page 61. Whatever the reafon may be
which dictates the particular arrangement chofen, it will generally
happen that the inftances near to the boundary poffefs the characteriftics of the name in a fmaller degree than thofe nearer the
middle. Let the contiguous arrangement be made of all the inftances of the name Y, the univerfe being U. Let another name
X begin to grow, commencing with one inftance, that is, being
applied to one of the objects in the univerfe U, be it a Y or not;
then to another contiguous, and fo on. We are to enumerate the
ways in which fuch changes, whether of increafe or diminution, may caufe one name to change its relation to another.
According as the change is made by acceffion or retrenchment,
it may be denoted by (+) or (−).

Let the name X begin within the limits of the name Y : its
initial relation to Y is then D_1 And the poffibility of the
following continuous changes is obvious :

$$D_1 (+) D (+) D' \qquad D_1 (+) P (+) C'$$
$$D_1 (+) P (+) D' \qquad D_1 (+) P (−) C_1$$

Hence D_1 may become D' through either D or P, but C_1 or C'
only through P. Next, let X begin without the limits of Y :
the initial relation is C_1. We may have then

$$C_1 (+) C (+) C' \qquad C_1 (+) P (+) D'$$
$$C_1 (+) P (+) C' \qquad C_1 (+) P (−) D_1$$

Let X begin both within and without Y : its initial relation is
then P. And we have

$$P (+) D', \quad P (+) C', \quad P (−) D_1, \quad P (−) C_1$$

But when (−) follows D_1 or D, C_1 or C, we have nothing
except

$$D_1 (−) D_1, \quad D (−) D_1, \quad C_1 (−) C_1, \quad C (−) C_1$$

If we begin at the other extreme, with the name U, we have

$$U \,(-)\, D' \qquad\qquad U \,(-)\, C'$$

Beginning from D' and C' we have

$$D' \,(-)\, D \,(-)\, D_{\scriptscriptstyle |} \qquad\qquad D' \,(-)\, P \,(-)\, C_{\scriptscriptstyle |}$$
$$D' \,(-)\, P \,(-)\, D_{\scriptscriptstyle |} \qquad\qquad D' \,(-)\, P \,(+)\, C'$$
$$C' \,(-)\, C \,(-)\, C_{\scriptscriptstyle |} \qquad\qquad C' \,(-)\, P \,(-)\, D_{\scriptscriptstyle |}$$
$$C' \,(-)\, P \,(-)\, C_{\scriptscriptstyle |} \qquad\qquad C' \,(-)\, P \,(+)\, D'$$

But when (+) follows D' or D, C' or C, we have only

$$D' \,(+)\, D' \quad D \,(+)\, D', \quad C' \,(+)\, C', \quad C \,(+)\, C'$$

From the above lift it appears that the tranfition which is accompanied by a change of prepofition only can be made either through the letter without prepofition or through P: and in all cafes with one continued mode of alteration. But when the tranfition involves change of letter, it can only be made through P: with continuation of the mode of alteration when the prepofitions are different, and change in the mode when they are the fame. The following fucceffions contain the arrangement of the refults.

With one altera- tion (+)	With one altera- tion (−)	With two altera- tions (+−)
$D_{\scriptscriptstyle |}$ D D'	D' D $D_{\scriptscriptstyle |}$	$D_{\scriptscriptstyle |}$ P $C_{\scriptscriptstyle |}$
$D_{\scriptscriptstyle |}$ P D'	D' P $D_{\scriptscriptstyle |}$	$C_{\scriptscriptstyle |}$ P $D_{\scriptscriptstyle |}$
$C_{\scriptscriptstyle |}$ C C'	C' C $C_{\scriptscriptstyle |}$	———
$C_{\scriptscriptstyle |}$ P C'	C' P $C_{\scriptscriptstyle |}$	(−+)
———	———	D' P C'
$D_{\scriptscriptstyle |}$ P C'	D' P $C_{\scriptscriptstyle |}$	C' P D'
$C_{\scriptscriptstyle |}$ P D'	C' P $D_{\scriptscriptstyle |}$	

The following confiderations will further ferve to illuftrate the want of the extenfion of the doctrine of propofitions made in this chapter, and alfo the completenefs of it. Among our moft fundamental diftinctions is that of *neceffity* and *fufficiency*; of what we *cannot do without*, and what we *can do with*; of that which *muft precede*, and that which *can follow*. The contraries of thefe are *non-neceffity* and *non-fufficiency*. In thefe four words, applied to both Y and y, we have the defcription of the eight re-

lations of X to Y. For inftance A₁ or X)Y tells us that to
have an X, we muft take a Y, or to be X, it is *neceffary* to be
Y. Treating all in the fame way, we have

A₁	X) Y	To take an X it is	*neceffary*	to take a	Y
A'	Y) X	. . . X . .	*fufficient*	. . .	Y
E₁	X . Y	. . . X . .	*neceffary*	. . .	y
E'	x . y	. . . X . .	*fufficient*	. . .	y
I₁	X Y	. . . X .	*not neceffary*	. .	y
I'	x y	. . . X .	*not fufficient*	. .	y
O₁	X : Y	. . . X .	*not neceffary*	. .	Y
O'	Y : X	. . . X .	*not fufficient*	. .	Y

And the convertibility of the ordinary mode of defcription with
this new one may be eafily fhewn in any cafe. For example,
what can we mean by faying that to take a X, it is not fufficient
to take what is not Y ? Clearly that by taking not Y, or y, we
may at the fame time take a x, or that there are xs which are ys.
And fo on for the reft.

Of the four pairs XY, Xy, xy, xY, we know that each
propofition may be expreffed by three, and refufes to be expreffed
by one. If we now admit the two words *impoffible* and *contingent*,
meaning by the latter that which, as the cafe may be, is poffible
or impoffible, we fhall eafily fee the following table for the univerfals :

		XY	Xy	xy	xY
A₁	X) Y	N	I	S	C
E₁	X . Y	I	N	C	S
A'	Y) X	S	C	N	I
E'	x . y	C	S	I	N

The letters N, I, S, C, are the initials of neceffary, &c. And
we read in the firft line, that if X) Y, then to be X it is neceffary to be Y ; to be X, it is impoffible to be y ; to be x it is fufficient to be y ; and to be x, it is contingently poffible or impoffible
to be Y. Again, if by *n* and *s* we mean *not neceffary* and *not
fufficient*; by P, actually poffible ; and by C, as before (C being
its own contrary), we have the following table for the particulars :

		XY	Xy	xy	xY
O_{ι}	X : Y	n	P	s	C
I_{ι}	X Y	P	n	C	s
O'	Y : X	s	C	n	P
I'	xy	C	s	P	n

Of the four contrary pairs, n, P, s, C, are related to the particulars precifely as N, I, S, C, are to the univerfals. The interchange of Y and y is always accompanied by the interchange of N and I, S and C, n and P, s and C; the interchange of X and x is that of N and C, S and I, n and C, s and P; of both X and x, Y and y, is that of N and S, C and I, n and s, C and P.

The complex relations may be thus defcribed. According as X is fubidentical, identical, or fuperidentical of Y, to be X it is neceffary and not fufficient, neceffary and fufficient, or not neceffary and fufficient, to be Y : according as X is fubcontrary, contrary, or fupercontrary of Y, to be X it is neceffary and not fufficient, neceffary and fufficient, or not neceffary and fufficient, to be y.   Or, as in the following table :

	XY	Xy	xy	xY
D_{ι}	Ns	I	Sn	P
C_{ι}	I	Ns	P	Sn
D'	Sn	P	Ns	I
C'	P	Sn	I	Ns
D	NS	I	NS	I
C	I	NS	I	NS
P	nsP	nsP	nsP	nsP

Inftead of IC and PC, write I and C : for " impoffible, and poffible or impoffible as the cafe may be " is " impoffible " &c.

The names of the complex relations, fubidentity, identity, &c I fuppofe will be held tolerably fatisfaĉtory : thofe of the fimple relations fuggefted in page 68, fubaffirmative &c. have nothing in their favor except analogy with the former, and clofe connexion with the notation. A little praĉtice in their ufe might render thefe laft names available : but it will be advifable to connect them with names more defcriptive of the meaning, and to adopt thefe laft, whether we reject or maintain their fynonymes.

When X) Y, the relation of X to Y is well underftood as that of the *fpecies* to the *genus*. We may adopt thefe words, with the underftanding that the word fpecies includes the extreme cafe in which the fpecies is as extenfive as the genus. When X : Y, we may call X a *non-fpecies* of Y, and Y a *non-genus* of X. When X . Y we may call X an *exclufive* or *excludent* of Y, or elfe a *non-participant;* and alfo Y of X. When XY, we may fay that each is *participant*, or *non-exclufive*, of the other. When x . y, which means that X and Y together fill up, or more than fill up, the univere, we may fay that they are *complemental* names. When xy, which only means that X and Y do not between them contain the univerfe, we may call them *non-complemental*. We have then

Inconvertibles.   Name of X with refpect to Y.

A₁   X)Y   fpecies, or fubaffirmative.
O₁   X:Y   non-fpecies, or particular fubnegative.
A'   Y)X   genus, or fuperaffirmative.
O'   Y:X   non-genus, or particular fupernegative.

Convertibles.   Name of X and Y with refpect to each other.

E₁   X.Y   Exclufives, or non-participants, or fubnegatives.
I₁   XY   Non-exclufives, or participants, or particular fubaffir-
E'   x.y   Complements, or fupernegatives.   [matives.
I'   xy   Non-complements, or particular fuperaffirmatives.

The following exercifes in thefe terms, really contain the defcription of all the fyllogifms in the next chapter.

Inclufion in the fpecies is inclufion in the genus; and inclufion of the genus is inclufion of its parts (fpecies or not).

Exclufion from the genus is exclufion from the fpecies; and exclufion of the genus is exclufion of its parts (fpecies or not).

Inclufion or exclufion of the fpecies is part inclufion or exclufion of the genus.

When the fpecies is complemental, fo is the genus: and when the genus is not complemental, neither is the fpecies.

Exclufion from one complement is inclufion in the other.

Complements of the fame are participants.

Two species of one genus, are not complements ; neither are two exclusions from the same.

The complement of a genus is a non-species ; and the complement is a non-species of the non-complement.

CHAPTER V.

On the Syllogism.

A SYLLOGISM is the inference of the relation between two names from the relation of each of those names to a third. Three names therefore are involved, the two which appear in the conclusion, and the third or *middle term*, with which the names, or terms, of the conclusion are severally compared. The statements expressing the relations of the two *concluding* terms to the *middle* term, are the two *premises*. In this chapter, no ratio of quantities is considered except the definite *all* and the indefinite *some*.

A syllogism may be either *simple* or *complex*. A syllogism is *simple* when in it two simple propositions produce the affirmation or denial of a third : or the affirmation of a third, we may say, since every denial of one simple proposition is the affirmation of another. A *complex* syllogism is one in which two complex propositions produce the affirmation or denial of a third complex proposition.

It might be supposed that we ought to begin with the simple syllogism, and from thence proceed to the complex. On this point I have some remarks to offer, in justification of following precisely the reverse plan.

Hitherto the complex syllogism has never made its appearance in a work on logic, except in one particular case, in which it is allowed to be treated as a simple syllogism, though most obviously it is not so. I allude to the common *à fortiori* argument, as in 'A is greater than B, B is greater than C, therefore A is greater than C.' There is no middle term here : the predicate of the first proposition is 'a thing greater than B,' the subject of the second proposition is ' B.'

Admitting fully that the quality of the premises,—that which entitles the conclusion to be made, as it is said, *à fortiori*—marks this argument out as, if anything, stronger, clearer, and (could such a thing be) truer, than a simple syllogism ; yet it is plain that the very additional circumstance on which this additional clearness depends, takes the argument out of a syllogism, as defined by all writers. By beginning with the complex syllogism, and thence descending to the simple one, it will be seen that we begin with cases which present this *à fortiori* and clearer character. I think I shall shew that the complex syllogism is easier than the simple one.

Next, the syllogism hitherto considered has never involved any contrary terms ; the consequence of which has been that various legitimate modes of inference have been neglected. Moreover, several of the usual syllogisms are more strong than need be in the premises, in order to produce the conclusion. Thus Y)X and Y)Z being admitted as premises, the necessary conclusion is XZ. But if Y)X be weakened into YX, the same conclusion follows. If we call a syllogism *fundamental*, when neither of its premises are stronger than is necessary to produce the conclusion, it is obvious that every fundamental syllogism which has a particular premise, gives at least as strong a conclusion when that particular is strengthened into a universal. But, except when strengthening the premise also enables us to strengthen the conclusion, in which case we have a new and different syllogism, it seems hardly systematic to mix with fundamental arguments syllogisms which have quality or quantity more than is necessary for the conclusion.

The use of the complex syllogism will, as we shall see, give an independent and systematic derivation to these strengthened syllogisms, as well as to the rest.

Let X and Z be the terms of the conclusion ; and let Y be the middle term. Let the premise in which X and Y are compared come first of the two. Let the order of reference in each case be that of the alphabet

$$XY \qquad YZ \qquad XZ$$

So that by stating what X is with respect to Y, and what Y is with respect to Z, our syllogism involves the statement of what X therefore must be, or therefore cannot be, with respect to Z. We can, in every case, express the result in simple words. Thus, one of our fyllogifms being what I fhall reprefent by $D_iD_iD_i$ is as follows. If X be a fubidentical of Y, and Y a fubidentical of Z, then X is a fubidentical of Z. But all this merely amounts to the following 'A fubidentical of a fubidentical is a fubidentical.'

We have then to examine every way in which D_i or D' or C_i or C' can be combined with D_i or D' or C_i or C', giving fixteen cafes in all, and all conclufive in one way or the other. Inftead of taking an accidental order, and afterwards claffifying the refults, it will be better to predict the order which will give claffification. That order will be to take 1. a D followed by another of the fame prepofition 2. a C followed by another of different prepofition 3. a D followed by another of a different prepofition. 4. a C followed by another of a like prepofition. This arrangement gives us

1.	D_iD_i	$D'D'$	D_iC_i	$D'C'$	3.	D_iD'	$D'D_i$	D_iC'	$D'C_i$
2.	C_iD'	$C'D_i$	C_iC'	$C'C_i$	4.	C_iD_i	$C'D'$	C_iC_i	$C'C'$

Each of thefe cafes will be examined by a method fimilar to that propofed in page 61. But a clear perception of the meaning of the words will at once dictate the fixteen refults, which are as follows, preceded by the mode in which the fyllogifms are to be expreffed.

$D_iD_iD_i$	Subidentical of fubidentical is fubidentical.
$D'D'D'$	Superidentical of fuperidentical is fuperidentical.
$D_iC_iC_i$	Subidentical of fubcontrary is fubcontrary.
$D'C'C'$	Superidentical of fupercontrary is fupercontrary.
$C_iD'C_i$	Subcontrary of fuperidentical is fubcontrary.
$C'D_iC'$	Supercontrary of fubidentical is fupercontrary.
$C_iC'D_i$	Subcontrary of fupercontrary is fubidentical.
$C'C_iD'$	Supercontrary of fubcontrary is fuperidentical.
$D_iD':C'$	Subidentical of fuperidentical is *not* fupercontrary.
$D'D_i:C_i$	Superidentical of fubidentical is not fubcontrary.
$D_iC':D'$	Subidentical of fupercontrary is not fuperidentical.
$D'C_i:D_i$	Superidentical of fubcontrary is not fubidentical.
$C_iD_i:D'$	Subcontrary of fubidentical is *not* fuperidentical.
$C'D':D_i$	Supercontrary of fuperidentical is not fubidentical.
$C_iC_i:C'$	Subcontrary of fubcontrary is not fupercontrary.
$C'C':C_i$	Supercontrary of fupercontrary is not fubcontrary.

In the denials, the extreme limit is included: in the affirmations it is not. Thus ' not fuperidentical' and ' not fubidentical' both include ' not identical;' and the fame of contraries. In the affirmations, extreme limitation of *one* premife does not alter the conclufion: but that of *both* reduces the conclufion to its extreme limit. Thus

> Subcontrary of identical is fubcontrary.
> Contrary of fuperidentical is fubcontrary.
> Contrary of identical is contrary.

and fo on. The rules of this fpecies of fyllogifm are as follows. *For affirmatory conclufions ;*—(1.) Like names in the premifes give D in the conclufion, and unlike names C. (2.) D in the firft premife requires premifes of the fame prepofition; C in the firft premife, of different prepofitions. (3.) The prepofition of the conclufion agrees with that of the firft premife. *For negatory conclufions*, the preceding rules are reverfed. Thefe rules will do for the prefent, but they afterwards merge in others.

The fixteen forms of complex conclufion above given are of the clearnefs of axioms, as foon as the terms are diftinctly apprehended. The following diagrams will affift, and fhould be ufed until the propofitions fuggeft their own meaning. Though there be four, yet thefe four are really but one, as will be fhown.

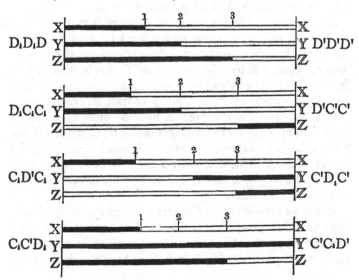

In each diagram are three lines, partly thick and partly open : thefe are meant to be laid over one another, but are kept feparate for diftinctnefs. A point on the firft line fignifies a X or a x; and one on the fecond or third, a Y or a y, and a Z or a z. The univerfe of the propofitions is fuppofed to be the whole breadth. Points which come under one another are fuppofed to reprefent the fame object of thought, varioufly named. Thus in the firft diagram, when the *thick* lines contain the points named X, Y, and Z, it is fhown that we mean to fay there are objects to which all the three names apply : for there are points under one another in the thick part of all the three lines.

When we read by the letters on the left, the thick lines are meant to reprefent the parts in which the Xs, Ys, and Zs muft be placed : and when by thofe on the right, the open lines. Accordingly, looking at the third diagram, and at the left, we fee C_1 D' C_1 : while in the diagram, it is clear that X is a fubcontrary of Y, or that X . Y and x y ; and that Y is a fuperidentical of Z, or that Z) Y and Y : Z. And the conclufion is equally manifeft, namely, that X is a fubcontrary of Z. But, looking at the left, and feeing C' D_1 C', we take the open parts to reprefent the fpaces in which Xs, Ys, and Zs are found, and the thick parts for thofe in which xs, ys, and zs are found. Here then we fee that X is a fupercontrary of Y, that Y is a fubidentical of Z, and that, *confequently*, X is a fupercontrary of Z.

Some attempts at laying down the premifes fo as to evade the conclufions, will be inftructive to any one who does not imme- diately fee the latter. And formal demonftration is always prac- ticable. Thus if X be a fubcontrary of Y, that is, if X and Y do not fill the univerfe, and have nothing in common ; and if Y be a fuperidentical of Z, or entirely contain Z, without being filled by it : then it is clear that X muft be more a fubcontrary of Z than of Y, by all the inftances which there are of a Y not being a Z. The diagram, however, is fo much clearer than this fort of demonftration, that the reader, until he has great com- mand of the language, may as well look to the former to fee that he is right in the latter.

It may be convenient, as a matter of language, to fpeak of a name as a kind of collective whole, confifting of inftances. And thus we may talk of one name being entirely in another, or partly in and partly out &c, as in fact we have already done.

All the complex fyllogifms which conclude by affirmation are obvioufly of the *à fortiori* character: I fhould rather fay, thofe of the firft three diagrams properly and obvioufly, thofe of the fourth by an eafy extenfion of language. The marks 1 2 3 in the middle of the diagrams fhow how this is. In the firft, on the left, X is more of a fubidentical of Z than it is of Y: the inftances in which its *fub*-identity appears confift of all thofe which prove the fubidentity of X to Y, *together with* all thofe which prove the fubidentity of Y to Z. In the third, read from the right, X is more fupercontrary to Z than it is to Y, by all the inftances which fhow the fubidentity of Y to Z. In the fourth diagram (from the left) we cannot fay that X is more fubidentical of Z than of fomething elfe, fimply becaufe there is no previous fubidentity among the relations. But ftill the diftinguifhing characteriftic of the conclufion takes its quantity from the addition of thofe of both the premifes.

If either of the premifes be brought to the limit which feparates it from the relation of an oppofite prepofition; that is, if C' or C_i be changed into C, or elfe D' or D_i into D: the nature of the conclufion is not altered, except by the lofs of the *à fortiori* character. One of the quantities which have hitherto contributed to the quantity of the conclufion, now difappears. Thus C_i D gives C_i as well as C_i D'; and C D' gives C_i as well as C_i D'; C_i C gives D_i as well as C_i C'.

Let one of the premifes pafs over the limit, and take the oppofite prepofition. Choofe C_i D', which gives C_i, and continues to give it, though weakened, when the firft C_i becomes C. Then let C_i become C': fo that our premifes are C' D'. The diagram is then as follows

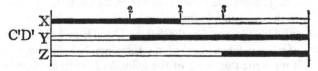

The *quantity* of the conclufion now depends upon the *difference* between the number of inftances in (12) and (23) and its *quality* upon whether (12) has fewer inftances than (23), or the fame number, or more. As I have drawn it, C_i is the conclufion, ftill: ftrengthen the firft premife ftill more, and the conclufion will pafs through C into C' or elfe into P, and in the fecond cafe may pafs into D', as in the following diagram

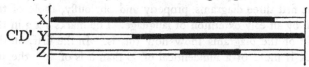

Nothing is impoffible except D_i or D. Hence C' D' enables us only to deny D_i and its limit D. Treat the other cafes in the fame manner, and, remembering that denial is to include denial up to the limit (while affirmation only affirms to any thing fhort of the limit) we have

D_i D' denies C'		D' D_i denies C_i
D_i C' . . D'		D' C_i . . D_i
C_i D_i . . D'		C' D' . . D_i
C_i C_i . . C'		C' C' . . C_i

The rules given above in page 79 may be collected from the inftances.

As long as we keep contraries out of view, the ultimate element of inference is of a twofold character. It is either ' X and Z are both Y ' ; therefore X is Z ' or elfe ' X is Y and Z is not Y ; therefore X is not Z ' : X, Y, Z, being fingle inftances of three names ; and Y the fame inftance in both premifes. But the ufe of contraries enables us to give an affirmative form to the latter cafe. It is ' X is Y, and not-Z is Y ' ; therefore ' X is not-Z '.

Connected with this change of expreffion is the following theorem : that all the eight affirmatory complex fyllogifms are reducible to any one among them : and the fame of the negatory ones. The reader may trace this theorem to the order of the figures 1, 2, 3, being the fame in all the four diagrams. Taking $D_i D_i D_i$ as the moft fimple and natural form, and looking at the diagram of $C_i D'C_i$, we fee the laft as $D_i D_i D_i$ in ' X is fubidentical of y ; y is fubidentical of z ; therefore X is fubidentical of z.' If we write the terms of the fyllogifm after its defcriptive letters, as in $D_i D_i D_i$ (XYZ) we have the following refults ;—

$D_i D_i D_i$ (XYZ) $= D_i D_i D_i$ (XYZ)	D'D'D' (XYZ) $= D_i D_i D_i$ (xyz)
$D_i C_i C_i$ (XYZ) $= D_i D_i D_i$ (XYz)	D'C'C' (XYZ) $= D_i D_i D_i$ (xyZ)
$C_i D'C_i$ (XYZ) $= D_i D_i D_i$ (Xyz)	C'D_iC' (XYZ) $= D_i D_i D_i$ (xYZ)
C_iC'D_i (XYZ) $= D_i D_i D_i$ (XyZ)	C'C_iD' (XYZ) $= D_i D_i D_i$ (xYz)

Thinking of the firft defcription only as to relations, and of the fecond only as to terms, we fee the following rules of connexion. In the firft and fecond premifes and terms, there are X and Y in the terms, or their contraries, according as there are fub-accents or fuperaccents in the relations. But in the conclufion, the term is Z for D_1 and C', z for D' and C_1. And we may thus reduce any fyllogifm involving any one of the eight varieties of relation combined with any one of the varieties of terms, either to $D_1D_1D_1$ or to XYZ. Thus $C_1D'C_1$ (XyZ) is $D_1D_1D_1$ (XYz), or $D_1C_1C_1$ (XYZ). Not to load the fubject with de-monftration of forms, I will give at once the general rules by which changes of accent and letter are governed: remarking that they apply throughout the whole of my fyftem.

The varieties in queftion are eight:

XYZ, xyz; xYZ, Xyz; XyZ, xYz; XYz, xyZ.

in which (thinking of XYZ) all are kept; or all changed; or one only kept; or one only changed. Learn to connect each letter with the propofitions in which it occurs; marking the pro-pofitions, premifes and conclufion, as 1, 2, 3. Connect X with 1,3; Y with 1,2; Z with 2,3. Keeping all, or changing all, makes no alteration of letters: keeping only one, or changing only one, alters the letters in the premifes in which that one occurs. Thus, be the accents what they may, if in DDD we change only the firft letter into its contrary, the fyllogifm becomes CDC; and the fame if we keep only the firft letter unchanged.

As to accents, remember that change of Z produces no effect: look then only at X and Y. When either letter is changed into its contrary, change the accents belonging to the premifes in which that letter comes firft; 13 for X, 2 for Y, 123 for XY. For example, what is $C_1C'D_1$ (Xyz). Here, as to letters, X alone (1,3) is unchanged: then CCD becomes DCC. As to accents, Y is changed, which comes firft only in 2: change C' into C_1. Hence $C_1C'D_1$ (Xyz)=$D_1C_1C_1$ (XYZ). Here we have paffed from a fyllogifm in Xyz to the correfponding equi-valent in XYZ: the rules equally hold for the inverfe procefs, and for all combinations of letters. For the change of XYZ into Xyz, and that of Xyz into XYZ, have only one defcription: the firft only left unchanged. Now fuppofe it required to know what fyllogifm in xYz anfwers to $D_1C_1C_1(Xyz)$. The key words are, *the third only unchanged.* Alter then DCC into DDD by the firft rule, and change all the accents. Thus $D_1C_1C_1(Xyz) = D'D'D'(xYz)$. The independent rules are that change of fubject only, changes *both* letter and accent; predicate only, letter; fubject and predicate, accent. Thus to find what $D'C'C'(xYz)$ is, expreffed in XYz, the changes are, in the three premifes S, neither, S, and $D'C'C'(xYz) = C_1C'D_1(XYz)$. The following table may be verified for exercife: it fhows the effect of all changes except that of the middle term.

XYZ	xYZ	XYz	xYz
$D_1D_1D_1$	$C'D_1C'$	$D_1C_1C_1$	$C'C_1D'$
$C'D_1C'$	$D_1D_1D_1$	$C'C_1D'$	$D_1C_1C_1$
$D_1C_1C_1$	$C'C_1D'$	$D_1D_1D_1$	$C'D_1C'$
$C'C_1D'$	$D_1C_1C_1$	$C'D_1C'$	$D_1D_1D_1$

Similarly, $D'D'D'$ would have $C_1D'C_1$ $D'C'C'$ &c. When the middle term only is changed, the table may ftand thus;—

XYZ	$D_1D_1D_1$	$C'D_1C$	$D_1C_1C_1$	$C'C_1D'$
XyZ	$C_1C'D_1$	$D'C'C'$	$C_1D'C_1$	$D'D'D'$

It will of courfe have been obferved that the eight fyllogifms go in pairs, each one of a pair differing from the other in accentuation, and nothing elfe. When we take fets of four, the ones put together fhould be thofe in which the firft premife, or the fecond, or the conclufion (whichever we take for a ftandard) has D_1 and C', or elfe has D' and C_1.

The fame rules of transformation apply to negatory complex fyllogifms; thus $D'D_1:C_1(XYZ)$ is $C'D':D_1(Xyz)$. In fact thefe rules do not depend upon the character of the inference, nor even upon its validity, but merely on the effects produced in the fingle propofitions by changes of term. Thus the ftatement $D'D_1C_1(XYZ)$, an invalid inference, is the fame ftatement (equally invalid of courfe) as is expreffed in $D_1C'D'(xyZ)$.

An examination of the complex particular relation $P = I_1 + I' + O_1 + O'$, whether by the diagram or by unaffifted thought, will fhow that when this relation exifts between X and Y, it alfo exifts between x and Y, X and y, x and y. Hence PC, CP, PD, DP, give P. Moreover, two complex particulars give no poffibility of any conclufion, all being equally poffible. Thus PP may give C_i or C or C', or D_i or D or D'.

Now combine one of the others, as D_i, with P : examine PD_i and D_iP. It will be found that the complex particular of a fubidentical may be either complex particular, fubidentical, or fupercontrary; or that PD_i may be either P, D_i or C'. Examine all the cafes, and the rules will be found in

$$(D_iC_i)P \qquad P(D_iC')$$
$$(D'C')P \qquad P(C_iD')$$

thus interpreted. Either premife from between the parenthefes, with P, in order as written, may have either, and muft have one, of the three for its conclufion. That D_iP muft give either D_i C_i or P, and fo muft C_iP : but PC_i muft have either P, C_i, or D'.

Before proceeding to the fimple fyllogifm, as I have called it, I will ftate that I much doubt the propriety of the terms *fimple* and *complex*. Undoubtedly the phrafes are hiftorically juft, for each of the fyllogifms which I propofe to call *complex* is, as we fhall fee, neceffarily compofed of three of thofe which are always called *fimple*. But in another point of view, the phrafeology ought to be reverfed; the fimple fyllogifm is the affirmation of the exiftence of one out of feveral of the complex ones. Thus X)Y+Y)Z=X)Z, or $A_iA_iA_i$, is really (D_i or D, not known which) (D_i or D, not known which) (D_i or D, not known which) and afferts that there is either $D_iD_iD_i$ or D_iDD_i or DD_iD_i or DDD.

But it will be faid, furely the complex propofition requires the *conjunctive* exiftence of two fimple ones: $D_i=A_i+O'$; and is therefore *compound* at leaft. I anfwer that, on the other hand, the fimple propofition requires the *disjunctive* exiftence of two complex ones: as $A_i=D_i$ or D. Which is moft fimple, *both*, or *one or the other?* to me, I think, the firft. Certainly the fyllogifm $D_iD_iD_i$ is one which I more readily apprehend than $A_iA_iA_i$. Indeed, to moft minds, the latter is the former, if they are left to themfelves: and the cafes D_iDD_i, &c. are only admitted when produced and infifted on.

But further, is the fimple propofition properly called *fimple?* Is there in it but one affertion to deny or admit? Is but one

queftion anfwered? When I affirm ' Every X is Y,' I affirm
1. Comparifon of X and Y. 2. Coincidences. 3. The greateft
poffible amount of them. 4. Thàt *every* X has been ufed in obtaining them. In ' Some Xs are Ys the firft two of the preceding are employed. In ' No X is Y,' we have, 1. Comparifon
of Xs and Ys. 2. Exclufions. 3. The greateft amount. 4. The
comparifon of *every* X with every Y. And ' Some Xs are not
Ys' omits the third, and fubftitutes Xs for every X in the fourth.

Now the fubidentical, for inftance, only contains, befides what
is in the fubaffirmative, the notion that there are more Ys than
Xs in exiftence. The fubcontrary confifts, over and above what
is in the fubnegative, in that Xs and Ys are not every thing that
the propofition might have applied to : and fo on. On thefe
confiderations, I think it may be allowed to treat the words fimple and complex as only of hiftorical reference, and to confider
the firft as disjunctively connected with the fecond, the fecond
as conjunctively connected with the firft, in the manner above
noted. I think I fhall make it clear enough, that the paffage
from the conjunctions to the disjunctions is better fuited to a
demonftrative fyftem than the converfe. If the plan which I
propofe fhould gain any reception, I fhould imagine that *disjunctive* and *conjunctive* would be the names given to the claffes
which I have called fimple and complex : the conjunctive compofed of feveral of the disjunctive, the disjunctive confifting of
one or the other out of feveral of the conjunctive.

When a propofition R, is the neceffary confequence of two
others, P and Q, it neceffarily follows that the denial of R, muft
be the denial of one at leaft of P and Q. For every propofition
admits but of affirmation or denial : and he who affirms *both* P
and Q muft *affirm* R. If then P be affirmed and R denied, the
denial of Q muft follow : if Q be affirmed and R denied, the
denial of P muft follow.

A *fimple fyllogifm* is one, the two premifes and conclufion of
which are to be found among the fimple propofitions A_1, E_1, I_1,
O_1, A', E', I', O'. Thus we have $A_1E_1E_1$ or $X)Y + Y.Z =$
X.Z, as an inftance. The order of reference is always XY,
YZ, XZ.

The following theorems will be neceffary;—1. *A particular
premife cannot be followed by a univerfal conclufion.*

If poſſible, let A_1I_1 for example, have a univerſal concluſion. Take the complex premiſes D_1P or $(A_1 + O')(I_1 + I' + O_1 + O')$. All that can be inferred is that one of *three* concluſions (page 85) is valid, and neither D nor C: either D_1 or P or C_1. But if a univerſal be true, one of *two* concluſions muſt be valid (page 69) and one of them D or C. If then A_1 and I_1 alone yielded a univerſal concluſion, quite as much muſt D_1P: or a form which is indifferent to three concluſions, and not having D nor C, is neceſſarily productive of one of two concluſions, one of which is D or C. This contradiction cannot exiſt: or A_1I_1 cannot yield a univerſal concluſion.

2. *From two particular premiſes no concluſion can follow.*

If poſſible, let I_1I_1 yield a concluſion; which by the laſt theorem, muſt be only particular. Now PP or $(I_1 + I' + O_1 + O')$ $(I_1 + I' + O_1 + O')$ is indifferent to all complex concluſions: quite as much is I_1I_1. But if theſe premiſes yield a particular concluſion, two complex concluſions are denied (page 69). This contradiction cannot exiſt: or particular premiſes can yield no concluſion.

Let a ſimple ſyllogiſm with premiſes and concluſion all univerſal, be called *univerſal*: and with either premiſe (and therefore the concluſion) particular, be called *particular*. Then every univerſal ſyllogiſm has two particular ſyllogiſms deducible from it. Thus if $A_1E_1E_1$ be valid, then A_1 joined with the denial of E_1 gives the denial of E_1: or $A_1I_1I_1$ *ſeems to be valid*. But the alteration of the places of the propoſitions requires us to ſay that it is $A'I_1I_1$ which is valid: and this point requires cloſe attention.

Take $A_1E_1E_1$ or $X)Y + Y.Z = X.Z$. Then $X)Y$ with the denial of $X.Z(or XZ)$ gives the denial of $Y.Z(or YZ)$; and we have

$$X)Y + XZ = YZ$$

This is valid, if the firſt be (as it is) valid: but its ſymbol is not $A_1I_1I_1$. For the middle term is, in our notation, made *middle* in the order of reference, which is therefore YX, XZ, YZ: and the ſyllogiſm is $A'I_1I_1$. Similarly we have

$$XZ + Y.Z = X:Y$$

produced by coupling the denial of X.Z with Y.Z. But this is $I_1E_1O_1$: for the order of reference is now XZ, ZY, XY, and

E_i is not changed by change of order. The rule is as follows. When the denials of the conclusion and of a premise are made to take the places of that premise and the conclusion, the order of reference remains undisturbed as to the transposed terms, and is changed as to the standing term. This last must therefore have the preposition of the inconvertible proposition changed; but not that of the convertible proposition.

Thus $E'A_iE'$, if valid, gives $E'I'O_i$ and $I'A'I'$. Again, in a similar way it may be shown that from each particular syllogism follows a universal: thus $I_iE'O'$, if valid, shows that denial of O', and E', give denial of I_i or $A'E'E_i$. In this case neither is valid. And $E'I'O_i$, besides $E'A_iE'$, also gives $A_iI'I'$.

Such classification of these *opponent forms* as is useful, will presently be given.

Since there are eight forms of assertion, with reference to each of the orders XY YZ, it follows that there are sixty-four combinations of a pair of premises each. But of these the only ones which have a chance of yielding a conclusion are, 1. sixteen with premises both universal; 2. thirty-two with one universal and one particular. If, for a moment, U stand for universal and P for particular, the form of a syllogism is either UUU, PUP, UPP, or UUP. Of these, the first, second, and third are so related that each form has the other two for its opponents: but the fourth has its own form in each of its opponents.

Now examine one of the complex affirmative syllogisms, say $D_iD_iD_i$, by the diagram in page 79. The premises are $A_i + O'$ and $A_i + O'$, giving the four combinations A_iA_i, A_iO', $O'A_i$ and $O'O'$. The conclusion is $A_i + O'$: but it is not merely twofold, but threefold: for the *à fortiori* character explained in page 81, shows that O' is obtainable on two different grounds, and is the sum, as it were, of two different and necessary parts of the conclusion. That every X is Z, follows from X)Y and Y)Z, or we have the syllogism.

$$A_iA_iA_i \quad X)Y + Y)Z = X)Z$$

But as far as the Zs which are below (12) are concerned, it follows that they are not Xs because they are the Ys which are not Xs: or we have

$$O'A_iO' \quad Y:X + Y)Z = Z:X$$

and as to the Zs below (23) they are not Xs becaufe they are not Ys, among which are all the Xs. Accordingly we have

$$A_1O'O' \qquad X)Y+Z{:}Y=Z{:}X$$

or $D_1D_1D_1$ requires the coexiftence of $A_1A_1A_1$, $O'A_1O'$, $A_1O'O'$. Apply this reafoning to the contraries x, y, z, or elfe examine $D'D'D'$ in the fame way, and we find that $D'D'D'$ requires the coexiftence of $A'A'A'$, $O_1A'O_1$, $A'O_1O_1$.

By applying the preceding refults to x, Y, Z, &c. as in page 82, or, as is better at firft, by examining all the cafes of the diagram in page 79, we get the following table of derivations from the eight affirmatory complex fyllogifms. The firft column fhews the terms which muft be ufed, to deduce all from $D_1D_1D_1$.

XYZ	$D_1D_1D_1$ ..	$A_1A_1A_1$ $X)Y+Y)Z=X)Z$	
		$O'A_1O'$ $Y{:}X+Y)Z=Z{:}X$	(12)
		$A_1O'O'$ $X)Y+Z{:}Y=Z{:}X$	(23)
x y z	$D'D'D'$ ..	$A'A'A'$ $Y)X+Z)Y=Z)X$	
		$O_1A'O_1$ $X{:}Y+Z)Y=X{:}Z$	(12)
		$A'O_1O_1$ $Y)X+Y{:}Z=X{:}Z$	(23)
x YZ	$C'D_1C'$ ..	$E'A_1E'$ $x.y +Y)Z= x.z$	
		$I_1A_1I_1$ $XY +Y)Z=XZ$	(12)
		$E'O'I_1$ $x.y +Z{:}Y=XZ$	(23)
X y z	$C_1D'C_1$ ..	$E_1A'E_1$ $X.Y+Z)Y=X.Z$	
		$I'A'I'$ $xy +Z)Y=xz$	(12)
		E_1O_1I' $X.Y+Y{:}Z=xz$	(23)
XYz	$D_1C_1C_1$ ..	$A_1E_1E_1$ $X)Y+Y.Z=X.Z$	
		$O'E_1I'$ $Y{:}X+Y.Z=xz$	(12)
		$A_1I'I'$ $X)Y+yz =xz$	(23)
x y Z	$D'C'C'$ ..	$A'E'E'$ $Y)X+y.z =x.z$	
		$O_1E'I_1$ $X{:}Y+y.z =XZ$	(12)
		$A'I_1I_1$ $Y)X+YZ =XZ$	(23)
x Y z	$C'C_1D'$ ..	$E'E_1A'$ $x.y +Y.Z=Z)X$	
		$I_1E_1O_1$ $XY +Y.Z=X{:}Z$	(12)
		$E'I'O_1$ $x.y +yz =X{:}Z$	(23)
X y Z	$C_1C'D_1$ ..	$E_1E'A_1$ $X.Y+y.z =X)Z$	
		$I'E'O'$ $xy +y.z =Z{:}X$	(12)
		E_1I_1O' $X.Y+YZ =Z{:}X$	(23)

Before forming any rule, or making any remark, I proceed to

collect the refults of the remaining cafes. And firft, let a premife be brought to its limit, D or C: fay that $D_1D_1D_1$ becomes
DD_1D_1. In the diagram it immediately appears that one of the
particular conclufions is loft; not contradicted, but nullified: for
(12) difappears, becaufe X and Y are identical names. That is,
$A_1A_1A_1$ remains, and $A_1O'O'$: but the conclufion of $O'A_1O'$ is
nullified. But this very circumftance creates, not a new conclufion, for it is only a part of one already exifting, but a new form
of deduction. The premifes are now $A_1 + A'$ and $A_1 + O'$, and
the conclufion is $A_1 + O'$. The fyllogifms $A_1A_1A_1$ and $A_1O'O'$
are as before, and for the fame reafons: but there is now the
combination $A'A_1$ among the premifes, which produces the conclufion I_1, and we have

$$A'A_1I_1 \quad Y)X + Y)Z = XZ$$

This fyllogifm, though new as far as $D_1D_1D_1$ is concerned, is
only a ftrengthened form of $I_1A_1I_1$, a concomitant of $E'A_1E'$.
For (page 65) I_1 is true whenever A' is true, fo that $A'A_1$ includes I_1A_1 and its neceffary confequence I_1. But if I_1 had been
ftrengthened into A_1 inftead of A', we fhould have had $A_1A_1I_1$
which though perfectly valid, yet admits of a ftronger conclufion, as feen in $A_1A_1A_1$.

Of the two modes of ftrengthening a particular propofition
(as I_1 into A_1 or A') there is one which ftrengthens the quantity
of the firft form of the propofition, and another that of the fecond.
Thus XY or I_1 becomes X)Y or A_1 when the firft form, and
Y)X or A', when the fecond form, is ftrengthened. Similarly
O_1 or X:Y becomes X.Y or E_1, and y.x or E', according as
the form ftrengthened is X:Y or y:x. The prepofition remains
the fame, or changes, according as the firft or fecond form is
ftrengthened. If the *firft* form of the *fecond* premife of a fyllogifm, or the *fecond* form of the *firft* premife, be ftrengthened, no
ftrength is added to the conclufion. Thus, as far as the fyllogifms in this chapter are concerned, I_1A_1 gives as much as $A'A_1$,
and E_1O_1 as E_1E_1. But if the firft form of the firft premif, or
the fecond form of the fecond, be ftrengthened, the conclufion
has its firft form ftrengthened.

A very fimple and obvious theorem contains all thefe refults.
The concluding terms are, in our order of reference, the firft

term of the firſt premiſe and the ſecond term of the ſecond. The concluſion is never ſtrengthened by augmenting the quantity of the middle term, nor *only* weakened (it may be altogether deſtroyed) by weakening the middle term. A wider field of compariſon does not by itſelf give more compariſons: nor can more compariſons ariſe except by augmenting the number of things compared in that field. Since the concluſion can obviouſly ſpeak of no more than was in the premiſes, no term of that concluſion can be augmented in quantity, until the ſame thing has taken place in its premiſe. But no ſtrengthening of a propoſition ſtrengthens both terms: conſequently, to make ſuch a thing effective, it muſt be the concluding, and not the middle, term which is ſtrengthened.

The following table is only worth inſerting as a collection of exerciſes. The fourth column ſhows the eight *ſtrengthened particular ſyllogiſms*, as I will call them, having univerſal premiſes but only a particular concluſion, not ſtronger than might have been inferred from the particular ſyllogiſm itſelf.

Alteration of	into	removes	and ſubſtitutes	ſtrengthened from	occurring in
$D_iD_iD_i$	DD_iD_i	$O'A_iO'$ ⎫	$A'A_iI_i$	$I_iA_iI_i$ ⎫	$C'D_iC'$
$D'D'D'$	$D'DD'$	$A'O_iO_i$ ⎬		$A'I_iI_i$ ⎭	$D'C'C'$
$D_iD_iD_i$	D_iDD_i	$A_iO'O'$ ⎫	$A_iA'I'$	$I'A'I'$ ⎫	$C_iD'C_i$
$D'D'D'$	$DD'D'$	$O_iA'O_i$ ⎬		$A_iI'I'$ ⎭	$D_iC_iC_i$
$D_iC_iC_i$	DC_iC_i	$O'E_iI'$ ⎫	$A'E_iO_i$	$I_iE_iO_i$ ⎫	$C'C_iD'$
$D'C'C'$	$D'CC'$	$A'I_iI_i$ ⎬		$A'O_iO_i$ ⎭	$D'D'D'$
$D_iC_iC_i$	D_iCC_i	$A_iI'I'$ ⎫	$A_iE'O'$	$I'E'O'$ ⎫	$C_iC'D_i$
$D'C'C'$	$DC'C'$	$O_iE'I_i$ ⎬		$A_iO'O'$ ⎭	$D_iD_iD_i$
$C_iD'C_i$	$CD'C_i$	$I'A'I'$ ⎫	$E'A'O_i$	$O_iA'O_i$ ⎫	$D'D'D'$
$C'D_iC'$	$C'DC'$	$E'O'I_i$ ⎬		$E'I'O_i$ ⎭	$C'C_iD'$
$C_iD'C_i$	C_iDC_i	E_iO_iI' ⎫	E_iA_iO'	$O'A_iO'$ ⎫	$D_iD_iD_i$
$C'D_iC'$	CD_iC'	$I_iA_iI_i$ ⎬		E_iI_iO' ⎭	$C_iC'D_i$
$C_iC'D_i$	$CC'D_i$	$I'E'O'$ ⎫	$E'E_iI_i$	$O_iE'I_i$ ⎫	$D'C'C'$
$C'C_iD'$	$C'CD'$	$E'I'O_i$ ⎬		$E'O'I_i$ ⎭	$C'D_iC'$
$C_iC'D_i$	C_iCD_i	E_iI_iO' ⎫	E_iE_iI'	$O'E_iI'$ ⎫	$D_iC_iC_i$
$C'C_iD'$	CC_iD'	$I_iE_iO_i$ ⎬		E_iO_iI' ⎭	$C_iD'C_i$

I will now examine the negatory complex ſyllogiſms, premiſing however than we cannot get any new concluſions from them.

For we have now got all the fixteen cafes in which both premifes are univerfal: and we know that there can be no fyllogifm with a particular premife, except it have one of thofe with univerfal premifes for its opponents.

Take $D_1D':C'$ or A_1+O' and $A'+O_1$ together deny $E'+I_1$, that is, deny the coexiftence of E' and I_1, that is, deny either E' or I_1, that is, affert either I' or E_1. This fyllogifm then may be written thus,

$$(A_1+O')\ (A'+O_1)\ (\text{either } E_1 \text{ or } I')$$

Now the fact is that this disjunction is fuperfluous; it is I' which is always afferted, and E_1 is never a neceffary confequence of D_1D'. For A_1A' gives I' as already fhown, and A_1O_1 and $O'A'$ are inconclufive (and $O'O_1$ of courfe). And the rationale of the inference is as follows: fince X is a fubidentical of Y, and Y a fuperidentical of Z, it follows that Y is fuperidentical both of X and Z; confequently, Y not filling the univerfe (our fuppofition throughout) it follows that there are things which are neither Xs nor Zs, namely, all which are not Ys. Again, in $C_1C_1:C'$, which the fame reafoning fhows to be only C_1C_1I', none either of X or of Z is in Y, therefore every inftance in Y is both x and z. And thus it will appear that in every negatory complex conclufion *the whole middle term*, or the *whole of its contrary*, makes the fubject matter of the ftrengthened particular fyllogifm which is all that can be collected.

Our conclufion is that no negatory complex fyllogifm is of any more logical effect than the ftrengthened particular derived from it. Thus we may fay that, fo far as the extent and character of the inference is concerned, the former is the latter.

I will now pafs to the general rules of the complete fyftem of fyllogifms;—

The reader muft take pains to remember two rules of formation, perfect contraries of each other, for the dependence of the *accents* (or *prepofitions*) on the *fign* (affirmative or negative character) of the firft premife. I exprefs them in the briefeft way poffible.

Direct Rule. Affirmation (in the firft premife) makes the fecond *premife* agree with both the other propofitions, or *ifolates* nothing: negation makes the fecond premife differ from both the others, or ifolates the fecond premife. *Inverfe rule.* Affirmation *ifolates the firft premife*, makes the firft premife differ from both the others in prepofition : negation *ifolates the conclufion*, makes the conclufion differ from both the others. Thefe rules might be expreffed fo as to make their contrariety more complete. Thus in the $\frac{\text{direct}}{\text{inverfe}}$ rule, affirmative commencement fhows $\frac{\text{like}}{\text{unlike}}$ prepofitions in the two premifes, and the conclufion $\frac{\text{agreeing}}{\text{differing}}$ $\frac{\text{with}}{\text{from}}$ the firft premife in prepofition : but negative commencement fhows $\frac{\text{unlike}}{\text{like}}$ prepofitions in the two premifes, and the conclufion $\frac{\text{agreeing with}}{\text{differing from}}$ the firft premife in prepofition.

The fubjects of the following rules are,

1. The eight affirmatory complex fyllogifms.
2. The eight univerfal fimple fyllogifms.
3. The eight ftrengthened particular fimple fyllogifms.
4. The fixteen particular fimple fyllogifms.

Omit the negatory complex fyllogifms, as fully contained in the third of this enumeration, and the complex fyllogifms which contain the unaccented D or C, as carrying a momentary accent for the rule, to be expunged when the formation is completed. Confider $D_i, D, D', A_i, A', I_i, I'$, as of the affirmative figns, and $C_i, C, C', E_i, E', O_i, O'$, as negative.

Rule 1. In the complex fyllogifm all parts are complex ; in the univerfal fimple fyllogifm all parts are univerfal ; in the ftrengthened particular only the conclufion is particular ; in the particular only a premife is univerfal.

Rule 2. Premifes of like fign have an affirmative conclufion ; of unlike fign, a negative.

Rule 3. The complex, the univerfal, the particulars which begin with a particular, follow the direct rule ; the ftrengthened particulars, and the particulars which begin with a univerfal (all that commence with a univerfal, and conclude with a particular) follow the inverfe rule. [Or thus ; all which begin and end alike, follow the direct rule ; all which begin and end differently, the inverfe.]

The complex fyllogifms and univerfals are eafily remembered

by rule: the particulars almoſt as eaſily. The following ſub-rules may be noted, as far as theſe laſt are concerned.

Sub-rule 1. *Firſt and ſecond premiſes.* A and O in the firſt premiſe demand unlike prepoſitions in the two premiſes: E and I demand like prepoſitions. Thus A_1O_1 muſt be inconcluſive: A_1O' muſt be concluſive. But E_1O_1 muſt be concluſive: and E_1O' muſt be inconcluſive.

Sub-rule 2. *Firſt premiſe and concluſion.* A univerſal in the firſt premiſe demands an unlike prepoſition in the concluſion: a particular firſt premiſe, a like prepoſition in the concluſion.

Sub-rule 3. *Second premiſe and concluſion.* Every ſecond premiſe demands its own prepoſition in a concluſion of like ſign: and the other prepoſition in a concluſion of unlike ſign.

As far as the four ſpecies are concerned, every ſyllogiſm formed according to the three rules is valid; and every one not ſo formed is invalid. The following remarks are partly recapitulatory, partly new.

Remark 1. Every complex ſyllogiſm gives one univerſal ſyllogiſm * and two particular ones, its concomitants: and the concomitants are formed by changing one of the premiſes of the univerſal and the concluſion, into their particular concomitant propoſitions (page 63.)

Remark 2. Every ſyllogiſm has its *contranominal*, which aſſerts of the contraries in the ſame manner as the firſt does of the direct terms: and contranominals have all their accents different, as in $O'A_1O'$ and $O_1A'O_1$ (page 62.)

Remark 3. Every ſyllogiſm has *two* opponents, made by interchanging the contradictories of one premiſe and of the concluſion, and altering the accent of the remaining premiſe, if inconvertible (A or O) (page 88.)

Remark 4. Every complex ſyllogiſm has two ſuch opponents formed in the ſame way, the Ds being the inconvertibles, the Cs the convertibles. Thus (:) meaning *denial of*, the opponents of $C_1D'C_1$ are $C_1:C_1:D'$ and $:C_1D_1:C_1$. The firſt of theſe is

$$(E_1 + I') \; (I_1 \; or \; E') \; (O' \; or \; A_1)$$

containing the valid ſyllogiſms $E_1E'A_1$, E_1I_1O', $I'E'O'$; being

---

* *Syllogiſm*, not preceded by *complex*, means ſimple ſyllogiſm.

$E_iE'A_i$ and its concomitants. And $:C_iD_i:C_i$ gives $E'A_iE'$ (the contranominal of $E_iA'E_i$) and its concomitants. And the fame of the reft.

Remark 5. Each univerfal fyllogifm has two weakened forms, made by weakening one premife and the conclufion. When the *firft* premife is weakened, it is without change of prepofition : but when the fecond, with change. Thus the weakened forms of $E_iA'E_i$ are $O_iA'O_i$ and E_iI_iO'.

Remark 6. Each particular fyllogifm has two ftrengthened forms, one of which is a univerfal, the other only a ftrengthened particular. Thus the ftrengthened forms of $O_iA'O_i$ are $E_iA'E_i$ and $E'A'O_i$.

Remark 7. In every fyllogifm except the ftrengthened particular, the middle term is univerfal in one premife, and particular in the other : and its contrary is therefore the fame. But in the ftrengthened particular, the middle term is univerfal in both premifes, or particular in both. This affords a complete criterion of fyllogifm, as will be noticed hereafter : in fact, the completenefs of this fyftem crowds us with relations, from many of which general rules might be deduced, though they need only appear here by cafual remark.

In $O'A_iO'$, $A'O_iO_i$, $I_iA_iI_i$, E_iO_iI', $O'E_iI'$, $A'I_iI_i$, $I_iE_iO_i$, E_iI_iO', the middle term enters univerfally in the univerfal, and particularly in the particular. In all the others it enters particularly in the univerfal, and univerfally in the particular. In the firft fet, the convertible premifes are all *fubs*, the inconvertibles are *fubs* in the fecond premife, and *fupers* in the firft. In the fecond fet, thefe rules are inverted.

Remark 8. Of the twelve poffible pairs of premifes AA, AE, AI, AO, EA, EE, EI, EO, IA, IE, OA, OE, which *can* give a conclufion, each one *will*, in two ways, which two ways are inverted in their accents. Thus EO appears in $E'O'I_i$ and E_iO_iI'. The two premife-letters and one accent dictate all the reft : thus $I'A$ can belong to nothing but $I'A'I'$. When the fyftem is well learnt, it will be found unneceffary to write more than $I'A$, for the fymbol of $I'A'I'$. I now fpeak only of fundamental fyllogifms : the ftrengthened fyllogifm $A_iA'I'$ might be fignified by $A'A'$.

Remark 9. The fyllogifms of the three firft claffes are all really specimens of one, those of the fourth of two, among them, with the eight variations XYZ, xYZ, XYz, xYz, XyZ, xyZ, Xyz, xyz. The rules for conducting these changes are

Change of subject is change of both accent and letter.
Change of predicate is change of   letter.
Change of both is change of   accent.

thus to pass from $E'E_1A'$ to $A_1E_1E_1$ we note in XY change of subject, in YZ change of neither, in XZ change of subject: therefore xYZ is the set of terms into which XYZ must be changed: and the $E'E_1A'$ syllogism of either set is the $A_1E_1E_1$ syllogism of the other.

The 24 syllogisms, which are 24 with reference to the order XY, YZ, XZ, are only 12 if the order ZY, YX, ZX, be allowed. Thus $A_1I'I'$ of the first is the $I'A'I'$ of the second. These syllogisms are essentially the same in the mode of inference they afford. To change a syllogism into another of the same mode of inference, invert the premises and change the preposition of all the inconvertibles. Thus $A'O_1O_1$ and $O'A_1O'$ are of the same inference. The pairs which in this point of view are identical are

$A_1A_1A_1 = A'A'A'$	$E'A_1E' = A'E'E'$	$E_1A'E_1 = A_1E_1E_1$	$E'E_1A' = E_1E'A_1$
$O'A_1O' = A'O_1O_1$	$I_1A_1I_1 = A' I_1 I_1$	$I'A'I' = A_1I'I'$	$I_1E_1O_1 = E_1I_1O'$
$A_1O'O' = O_1A'O_1$	$E'O'I_1 = O_1E'I_1$	$E_1O_1I' = O'E_1I'$	$E'I'O_1 = I'E'O'$

The ninth remark admits of considerable extension. The '*some*' of a logical proposition may have a much more definite character in some cases than in others. It may be a selected, or at least a distinguishable *some*, which want nothing but a nominal distinction to make the particular proposition easily and usefully universal. Whether it can be done more or less easily, and more or less usefully, is no question of formal logic. If it be supposed done, the particular is converted into a universal. In 'some Xs are Ys,' if we make a name for every X which is Y, say M, we have then 'Every M is Y'. This proposition may be purely identical, or it may not. If we call every X which is Y by the name M merely because it is Y, then our universal is only 'Every *X which is Y* is Y'. But if the name M be conferred from any other circumstance, which distinguishes the Xs that are Ys from other Xs, then the change from the particular to the universal by means of the new reſtriction impoſed by the new name, is the expreſſion of new knowledge.

The quantities in the concluſion are of two kinds. There are thoſe which are brought in with the terms, and which continue in the concluſion ſuch as they were introduced in the premiſes : and there are thoſe which depend on the union of the premiſes, and which are what they are only in virtue of the joint exiſtence of the premiſes. For example, in $I_1A_1I_1$ we have ' ſome Xs are Ys, but every Y is Z, therefore ſome Xs are Zs' : if we aſk, what Xs are Zs, the anſwer is, thoſe which are Ys, and no others, ſo far as this concluſion affirms. But when we look at $O'A_1O'$ or 'ſome Ys are not Xs, and every Y is Z ; therefore ſome Zs are not Xs : and if we then aſk *what* Zs are not Xs ; the anſwer is, that this quantity does not enter with Z, but depends upon the other premiſe, namely, upon the number of Ys which are not Xs. In a particular ſyllogiſm, let us call the quantity of the ſubject in the concluſion *intrinſic* or *extrinſic* according as it is that of the premiſe which introduces that ſubject, or of the other premiſe. Examination will ſhow that in every particular ſyllogiſm which concludes in I_1 or I', in which both terms are particular, the quantities of the terms are, of the one intrinſic, of the other extrinſic : but that where the concluſion is in O_1 or O', either the quantity of the ſubject is intrinſic and that of the contrary of the predicate extrinſic, or *vice verſâ*.

When the quantity of a particular term in the concluſion is intrinſic, the invention of a name will convert the ſyllogiſm into a univerſal. Thus $I_1A_1A_1$ or XY + Y)Z = XZ, if M be taken to repreſent all thoſe Xs which are Ys, and nothing elſe, becomes M)Y + Y)Z = M)Y, of the form $A_1A_1A_1$. Again, $O'A_1O'$ or Y:X + Y)Z = Z:X, thrown into the form x:y + z)y = x:z, becomes m.y + z)y = m.z, of the form $E_1A_1E_1$, when the xs which are ys are diſtinguiſhed from the reſt of the univerſe by the name m. There is nothing either illegitimate or uncommon in diſtinguiſhing by a peculiar name *certain ſome* (or even *uncertain ſome*, if *certainly always the ſame ſome*) of another name. Again, ſince we know that every univerſal ſyllogiſm is reducible to the form $A_1A_1A_1$ by uſe of contraries, we have now reaſon to know that there is no fundamental inference, of the kind treated in this chapter, which is any other than that in $A_1A_1A_1$, or, the *contained of the contained is contained.* And there is no better exerciſe than learning to read off each of the ſyllogiſms, univerſal and particular, into this one form, by perception, and without uſe of rules. Take as an inſtance $X:Y+y.z=XZ$: what is the container, what is the contained, and what is the middle container of one and contained of the other. It is a parcel of Xs which are contained in y, all y in Z, and therefore *that parcel* of Xs in Z.

This general principle ſuggeſts a notation for all the complex, univerſal, and .fundamental particular, ſyllogiſms. If we abbreviate $X)Y+Y)Z=X)Z$ into $XYZ)$, and if we denote by XYZ, without), that it is only a parcel of Xs (all or ſome, defined or undefined, but always the ſame), we have the following,

For $A_iA_iA_i$ read XYZ) or zyx)　　For A'A'A' read xyz) or ZYX)
— $O'A_iO'$ — xYZ　　　　　　— $O_iA'O_i$ — Xyz
— $A_iO'O'$ — Zyx　　　　　　— $A'O_iO_i$ — zYX

For $E'A_iE'$ read xYZ) or zyX)　　For $E_iA'E_i$ read Xyz) or ZYx)
— $I_iA_iI_i$ — XYZ　　　　　　— I'A'I' — xyz
— $E'O'I_i$ — ZyX　　　　　　— E_iO_iI' — zYx

For $A_iE_iE_i$ read XYz) or Zyx)　　For A'E'E' read xyZ) or zYX)
— $O'E_iI'$ — xYz　　　　　　— $O_iE'I_i$ — XyZ
— $A_iI'I'$ — zyx　　　　　　— $A'I_iI_i$ — ZYX

For $E'E_iA'$ read xYz) or ZyX)　　For $E_iE'A_i$ read XyZ) or zYx)
— $I_iE_iO_i$ — XYz　　　　　　— I'E'O' — xyZ
— $E'I'O_i$ — zyX　　　　　　— E_iI_iO' — ZYx

Here, uſing P,Q,R, as general terms, PQR) denotes that all Ps are Qs, and all Qs are Rs, whence all Ps are Rs : while PQR only denotes that there is a parcel of Ps among the Qs, and all Qs are among the Rs, whence that parcel of Ps is among the Rs.

The rules for the connection of theſe ſyſtems are not complicated, conſidering the extent of the caſes they are to include. Let the letters A,E, &c. be called *proponents* ; X,Y,Z, *nominals:* and by the *order* of the nominals we always mean that X is firſt, &c. both in XYZ, and ZYX. The nominals being *direct* (X,Y,Z) and *contrary* (x,y,z), remember that, *firſt,*

An affirmative $\begin{cases} \textit{firft} \\ \textit{fecond} \\ \textit{third} \end{cases}$ proponent denotes that the $\begin{cases} \textit{firft} \text{ and fecond} \\ \textit{fecond} \text{ and third} \\ \textit{third} \text{ and firft} \end{cases}$

nominals agree (are both direct or both contrary).

A negative $\begin{cases} \textit{firft} \\ \textit{fecond} \\ \textit{third} \end{cases}$ proponent denotes that the $\begin{cases} \textit{firft} \text{ and fecond} \\ \textit{fecond} \text{ and third} \\ \textit{third} \text{ and firft} \end{cases}$

nominals differ (are one direct, one contrary).

> Thus EIO muft give Xyz or xYZ or zyX or ZYx
> IEO muft give XYz or xyZ or zYX or Zyx

Secondly, whether the middle term be Y or y depends only on the accent of the middle proponent: a *fub*-accent gives Y, a *fuper*-accent gives y. In the univerfal fyllogifm however, either gives either.

Thirdly, the XYZ fyllogifms are the particulars which begin with a particular: and the ZYX fyllogifms are the particulars which begin with a univerfal.

For example, required $O_iE'I_i$. Seeing the particular O_i, at the beginning, take the order XYZ, feeing the fuperaccent in E' make it XyZ. Seeing the negative O_i, let the exifting difagreement of the firft and fecond nominals continue; and the fame of the fecond and third from the negative E_i. Confequently XyZ is the fyllogifm expreffed in nominals. Or the rationale of the inference in $O_iE'I_i$ is that a parcel of Xs are among the Zs becaufe among the ys which are all among the Zs.

Again, required the nominal mode of expreffing $E'I'O_i$. Seeing the univerfal E at the beginning, write down ZYX; for the fuperaccent in I', write down ZyX; for the negative in E', continue yX; for the affirmative in I', write zy: hence zyX is the nominal form of $E'I'O_i$.

Required the proponent mode of expreffing xYz. Here xY, Yz, fhow us that the premifes are negative forms, and the direction of the order x, Y, z, that the firft premife is particular. Then OE are the premifes, and I the conclufion. And Y tells us that the middle proponent has a fubaccent. Whence OE_iI is, fo far as it goes, the proponent expreffion. And, by the laws of form, the other accents muft be as in $O'E_iI'$, fince the fyllogifm follows the direct rule (page 93).

Required the proponent mode of expressing ZYx. Here we note in succession—universal commencement—first premise negative—second, affirmative—middle accent sub. This gives EI_1O of the inverse rule, or E_1I_1O'.

Required the proponent notation for the universal xYZ) or zyX). We see at once EA_1E, or $E'A_1E'$.

The concomitants of a universal are found by changing the first nominal into the contrary, in each of the forms, and throwing away the sign of universality [)] . Thus the concomitants of XyZ) or zYx) are xyZ and ZYx.

The weakened forms of a universal are found by merely throwing away the symbol of universality [)] from the two forms of the universal. Thus the weakened forms of XYZ) which is also zyx) are XYZ and zyx.

But we have not yet reached the climax of symbolic simplicity in the mere representation of syllogisms. An algebraist would say that the structure of the inference, as now considered, does not depend upon the names; but only upon their reference to the names in the fundamental form XYZ). He would therefore propose a simple symbol to represent *letting alone*, and another to represent *changing into the contrary*. These, with a sign of complete universality, and another of inversion of order, are all that he would find necessary. Let o and 1 signify letting alone and changing into the contrary : let the terminal parenthesis denote complete universality, as before, and let inversion of order be denoted by a negative sign prefixed. Thus XYZ or $I_1A_1I_1$, would be denoted by ooo ; Zyx or $A_1O'O'$ by—o11 ; $A_1E_1E_1$ or XYz) by oo1) or its equivalent—o11. Thus—o11 tells us that some of the Zs are ys, all the ys are xs, whence some of the Zs are xs. To write its proponent form, observe that — instructs us to write a universal first ; 11 to make it affirmative ; 1 in the middle to superaccent the middle proposition ; o1 to make the second premise negative. We have then $A_1O'O'$ or X)Y+ Z:Y=Z:X which is Zy+y)x=Zx, as asserted.

All that relates to universals in the preceding, applies to the complex syllogisms. Let a couple of parentheses imply a complex syllogism : thus $D_1D_1D_1$ may be (XYZ) or (ooo). Then in (o1o) or (XyZ), we are to see that X is a subidentical of y, and y of Z, whence X is the same of Z. But Xy and yZ warn us to write *contraries* for the firſt and ſecond premiſes and y to ſuperaccent the middle letter: whence $C_iC'D_i$ is the ſyllogiſm expreſſed by the names XYZ. The equivalent forms—(101) and (zYx) expreſs it by ſaying that z is a ſubidentical of Y and Y of x, whence z is a ſubidentical of x.

I now look at the ſtrengthened particular ſyllogiſms. All inference which is fundamental, that is, which will come from nothing weaker than the premiſes given, has been reduced to the one eaſy caſe of ' the contained of the contained is contained.' The ſtrengthened particular, the type of which is $A'A_iI_i$, obeying the inverſe rule of formation, and written at more length in Y)X + Y)Z $=$ XZ, may be ſtated thus ' all names are common as to what they contain in common.' If we denote this ſtrengthened ſyllogiſm by XYZ|, a ſymbol intended to imply ſomething between XYZ and XYZ) in the amounts of quantity introduced, we ſhall find that the eight ſtrengthened ſyllogiſms muſt be repreſented by

$$A'A_iI_i = XYZ| \qquad A_iA'I' = xyz|$$
$$A'E_iO_i = XYz| \qquad A_iE'O' = xyZ|$$
$$E'A'O_i = Xyz| \qquad E_iA_iO' = xYZ|$$
$$E'E'I_i = XyZ| \qquad E_iE_iI' = xYz|$$

The rules of connexion are preciſely thoſe for the particular ſyllogiſms: and inverſion is abſolutely ineffective. Thus XYZ| $=$ ZYX|.

A few words will ſerve to diſpoſe of the *mixed complex ſyllogiſms* in which a complex premiſe is combined with a ſimple one, univerſal or particular. Firſt, when a complex and a univerſal are premiſed, and ſigns and accents are as in the *direct rule* (page 92), the concluſion is as it would be if the A were heightened into D, or E into C. Thus E_iD' gives C_i, the ſame as C_iD'. For E_i is C or C_i, and both CD' and C_iD' give C_i, but with different quantities. But if the premiſes be conſtructed on the inverſe rule, there is no more inference than can be obtained when the complex premiſe is lowered into a univerſal: or we have only a ſtrengthened particular. Thus in D_iE' or $(A_i + O')E'$, A_iE' gives the ſtrengthened particular $A_iE'O'$, and O'E' is inconcluſive. And when the complex premiſe is combined with a particular, we have only what would follow if the complex premiſe were lowered into a univerſal. Thus D_iI', or

$(A_{i}+O')I'$ can only give $A_{i}I'I'$; and $D'I'$ or $(A'+O_{i})I'$ gives no conclufion, for $A'I'$ is inconclufive.

The claffification of opponent forms may be thus treated. We know that opponent forms of AEE, for inftance, be it $A_{i}E_{i}E_{i}$ or $A'E'E'$, muft be IEO and AII. Now whether $A_{i}E_{i}E_{i}$ fhall have $I_{i}E_{i}O_{i}$ or $E_{i}I_{i}O'$, whether $A'I_{i}I_{i}$ or $I_{i}A_{i}I_{i}$, depends upon the introduction of a new and arbitrary notion of the order to be adopted. Our firft fyllogifm being defcribed by XY, YZ, XZ, the opponent which ends in the contradiction of the firft premife is in XZ, YZ, XY; which, keeping Z middle, is either to be defcribed with reference to XZ, ZY, XY, or to YZ, ZX, YX. Now in adopting the firft of thefe three orders, there is nothing which compels us therefore to prefer the fecond to the third, or *vice verfâ.*

The effect of the change of order which confifts in the interchange of Z and X is as follows. The premifes change places; A and O with altered accents, altered alfo in the conclufion, E and I with unaltered accents. Thus $A_{i}I'I'$ becomes $I'A'I'$; $E'O'I_{i}$ becomes $O_{i}E'I_{i}$. Accordingly, it is matter of new arrangement whether for inftance, $I_{i}E_{i}O_{i}$ or $E_{i}I_{i}O'$ fhall be called the opponent of $A_{i}E_{i}E_{i}$; and I prefer to give the name to both. The confequence is, the following diftribution of opponents;—

$$\left.\begin{matrix} AA & AO & \dot{O}A \\ EE & EO & OE \end{matrix}\right\| \quad AE \quad EA \quad AI \quad IA \quad EI \quad IE.$$

The three fets reprefent letters combined in reprefentation of premifes: the firft two containing fix fyllogifms each, the third twelve. The third muft be divided into two fets of fix each, in one of which the fubaccents are in greater number, in the other the fuperaccents. There are then four fets in all. Pick any two out of a fet, which only differ in change of order: thefe two have the fame opponent forms, namely, the other four of the fet. For inftance, $A'I_{i}I_{i}$ and $I_{i}A'I_{i}$, in which fubaccents predominate. Take AE, EA, EI, IE, and complete fyllogifms in fuch manner as to make fubaccents predominate: giving $A_{i}E_{i}E_{i}$, $E_{i}A'E_{i}$, $E_{i}I_{i}O'$ $I_{i}E_{i}O_{i}$. The laft four are the opponents of the firft two.

In the fet of ftrengthened particulars the opponent forms will be found to be univerfals weakened in the conclufion without being weakened in the premifes. Thus $A_1A'I'$ has $A'E'O'$ for one of its opponents : but $A'E'$ may produce the univerfal con-clufion E' as well as its weaker form O'.

Some readers, particularly thofe who have a tincture of algebra, are more helped by fymbolic notation than by language : with others it is the converfe. To fuit the latter, obferve that the language of page 78 may eafily be adapted to fimple fyllogifms. Thus A_1 being fubaffirmation, I_1 may be fome fubaffirmation, O' may be fome fupernegation ; and fo on. Thus inftead of $E'I'O_1$ we may fay that ' fupernegation of fome fuperaffirmation gives fome fubnegation.' Practice in this language would make the phrafe fuggeft fomething more than the notation it is derived from. The phrafe refers to Z : there is a term partially fuperaffirmed of Z, namely Y ; and a complete fubnegative of Y, namely X. The partial fubaffirmation declares fome things neither Y nor Z ; the complete fupernegation declares that whatever is not Y is X. Confequently there are fome Xs which are not Zs : or X is a partial fubnegative of Z. This fubject will be refumed.

In what precedes are two views of the deduction of all the varieties of fyllogifm. The firft, taking the complex fyllogifm as the fource, connects the ftrengthened fyllogifms and the parti-cular ones with the univerfals, and thus in fact reduces every thing to the conftituents of $D_1D_1D_1$ or DD_1D_1. The fecond pro-ceeds from $A_1A_1A_1$, $A'A_1I_1$, $A_1I'I'$, and $I_1A_1I_1$, and forms the claffes of univerfal, ftrengthened, and particular, fyllogifms by fubfti-tuting contraries in every way in which it can be done. Thefe two fyftems have clofe connexion, but not fo clofe as might perhaps be thought : for $I_1A_1I_1$ is not one of thofe which are connected with $A_1A_1A_1$ in the formation of a complex fyllo-gifm.

The two new views which I now proceed to give are alfo clofely connected, and different from the former ones, in which we held it equally admiffible to refer one of the concluding terms to the middle, as in $X)Y$, or the middle to one of the concluding terms, as in $Y)X$. But now I afk whether it be not poffible fo to conftruct the fyftem, that we may firft lay down the middle term and its contrary, as conftituting the univerfe of the fyllogifm, and then complete the premifes and their conclufion, by properly laying down the concluding terms in their places. We may fucceed, if, in the firft inftance, we confider none but convertible propofitions. And this we can do; for univerfal exclufion and particular inclufion comprehend all affertion. Thus univerfal inclufion is only univerfal exclufion from the contrary, and particular exclufion is only particular inclufion in the contrary.

Setting out then with the middle term and its contrary, and reftricting ourfelves to E and I, let E fignify (univerfal) exclufion from the middle term, and e from its contrary; let I fignify (particular) inclufion in the middle term, and i in its contrary. Choofing a pair of concluding terms, we reject II, Ii, and ii on grounds already demonftrated, and very eafily feen in this view, and proceed to confider Ee, EE and ee, EI and ei, Ei and eI.

Ee. From this a univerfal conclufion muft follow. If one term be completely excluded from the middle and the other from its contrary, the terms are completely excluded each from the other. The fundamental forms are,

$$E_{\iota}A'E_{\iota}, \ X.Y+Z.y=X.Z \ ; \ A_{\iota}E_{\iota}E_{\iota}, \ X.y+Z.Y=X.Z$$

and by ufe of XZ, Xz, xZ, xz, we thus bring out the eight univerfal fyllogifms.

EE and ee. From thefe a particular inclufion muft follow. Exclufion of both terms from a third, gives partial inclufion of their contraries in each other: for all that third term belongs to the contraries of the other two. The fundamental forms are,

$$E_{\iota}E_{\iota}I', \ X.Y+Z.Y=xz \ ; \ A_{\iota}A'I' \ X.y+Z.y=xz$$

from which, as before, the eight ftrengthened fyllogifms are deduced.

EI and ei. From thefe a particular inclufion muft follow. The exclufion of one term from a third, and the inclufion of part of a fecond term in that third, tell us that part of the particularized term is in the contrary of the univerfalized term. The fundamental forms are,

$$E_\iota I_\iota O', \ X.Y + ZY = Zx \ ; \ A_\iota O'O', \ X.y + Zy = Zx$$
$$I_\iota E_\iota O_\iota, \ XY. + Z.Y = Xz \ ; \ O_\iota A'O_\iota, \ Xy + Z.y = Xz$$

from which the fixteen particular fyllogifms are deduced.

Ei and eI. From thefe no conclufion can be drawn. All that is fignified is that one concluding term is wholly excluded from a third, and the fecond partially excluded (or included in the contrary).

It thus appears that a fyllogifm with one particular premife is valid when the premifes reduced to convertible forms, fhow the middle term in both or the contrary of it in both; otherwife, invalid. Alfo, that the conclufion in its convertible form, takes directly from the particular premife and contrariwife from the univerfal.

It alfo appears that a fyllogifm with both premifes univerfal is always valid; with a univerfal conclufion when the premifes (made convertible) fhow one the middle term and the other its contrary; with a particular conclufion when both fhow the middle term or both its contrary. And the convertible form of the conclufion takes directly from both in the firft cafe, and contrariwife from both in the fecond.

The other view which I here propofe is really a different mode of looking at that juft given. By the time we have made every name carry its contrary, as a matter of courfe, we become prepared to take the following view of the nature of a propofition. A name by itfelf is a found or a fymbol : its relation to things (be they objects or ideas) is twofold. There may be *in rerum natura* that to which the name applies, or there may not. I do not here fpeak of how many things there may be to which a name applies : it is not effential to know whether they be more or fewer, either abfolutely or relatively. The introduction of *contraries* may be made the expulfion of *quantity.* With reference to application, then, let a name be called *poffible* or *impoffible* according as the thing to which it applies can be found or not.

A name may be compounded of others ; the compound name being that of everything to which all the components apply. Thus *wild animal* is the name of all things to which both the names *wild* and *animal* apply. To call this compound name impossible is to say that there is not such a thing as a wild animal: to call it possible is to say that there is such a thing.

X and Y being two names, the compound name may be represented by XY when possible, and by XY) when impossible. This does not alter the meaning of our symbol XY, as hitherto used: as yet it has been ' there are Xs which are Ys' and now it is ' XY, the name of that which is both X and Y, is the name of some thing or things;' and these two are the same in meaning, so far as their use in inference is concerned. Nor need XY), as just defined, be treated as a departure from, otherwise than as an extension of, the use of X)Y. In X)Y, we assert that X is something, namely Y: in X) we assert that X is *nothing whatever*. The proper notation, however, for indicating that the name X has no application, is X)u, u being the contrary of U, which last includes everything in the universe spoken of; so that u may denote nonexistence.

The proposition ' Every X is Y' asserts that Xy is the name of nothing, or X)Y=Xy). Similarly ' No X is Y' asserts that XY is the name of nothing, or X.Y=XY). But ' Some Xs are Ys' and ' Some Xs are not Ys' merely assert the possibility of the names XY and Xy.

A syllogism, then, is the assertion that from the possibility or impossibility of the names produced by compounding X or x, Z or z, each with Y or y, may be inferred the possibility or impossibility of a name compounded of X or x with Z or z. The rules of the last system are now so easily changed into the language of the present one, that it is hardly worth while to state more than one for example. Thus, if X compounded with Y, and Z compounded with y, both give impossible names, then X compounded with Z gives an impossible name. This is XY)+ Zy)=ZX) or X.Y+Z.y=Z.X, or $E_1A'E_1$

The view here taken of compound names will be extended in the next chapter.

CHAPTER VI.

On the Syllogifm.

WHEN the premifes of a fyllogifm are true, the conclufion is alfo true, and when the conclufion is falfe, one or both of the premifes are falfe. There are two kinds of modifications which it may be ufeful to confider: thofe which concern the entrance of the propofition into the argument; and thofe which affect the connexion of the fubject and predicate.

As to the propofition itfelf, it may be true or falfe abfolutely, or it may have any degree of truth, credibility, or probability. This relation will be hereafter confidered; and, according to the principles of Chapter IX. fo far as the propofition is probable it is credible, and fo far as it is credible, it is true. But as to other modes of looking at the fyllogifm, are we entitled to fay that every thing which can be announced as to the premifes may be announced in the fame fenfe as to the conclufion? The anfwer is, that we cannot make fuch announcement abfolutely; but of the premifes *as derived from that conclufion* we can make it. In what manner foever two premifes are applicable, their conclufion as from thofe premifes is alfo applicable: becaufe the conclufion is in the premifes. For inftance, in the fyllogifm 'all men are trees, all trees are rational, therefore all men are rational,' the premifes are abfurd and falfe, and the conclufion taken independently is rational and true: but that conclufion, as from thofe premifes, is as abfurd as the premifes themfelves. Again, in 'all pirates are convicted, all convicts are punifhed, therefore all pirates are punifhed,' the premifes are *defirable*, and fo is the conclufion with thofe premifes. But the conclufion is not defirable in itfelf: as that pirates fhould be punifhed with or without trial. Neither may we fay 'X ought to be Y and Y ought to be Z, therefore X ought to be Z' except in this manner, that we affirm X ought to be Z in a particular way. We may not even fay that when 'X ought to be Y, and Y *is* Z' it follows that 'X ought to be Z,' for it may be that Y ought not to be Z. Thus a royalift, in 1655, would fay that the hundred excluded members of Cromwell's parliament ought to be allowed to take their feats, and alſo that all who took any feats in that parliament were rebels; but he would not infer that the hundred members ought to be rebels.   There is nothing which, being the property of the premiſes, is neceſſarily the independent property of the concluſion, except *abſolute truth.*   It ſhould be noted that in common language and writing, the uſual meaning of concluſions is that they are ſtated as of their premiſes and to ſtand or fall with them, even as to truth.   Though a concluſion may be true when its premiſes are falſe, the proponent does not mean, for the moſt part, to claim more than his premiſes will give, nor that any thing ſhould ſtand longer than the premiſes ſtand.

Next, we are not to argue from what we may ſay of a propo-ſition to what we may ſay of the inſtances it contains, except as to what concerns the truth of thoſe inſtances, or elſe to what concerns the inſtances as parts of a whole.   If I ſay ' Every X is Y' I aſſert, no doubt, of each X independently of the reſt: that is, the truth of ' Every X is Y' involves the truth of ' this X is Y.'   But if, to take ſomething elſe, I maintain ' Every X is Y' to be a deſirable rule, I do not therefore aſſert ' this X is Y' to be a deſirable caſe, except upon an implied neceſſity that there ſhould be a rule.   And if I ſay that ' every X is Y' is unintelligible, I do not ſay that ' this X is Y' is unintelligible; and ſo on.   Thus, where there muſt be a rule, as in law, ' every man's houſe is his caſtle' is deſirable, becauſe there is but one alternative ' no man's houſe, &c.'   But the propoſition, by itſelf, may not be deſirable as to the inſtance of a generally reputed thief or receiver.

There is one caſe, however, in which a term cannot be ap-plied to the general propoſition, unleſs it can be applied in a higher degree to the inſtances.   The propoſition ' Every X is Y' cannot be announced as of any degree of probability, unleſs each inſtance has a much higher degree of probability.   If μ, ν, ρ, &c. be the probabilities of the ſeveral inſtances, ſuppoſed independent, that of the propoſition (Chapter IX.) is $\mu\nu\rho$ ... which product muſt be leſs than that of any one of the fractions of which it is formed.

I now come to the conſideration of circumſtances which mo-dify the internal ſtructure of the premiſes themſelves.   And firſt of conditions.

A *conditional* propofition is only a grammatical variation of the ordinary one; as in 'If it be X, then it is Y.' The common form of this, 'Every X is Y,' is called *categorical*, or *predicative*. Of the two forms, categorical and conditional, either may always be reduced to the other; as follows,

'Every X is Y' or 'If X, then it is Y'
'No   X is Y' or 'If X, then it is not Y'

The particular propofitions might be given conditionally in various ways, but the transformation is not fo common. Thus 'fome Xs are Ys' might be 'if X, then it may be Y' or 'if X, then Y muft not therefore be denied of it,' &c.

Of the two common fubject-matters of names, ideas and propofitions, it is moft common to apply the categorical form to the firft, and the conditional form to the fecond: in truth we might call the conditional form a grammatical convenience for the expreffion of dependence of propofitions on one another, and of names which require complicated forms of expreffion. Thus in pages 2 and 3, the conditional forms, containing *if*, are more fimple than the correfponding categorical forms.

A condition may be either *neceffary*, or *fufficient*, or both. A neceffary condition is that without which the thing cannot be; a fufficient condition is one with which the thing muft be. In pages 73, 74, I have fufficiently pointed out the completenefs of the connexion between the conditional and the categorical forms. In any one cafe the fufficient muft contain all that is neceffary, and may contain more.

After what is faid in page 23, it is not neceffary to dwell on the reduction of a conditional* fyllogifm to a categorical one. The premifes contain the conclufion: whatever gives us the premifes, gives us the conclufion. But I think that the reduction of conditional to categorical forms, though juft, and, for inference, complete, is not the reprefentation of the whole of what paffes in our minds.

As an example of what I mean, look forward to the numerical fyftem of Chapter VIII. Precedent to all propofitions,

---

* Wallis, as far as I know, was the firft who afferted that all fyllogifms are, or can be made, categorical. He did this in the fecond thefis attached to his logic, headed *Syllogifmi Hypothetici, aliique Compofiti, referendi funt omnes ad Ariftotelicos Categoricorum Modos.*

there are the numerical conditions which prescribe the limits of
the universe under consideration. Say there are 250 instances in
that universe: this is the first condition. Of these 100 are Xs and
200 are Ys; giving a second and third condition. If we take a
proposition, as 20XY, and ask whether it be spurious or not, we
have reference to the three conditions understood. But this is not
necessary: for it would be possible categorically to express these
conditions by "20Xs out of 100 in a universe of 250 instances
containing 200Ys are to be found among those 200 Ys? It is
of course the rule of brevity not to drag about these conditions
with every proposition which is employed, but rather to state
them once for all. There is however something more. The
conditions are a restriction upon the arguments intended to be
introduced, and a restriction throughout. The attachment of
them to each individual proposition does not express this: if they
be seen in twenty consecutive propositions, there is no more than
a presumption that they are to be seen in the twenty-first. It is
better that the limits allowed should be marked out by one boundary than that the several arguments should each have a description of the boundary to itself.

Just as a universe of names is defined by specifying one or
more names to constitute collectively the *summum genus*, or *universe*, so one of propositions may be defined by stating propositions which are to be true, or which are not to be contradicted, as
the case may be. These propositions may be conditions preceding all, or some only, of the premises which are used in argument; or some may precede some, and others others. In
analysing arguments, it would be found that many propositions
which enter as premises, enter each with a condition understood,
and well understood, to be granted. Whatever the conditions
may be, so long as the consequent propositions act logically together to produce the final result, then that same result depends at
last only on the conditions, and must be affirmed when the conditions, and their connexion with their consequents, are affirmed.
But then it must be understood that the result also stands upon
the conditions, and may fall with them. Let us now examine
the common syllogism, and see whether there be any preceding
conditions, on which the result depends.

On looking into any writer on logic, we shall see that *existence*

is claimed for the fignifications of all the names. Never, in the
ftatement of a propofition, do we find any room left for the alternative, *fuppofe there fhould be no fuch things*. Exiftence as objects, or exiftence as ideas, is tacitly claimed for the terms of
every fyllogifm.. The exiftence of an idea we muft grant whenever it is diftinctly apprehended, and (therefore) not felf-contradictory : we cannot for inftance admit the notion of a lamp
which is both metal and not metal; but, as an idea, we are at
liberty to figure to ourfelves fuch a lamp as that with which
Aladdin made his fortune. An attempt at a felf-contradicting
idea is no idea; we have not that apprehenfion of it in which an
idea confifts : but in no other way can we fay that the attempt
to produce an idea fails. It may then be more convenient here
to dwell on *objective* definition of terms, as more eafily conceived with relation to exiftence and non-exiftence. Accordingly,
let us take the propofitions X)Y and X.Y, of the character of
which the particulars muft partake, as to the point before us.
By the meaning of y, in relation to Y, it follows that every thing
is either Y or y : if we fay that Y does not exift, then every thing
is y. If then X exift, and Y do not, the propofition X)Y, or
X.y is falfe, and X)y, or X.Y is true. If neither X nor Y
exift, I will not fo far imitate fome of the queftions of the fchools
as to attempt to fettle what nonexifting things agree or difagree.
If Y exift, but not X, then y)x is certainly true, but not thence
X)Y, for when x is, as here, the whole univerfe, the proof of
y)x=X)Y fails to prefent intelligible ideas, that is, fails to be a
proof. But Y)x or Y.X is true.

If all my readers were mathematicians, I might purfue thefe
extreme cafes, as having intereft on account of their analogy
with the extreme cafes which the entrance of zero and of infinite
magnitude oblige him to confider. But as thofe who are not
mathematicians would not be interefted in the analogy, and thofe
who are can purfue the fubject for themfelves, I will go on to
fay that the preceding order is not the natural one. We cannot,
to ufeful purpofe, laying down the truth of the propofition, *firft*,
then proceed to enquire how the non-exiftence of one or both
terms affects the propofition. The exiftence of the terms muft
be firft fettled, and then the truth or falfehood of the propofition.
The affirmative propofition requires the exiftence of both terms :

the negative propofition, of one; being neceffarily true if the other term do not exift, and depending upon the matter, as ufual, if it do exift.

Let us make the exiftence of the terms to be preceding conditions of the propofitions. The fyllogifm $A_1A_1A_1$ is then as follows,

If X and Y both exift,	Every X is Y
If Z alfo exift	Every Y is Z
Therefore If X, Y, Z all exift	Every X is Z.

As to the concluding terms, X and Z, they remain, as it were, to tell their own ftory. Whatever conditions accompany their introduction unto the premifes, thefe fame conditions may be conceived to accompany them in the conclufion. But the middle term difappears: and, not fhowing itfelf in the conclufion, the conditions which accompany it muft be exprefsly preferved. The conclufion then is 'every X is Z, if Y exift' which may be thrown into theform of a dilemma, 'Either every X is Z, or Y does not exift'.

But taking X and Z to exift, let us confider the following fyllogifm, *as it appears to be,*

Every X is (Y, if Y exift)
Every (Y, if Y exift) is Z
Therefore Every X is Z.

If this be not a valid fyllogifm, what *expreffed* law of the ordinary treatifes does it break? The middle term, a curious one, is ftrictly middle: but there is no rule for excluding middle terms of a certain degree of fingularity. That it does break, and very obvioufly, an implied rule, I grant. And as to this work, the rule laid down in Chapter III. is broken in its fecond condition (page 50). The two ufes of the word *is* do not amount to one fuch ufe as is made in the conclufion. That X is (conditionally) Y which is (on the fame condition) Z, gives that X is (on the fame condition) Z. Accordingly, the abfolute conclufion is only true upon fuch conditions as give the middle term abfolute exiftence.

But it muft be particularly noted that it is enough if this exiftence be given to the middle term by the fulfilment of the conditions which precede the entrance of one of the concluding terms. The condition of the act of inference is, that the comparifon muft be really made, if the terms to be compared with the middle term really exift, or, which is the fame, if the conditions under which they are to enter be fatisfied. The other terms being ready, there muft *then* be a real middle term: and there will be, if the mere entrance of one of the concluding terms be proof of the exiftence of a middle term; while, if the other terms cannot be brought in, from nonexiftence, there is no occafion to inquire about a middle term, for it is otherwife known that the comparifon cannot be completed. I will take two concrete inftances, in the firft of which one of the concluding terms, if exifting, is held to furnifh a middle term as real as itfelf, and in the fecond of which no fuch fuppofition occurs. Of courfe I have nothing here to do with the truth of the premifes.

Philip Francis, (if the author of Junius), was an accufer whofe filence was fimultaneous with a government appointment: an accufer &c. reflects difgrace upon the government (if they knew that their nominee *was* the accufer): therefore Francis (if &c.) reflects difgrace upon government (if &c.).

Homer (if there were fuch a perfon) was a perfect poet (if ever there were one): a perfect poet (if &c.) is faultlefs in morals: therefore Homer (if &c.) was faultlefs in morals.

The firft inference is good, even though we grant that our only poffible mode of knowing of the exiftence of an accufer &c. is by eftablifhing that Francis was Junius: it is even good againft one who fhould affert that the accufer &c. is a contradiction in terms in every actual and imaginable cafe except that of Junius.

In the fecond cafe, we put it that the man Homer (if he ever exifted; fome critics having contended for the contrary) was a perfect poet, if ever there were one. There may never have been one; and then Homer (exiftent or nonexiftent) was not a perfect poet. There is no condition here, which being fulfilled, is held to amount to an affertion that the middle term muft have exifted: but the condition of the exiftence of the middle term is independent. Accordingly, the fecond inference is not good: it fhould be Homer (if &c.) was a perfect poet, if ever there were one: that is, or elfe there never was a perfect poet.

I

Thefe points refer to the matter of a fyllogifm, and not to the form ; or rather, perhaps, hold a kind of intermediate relation.

There is another procefs which is often neceffary, in the formation of the premifes of a fyllogifm, involving a transformation which is neither done by fyllogifm, nor immediately reducible to it. It is the fubftitution, in a compound phrafe, of the name of the genus for that of the fpecies, when the ufe of the name is particular. For example, ' man is animal, therefore the head of a man is the head of an animal' is inference, but not fyllo-gifm. And it is not mere fubftitution of identity, as would be ' the head of a man is the head of a *rational animal*' but a fubftitution of a larger term in a particular fenfe.

Perhaps fome readers may think they can reduce the above to a fyllogifm. If *man* and *head* were connected in a manner which could be made fubject and predicate, fomething of the fort might be done, but in appearance only. For example, ' Every man is an animal, therefore he who kills a man kills an animal.' It may be faid that this is equivalent to a ftatement that in ' Every man is an animal ; fome one kills a man ; therefore fome one kills an animal,' the firft premife, and the fecond premife *condi-tionally*, involve the conclufion as *conditionally*. This I admit : but the laft is not a fyllogifm : and involves the very difficulty in queftion. ' Every man *is* an animal ; fome one *is* the killer of a man' : here is no middle term. To bring the firft premife into ' Every killer of a man is the killer of an animal' is juft the thing wanted. By the principles of chapter III, undoubtedly the copula *is* might in certain inferences be combined with the copula *kills*, or with any verb. But fo fimple a cafe as the pre-ceding is not the whole difficulty. If any one fhould think he can fyllogize as to the inftances I have yet given, let him try the following. ' Certain *men*, upon the report of certain other *men* to a third fet of *men*, put a fourth fet of *men* at variance with a fifth fet of *men*.' Now every man is an animal : and therefore ' Certain *animals*, upon the report of certain other *animals*, &c.' Let the firft defcription be turned into the fecond, by any num-ber of fyllogifms, and by help of ' Every man is an animal.'

The truth is, that in the formation of premifes, as well as in their ufe, there is a poftulate which is conftantly applied, and there-fore of courfe conftantly demanded. And it fhould be demanded openly. It contains the *dictum de omni et nullo* (fee the next chapter), and it is as follows. For every term ufed univerfally *lefs* may be fubftituted, and for every term ufed particularly, *more.* The fpecies may take the place of the genus, when all the genus is fpoken of: the genus may take the place of the fpecies when fome of the fpecies is mentioned, or the genus, ufed particularly, may take the place of the fpecies ufed univerfally. Not only in fyllogifms, but in all the ramifications of the defcription of a complex term. Thus for 'men who are not Europeans' may be fubftituted 'animals who are not Englifh.' If this poftulate be applied to the unftrengthened forms of the Ariftotelian Syllogifm, (page 17) it will be feen that all which contain A are immediate applications of it, and all the others eafily derived.

I now pafs to the confideration of the invention of names, and of the diftinctions which are made to exift for the want of it.

Any one may invent a name, that is, may choofe a found or fymbol which is to apply to any clafs of ideas or of objects. The clafs fhould, no doubt, be well defined: but fmall caution is here neceffary, for invented words are generally much more definite than thofe which have undergone public ufage. They come from the coiner's hand as fharp at the edge as a new halfpenny: and in procefs of time we look in vain for any edge at all. The right of invention being unlimited, and the actual ftock having been got together without any uniform rule of formation, *there can be no reafon why we fhould admit any diftinction which can be abrogated by the invention of a name,* fo far as inference is concerned. I do not difpute that the modes of fupplying the want of names may be of importance in many points of view: what I deny is, that they create any peculiar modes of inference.

The invention of names muft either be by actually pointing out objects named, or by defcription in terms of other names. With the former mode of invention, as ' let this, that, &c (fhowing them) be called X' we can have nothing to do. As to the latter, we may make a fymbolic defcription of the procefs by joining together the names to be ufed, with a fymbol indicative of the mode of ufing them, in extenfion of the fyftem in page 106. Thus, P, Q, R, being certain names, if we wifh to give a name to everything which is all three, we may join them thus, PQR: if we wifh to give a name to every thing which is either of the three (one or more of them) we may write P,Q,R: if we want to fignify any thing that is either both P and Q, or R, we have

PQ,R. The contrary of PQR is p,q,r; that of P,Q,R is pqr; that of PQ,R is (p,q)r : in contraries, conjunction and disjunction change places. This notation would enable us to express any complication of the preceding conditions : thus, to name that which is one and one only of the three, we have Pqr, Qrp, Rpq ; for that which is two and two only, PQr, QRp, RPq. Thus, XY includes the instances common to X and Y ; but X,Y includes all X and all Y : accordingly X,Y is a wider term than XY, except when X and Y are identical. As in page 106, XY, the term, supposed to exist, is XY, the proposition of chapter IV ; if we wish to distinguish, we may make X-Y the term, and XY the proposition, the hyphen having its common grammatical use. Thus, X-Y P-Q tells us the same as XYP-Q, both meaning, *for inference*, no more than that there exist objects or ideas to which the four names are applicable. But the first tells it thus, some XYs are PQs ; and the second thus, some things are Xs, Ys, and PQs.

With respect to this and other cases of notation, repulsive as they may appear, the reader who refuses them is in one of two circumstances. Either he wants to give his assent or dissent to what is said of the form by means of the matter, which is easing the difficulty by avoiding it, and stepping out of logic : or else he desires to have it in a shape in which he may get that most futile of all acquisitions, called a *general idea*,* which is truly, to use the contrary adjective term as colloquially, *nothing particular*, a whole without parts.

If the difficulty of abstract assertion be to be got over, the easiest way is by first conquering that of abstract expression, to the extent of becoming able to make a little use of it.

Suppose we ask for the alternative of the following supposition, ' Both X, and either P, or Q and one of the two R or S.' This is no impossible complication : for instance, ' He was rich, and if not absolutely mad, was weakness itself subjected either to bad advice or to most unfavourable circumstances.' The representation of the complex term is X {P, Q(R,S)} ; of the contrary,

---

* " Je vous avoue, dit . . . ., que j'ai cru en deviner quelque chose, et que je n'ai pas entendu le reste. L'abbé de . . . . a ce discours, fit réflexion que c'était ainsi que lui-même avait toujours lu, et que la plupart des hommes ne lisaient guere autrement."

x, p(q,rs) or x,pq,prs. If not the above, he was either not rich, or both not mad and not very weak, or neither mad nor badly advised, nor unfavourably circumstanced.

When a name thus formed, whether conjunctively or disjunctively, enters a simple inference, it gives rise to what have been called the *copulative* syllogism, the *disjunctive* syllogism, and the *dilemma.* The two last are not well distinguished by their definitions as given : the disjunctive syllogism seems to be that in which *names* are considered disjunctively, the dilemma that in which *propositions* are so used. But a proposition entering as part of a proposition, enters merely as a name, the predicates being usually only *true* or *false*, or some equivalent terms. A proposition may only enter for its matter, or it may enter in such a way that its truth is the matter : in this last case it is only as a name that it is the subject of inference. Thus, 'It is true that he was fired at' is ' the assertion (that he was fired at) is a true assertion.' I believe the best way would be to apply the term *disjunctive* argument so as to include the dilemma, marking by the latter word (as a term rather of rhetoric than of logic) every argument in which the disjunctive proposition is meant to be a difficulty for the opponent on every case, or *horn*, of it.

Whatever has right to the name P, and also to the name Q, has right to the compound name PQ. This is an absolute identity, for by the name PQ we signify nothing but what has right to both names. According X)P + X)Q=X)PQ is not a syllogism, nor even an inference, but only the assertion of our right to use at our pleasure either one of two ways of saying the same thing instead of the other. But can we not effect the reduction syllogistically? Let Y be identical with PQ; we have then PQ)Y and Y)PQ, and also Y)P and Y)Q. Add to these X)P and X)Q, and we have all the propositions asserted. But we cannot deduce from them alone X)Y, the result wanted, by any syllogistic combination of the six. Nor must it be thought surprising that we cannot, by a train of argument, arrive at demonstration of it being allowable to give to anything which has right to two names, a third name invented expressly to signify that which has such right. We might as well attempt to syllogize into the result, that a person who sells the meat he has killed is a butcher.

I lay ftrefs upon this, to an extent which may for a moment appear like diligently grinding nothing in a mill which might be better employed, for two reafo is. Firft, the young mathematician is very apt to try, in algel ra, to make one principle deduce another by mere force of fymb ls: and the above attempt may fhow him what he is liable to. Secondly, I am inclined to fuppofe that the diftinction drawn between the claffes of fyllogifms to which I prefently come, and the ordinary categorical ones, is due to what muft be defcribed in my language as a want of perception of the abfolute, *lefs than inferential* (fo to fpeak) identity of X)P+X)Q and X)PQ. But all other propofitions of the kind, however fimple, may be made deductions. For inftance, 'if X be both P and Q, and if P be R, and Q be S, then X is both Q and S' is thus deduced: X)P+P)R=X)R, and X)Q +Q)S=X)S, and X)R+X)S is X)RS. Even P)R+Q)R= P,Q)R is deducible; being P)R+Q)R=r)p+r)q=r)pq= P,Q)R. Thus it is feen that, as foon as the *conjunctive* poftulate is laid down, the identity of the correfponding disjunctive poftulate with it may be fhown. Next, if X muft be either P or Q, or X)P,Q, and if P be always R, and Q be always S, then X)R,S may be *deduced* from the preceding.

Firft, that X)P and Y)Q give XY)PQ can be deduced; evident as it may be, it is a fucceffion of applications. XY)X+ X)P gives XY)P, and XY)Y+Y)Q gives XY)Q, and XY)P +XY)Q is XY)PQ by the poftulate. Next, X)P,Q is pq)x, and P)R is r)p, and Q)S is s)q, whence, as juft proved rs)pq. Now, rs)pq+pq)x=rs)x, which is X)R,S. It will be a good exercife for the reader to tranflate this proof into ordinary language.

I may now proceed to extend this idea and notation relative to propofitions of complex terms. The complexity confifts in the terms being conjunctively or disjunctively formed from other terms, as in PQ, that to which both the names P and Q belong conjunctively; and as in P,Q that to which one (or both) of the names P and Q belong disjunctively. The contrary of PQ is p,q; that of P,Q is pq. *Not both* is either not one or not the other, or not either. *Not either P nor Q* (which we might denote by :P,Q or .P,Q) is logically '*not P and not Q*' or pq: and this is then the contrary of P,Q.

The disjunctive name is of two very different characters, according as it appears in the univerfal or particular form : fo very different that it has really different names in the two cafes, *copulative* and *disjunctive*. This diftinction I here throw away : oppofing *disjunctive*, (having one or more of the names) to *conjunctive*, (having all the names). The disjunctive particle *or* has the fame meaning with the diftributive copulative *and*, when ufed in a univerfal. Thus, ' Every thing which is P or Q is R or S' means ' Every P *and* every Q is R *or* S.' But PQ is always ' both P and Q in one.' Accordingly

Conjunctive	PQR ufes *and* collectively.
Disjunctive	P,Q,R in a univerfal ufes *and* diftributively,
	P,Q,R in a particular ufes *or* disjunctively, in the common fenfe of that word.

' Either P or Q is true,' is an ambiguous phrafe, which is P,Q)T or T)P,Q according to the context.

The manner in which the component of a name enters, whether conjunctively or disjunctively, is to pafs as it were for a part of the quality of the name itfelf. Thus the contrary of P (conjunctive, as indicated by the abfence of the comma) is ,p (disjunctive, as indicated by the comma). To teft this affertion about the mode of making contraries, let us afk what is that of ' one only of the two P or Q?' We know it of courfe to be ' both or neither.' The name propofed is Pq, Qp and its contrary is (p,Q)(q,P), that is, one of the two p,Q, *and* one of the two q,P. It is then either pq, pP, qQ, or PQ: the fecond and third cannot exift, therefore it is pq, PQ, as already feen. I need hardly have remarked that (P,Q)(R,S) is PR, PS, QR, QS.

Obferve that though X)PQ gives X)P, and that XPQ gives XP, we may not fay that XY)P gives X)P, nor that X)P,Q gives X)P. But any disjunctive element may be rejected from a univerfal term, and any conjunctive element from a particular one. Thus P)QR gives P)Q and P,Q)R gives P)R. Alfo P.Q,R gives P.Q and PQ:R,S gives P:R. All thefe rules are really one, namely that PQ is of the fame extent at leaft as PQR. This will appear from our rules of tranfpofition prefently given.

Let change from one member of the proposition to the other be called *transposition*. I proceed to inquire how many transpositions the various forms will bear, and what they are. It will however be neceffary to complete our forms by the recognition, as a proposition, of the fimple affertion of exiftence or non-exiftence. By XU we mean that there are in the univerfe things to which the name X applies, and we fpeak only of fuch things under the name. Accordingly X)U and XU do not differ in meaning. By u, the contrary of U, we can only denote non-exiftence; thus X.U or X)u throws the name X out of confideration. Thus Y)X=U)X,y; Y.X=YX)u, &c. To fignify, for inftance, that X and Y are complements (contraries or fubcontraries, page 75) we have U)X,Y, which our rules will tranfpofe into xy)u, or x.y.

Having to confider fubject and predicate, conjunctive and difjunctive, affirmative and negative, univerfal and particular, we muft think of fixteen different forms. Thus the four forms of the univerfal affirmative are

$$XY)PQ; \quad X,Y)PQ; \quad XY)P,Q; \quad X,Y)P,Q$$

It will be beft here to neglect the contranominal converfes of A and O equally with the fimple converfes of E and I: thus XY)PQ may be read as identical with p,q)x,y. There is alfo one obvious tranfpofition which we muft not merely neglect but throw out; fince it does not give a refult identical with its predeceffor. I mean the tranfpofition of M)PQ into MP)Q: the fecond follows from the firft but not the firft from the fecond. Alfo the correfponding change of M.P,Q into Mp.Q, for the fame reafon.

This being premifed, the following are the rules ;—

Direct tranfpofition is the change from one member to the other without alteration of name or junction: *contrary*, with alteration of both.

The convertibles (E,I) allow direct tranfpofition of conjunctive elements either way, from fubject to predicate, or from predicate to fubject: and thefe are the only direct tranfpofitions. Thus X.YZ=XY.Z, and X-YZ=XY-Z.

The inconvertibles (A,O) allow contrary tranfpofition of conjunctive elements from fubject to predicate, and of disjunctive

elements from predicate to fubject: beft remembered by allowing SP to ftand for *conjunctive* and PS for *disjunctive*. And thefe
are the only contrary tranfpofitions. Thus XY)M=X)M,y
and M)X,Y=My)X.

An element that can be rejected cannot be tranfpofed, and
vice verfâ. Thus X,Y)M gives X)M, and Y cannot be tranfpofed.

The following table exhibits the varieties of the forms A and
E, equivalents being written under one another, and converfions,
contranominal or fimple, oppofite.

XY)P,Q	pq)x,y	XY.PQ	PQ.XY
Xp)Q,y	Yq)P,x	XP.QY	QY.XP
Xq)P,y	Yp)Q,x	XQ.PY	PY.XQ
X)P,Q,y	pqY)x	X.PQY	PQY.X
Y)P,Q,x	pqX)y	Y.PQX	PQX.Y
p)Q,x,y	XYq)P	P.QXY	QXY.P
q)P,x,y	XYp)Q	Q.PXY	PXY.Q
XYpq)u	U)P,Q,x,y	XYPQ.U	U.XYPQ

XY)PQ	p,q)x,y	XY.P,Q	P,Q.XY
X)PQ,y	[p,q]Y)x	X.[P,Q]Y	[P,Q]Y.X
Y)PQ,x	[p,q]X)y	Y.[P,Q]X	[P,Q]X.Y
XY[p,q])u	U)[x,y],PQ	XY[P,Q].U	U.XY[P,Q]

X,Y)P,Q	pq)xy	X,Y.PQ	PQ.X,Y
[X,Y]p)Q	q)xy,P	[X,Y]P.Q	Q.[X,Y]P
[X,Y]q)P	p)xy,Q	[X,Y]Q.P	P.[X,Y]Q
[X,Y]pq)u	U)xy,P,Q	[X,Y]PQ.U	U.[X,Y]PQ

X,Y)PQ	p,q)xy	X,Y.P,Q	P,Q.X,Y
[x,y][p,q])u	U)xy,PQ	[X,Y][P,Q].U	U.[X,Y][P,Q]

If for) we write (:) in the left hand divifions, and erafe the (.)
and ufe the hyphens of page 115, on the right, we have the tranfpofitions of O and I. And if we write p and q for P and Q on
the left, and change the form X)Y into X.y, we thereby change
the forms of A into thofe of E. If more than two elements
were ufed, the tranfpofitions would now be perfectly eafy.

It appears that there are no lefs than fixteen A forms into

which XY)P,Q may be varied: the reason is that both subject
and predicate are transposibly constructed. But XY)PQ shows
only a transposible subject; X,Y)P,Q only a transposible predicate: and these have only four forms each. Lastly, X,Y)PQ,
having neither transposible, has only two forms. By transposibly constructed, I mean capable of having the elements separated
by transposition. The whole term is always transposible: that is,
the complete subject, or the complete predicate, may be looked
on as conjunctive or disjunctive, at pleasure. Thus in X)Y, if
we consider this as XU)Y,u, we may make this yU)x,u or y)x.
So that the ordinary contranominal conversion may be considered
as a case of the more general rule. Just as, in arithmetic, a number, 5, may be made to obey the laws of $a + b$ as $0 + 5$, or of ab
as 1×5.

Syllogisms of complex terms might be widely varied, even if we
chose to consider only each first case of the preceding table as
fundamental. Thus

$$XY)P,Q + VW)P,Q = (x,y)\text{-}(v,w) \qquad A_1A'I'$$

would give sixty-four varieties of premises. I now proceed to
show that the ordinary disjunctive and dilemmatic forms are
really common syllogisms with complex terms, reducible to ordinary syllogisms by invention of names.

Example 1. Every S is either P, Q, R; no P is S; no Q is
S; therefore every S is R. Let S represent 'the true proposition' (singular), and let P, Q, R be names of propositions, and
this then represents a very common form, which would be expressed thus 'either A is B, or C is D, or E is F; but A is not
B, C is not D; therefore E is F.' I say that, where the necessary names exist, the final step of this could not be distinguished
from a common syllogism; which accordingly it becomes by invention of names.

We have S)P,Q,R, whence Spq)R. But S.P and S.Q or S)p
and S)q give S)pq, with which S)S combined gives S)pqS. And
S)pqS + pqS)R = S)R. Let M be the name of what is S and not P
and not Q, and the thing required is done. Here then is a syllogism
of the ordinary kind, to one premise of which we are led by a
use of the *conjunctive postulate* (page 116): the necessity for which
is the distinction between the class we are considering and others.
It happens here that two of the terms of our final syllogism are

identical: for Spq is of no greater extent than S. But the ufe made of S)S is perfectly legitimate.

Example 2. 'If A be B, E is F; and if C be D, E is F; but either A is B or C is D; therefore E is F.' This can be reduced to

$$P)R + Q)R + S)P, Q = S)R$$

which is immediately made a common fyllogifm by changing P)R + Q)R into P,Q)R.

Example 3. 'From P follows Q; and from R follows S; but Q and S cannot both be true; therefore P and R cannot both be true.' This may be reduced to

$$P)Q + R)S + T.QS = T.PR$$
$$\text{or } PR)QS \quad + T.QS = T.PR$$

Example 4. 'Every X is either P, Q, or R; but every P is M, every Q is M, every R is M; therefore every X is M.' This is a common form of the dilemma; it is obvioufly reducible to P,Q,R)M + X)P,Q,R = X)M.

Example 5. 'Every X is either P or Q, and every Q is X.' This is wholly inconclufive, and leads to an identical refult, as follows; X)P,Q gives Xp)Q, which with Q)X gives Xp)X, a neceffary propofition.

Example 6. If we throw X)R into the form X)R,R, we have Xr)R, or 'Every X which *is not* R *is* R,' a contradiction in terms. But it evidently implies that there can be no Xs which are not Rs; and thus alfo we return to X)R. Take 'every X is either P, Q, or R; every P is M; every Q is M; and every M is R.' Here X)P,Q,R = Xr)P,Q, which with P,Q)M gives Xr)M, which with M)R gives Xr)R or X)R.

Example 7. 'Every X is either P or Q, and only one.' This gives two propofitions, X)P,Q + X.PQ. Now X)XP,XQ is identical with X)P,Q, and this may be looked on as an extreme cafe of

$$X)P,Q + X)Y = X)PY; QY$$

but X.PQ gives XP)q and XQ)p, from which we can obtain

$$X)XP,XQ + XP)q + XQ)p = X)p,q$$

Hence X)P,Q + X)p,q = X)[P,Q,][p,q.]
$$= X)Pp,Pq,Qp,Qq = X)Pq,Qp$$

fince Pp and Qq are fubject to X.Pp and X.Qq. All this being

worked out in fyllogiftic detail, fhows us that the tranfition from
'Every X is P or Q, and no X is both' to 'Every X is either
P and not Q, or Q and not P' is capable of being made fyllogiftically.   The ftudent of logic may thus acquire the idea,
which fo foon becomes familiar to the ftudent of mathematics,
of perfectly felf-evident propofitions which are deducible from
one another, as diftinguifhed from thofe which are not.

Example 8. 'Every X is one only of the two, P or Q;
every Y is both P and Q, except when P is M, and then it is
neither; therefore no X is Y.'   Here is a cafe in which it is the
fact of the exception and not its nature which determines the
inference: M may be anything.   This ought to appear in our
reduction: and it does appear in this way.   From X)P,Q it is
obvious that X)P,Q,R,S, and fyllogiftically demonftrable from
X)P,Q, and Xrs)X.   Now in the fecond premife we have

$$Y)PQm,pqM, \text{ or } [p,q,M][P,Q,m])y$$
$$\text{or } pQ,Pq,PM,QM,pm,qm)y$$

from which, by rejection, follows pQ,Pq)y.   And the firft premife is X)Pq,Qp.   Whence X)y or X.Y.

It is not neceffary to multiply examples: I will conclude this
part of the fubject by pointing out that the ordinary propofitions
X)Y, &c. are, with reference to their inftances, disjunctively
compofed: the difference between the univerfal and particular
lying in the latter being indefinite in the number of its inftances.
Thus, if there be three Xs and four Ys, the four propofitions
are, applying the name to each inftance, as feen written at length in
X,X,X)Y,Y,Y,Y; X,X,X.Y,Y,Y,Y; (X,X,X)(Y,Y,Y,Y);
and (X,X,X).Y,Y,Y,Y.

The propofition in page 25, is a cafe of the preceding method.
I leave the reader to fhow it, and alfo that the hypothefis is
flightly overftated.

I now come to the *forites*, the *heap* or chain of fyllogifms, in
which the conclufion of the firft is a premife of the fecond, and
fo on.   Take a fet of terms, P, Q, R, S, &c. and let the order
of reference be PQ, QR, RS, &c.   Then $A_1A_1A_1A_1$ &c. is a
forites, and the only one ufually confidered: thus,

$$P)Q + Q)R + R)S + S)T = P)T$$

The firſt two links give P)R, which with the third gives P)S, which with the fourth gives P)T. Thus we have *links, intermediate concluſions,* and a *final concluſion.*

A great number of different ſorites may be formed, under the following conditions,

The firſt particular propoſition which occurs, be it link or concluſion, prevents any future link from being particular: for all the concluſions thence become particular.

Examine the caſes of ſyllogiſm which proceed by the firſt rule of accentuation (page 92), that is, which have beginning and ending both univerſal, or both particular: theſe only can occur in a ſorites, except at the end, or in the place where a particular propoſition firſt enters. It will be found that the concluſion, when the argument goes on, muſt come after ſomething connected with that which comes after it by the firſt rule of accentuation: except at the place where a particular concluſion comes in for the firſt time. For inſtance, E_1E' gives A_1, which, ſtill keeping concluſions univerſal, muſt be followed by A_1 or E_1, which follow E' by the firſt rule. Again, take O_1E', which gives I_1; this muſt be followed either by A_1 or E_1, which follow E' by the ſame rule: and ſo on. Accordingly,

Any chain of univerſals, in which affirmation is followed by a like prepoſition, and negation by a different one, as $A_1A_1E_1A'$ $E'A_1E_1E'$, &c. may be part of the chain of a ſorites. And the chain muſt be either of this kind wholly, or once only broken in one of two ways: either by the direct entrance of a particular propoſition, or by a breach of the rule. In a chain of this kind, unbroken, the concluſions are affirmative or negative, according as an even or odd number of negatives goes to the formation of them. All the concluſions have the ſame accent as the firſt link.

Let a particular premiſe be introduced, as in $A_1E_1E'I'$ &c. The accent of the particular introduced muſt be the ſame as or contrary to that of the firſt link, according as the preceding number of negatives is odd or even. For the accent of the firſt link remains as long as the concluſion is univerſal, and a ſyllogiſm with the ſecond premiſe particular follows the ſecond rule. Thus, inſerting the intermediate concluſions, the above is $A_1E_1(E_1)E'(A_1)I'(I')$. And after (I') muſt come A' or E', ſo that the firſt rule ſtill continues. But the accent of the concluſions changes.

Now let the rule of accentuation be broken. The *accent* of the conclufion ftill requires the firft rule to be refumed. Thus, E_iE'(rule unbroken) gives A_i, and E_iE_i (rule broken) gives I', and A_i requires A_i or E_i to follow E', while I' requires A' or E' to follow E_i. This one breach of rule only changes the conclufion from univerfal to particular. The accent of the conclufion changes as before.

The links of a forites, then, are either a chain of univerfals following the firft rule of accentuation, or fuch a chain with *one* breach of the rule, or fuch a chain with one particular inferted, of the fame or contrary accent to the firft link, according as the preceding negatives are odd or even, and made the commencement of the refumption of the rule (if broken). In all the cafes the conclufion is affirmative or negative according as the preceding negatives are even or odd in number : the unbroken chain has a univerfal conclufion with the accent of the firft link, and the broken one a particular with the contrary accent.

$A'E'E_iA'E'$	$E_iA'A_iE_iE'A_i$	$A_iE_iA'O_iA'E'$
$E'A'A'E'$	$E_iO'I'O'O'$	$E_iE_iI'I'O'$

Here are examples of the three kinds. The chain is in the firft row, the intermediate and final conclufions in the fecond. Thus the fecond example prefents the fyllogifms $E_iA'E_i$, E_iA_iO', $O'E_iI'$, $I'E'O'$, $O'A_iO'$; and at length is

$$P.Q + R)Q + R)S + S.T + t.u + U(V = V:P$$

The forites ufually confidered are only $A_iA_iA_i\ldots$ and $A'A'A'\ldots$ To thefe might be added without abandoning the Ariftotelian fyllogifm, fuch as $A_iE_iA'A'A'\ldots$, $A_iE_iA'A_iA_i \ldots$ But it would not be very eafy to follow the chain in thought without introducing the intermediate conclufions, and thus deftroying the fpecific charaƈter of the procefs.

And juft as the ordinary univerfal fyllogifm can be reduced to $A_iA_iA_i$, fo the univerfal forites can always be reduced to a chain of A_i. Thus $A'E'E_iA'E'$ or

$$Q)P + q.r + R.S + T)S + t.u = p.u$$
$$\text{is} \quad u)T + T)S + S)r + r)Q + Q)P = u)P$$

CHAPTER VII.

On the Aristotelian Syllogism.

FROM the time of Aristotle until now, the formal inference
has been a matter of study. In the writings of the great
philosopher, and in a somewhat scattered manner, are found the
materials out of which was constructed the system of syllogism
now and always prevalent: and two distinct principles of exclusion appear to be acted on. Perhaps it would be more correct
to say that the followers collected two distinct principles of exclusion from the writings of the master, by help of the assumption
that everything not used by the teacher was forbidden to the
learner. I cannot find that Aristotle either limits his reader in
this manner, or that he anywhere implies that he has exhausted
all possible modes of syllogizing. But whether these exclusions
are to be attributed to the followers alone, or whether those who
have more knowledge of his writings than myself can fix them
upon the leader, this much is certain, that they were adopted,
and have in all time dictated the limits of the syllogism. Of all
men, Aristotle is the one of whom his followers have worshipped
his defects as well as his excellencies: which is what he himself
never did to any man living or dead ; indeed, he has been accused
of the contrary fault.

The first of these exclusions is connected with the celebrated
dictum de omni et nullo, namely, that what is distributively affirmed
or denied of all, is distributively affirmed or denied of every some
which that all contains. It is there said that in every syllogism
the middle term must be universal in one of the premises, in order
that we may be sure that the affirmation or denial in the other
premise may be made of some or all of the things about which
affirmation or denial has been made in the first. This law, as
we shall see, is only a particular case of the truth: it is enough
that the two premises together affirm or deny of more than all
the instances of the middle term. If there be a hundred boxes,
into which a hundred *and one* articles of two different kinds are

to be put, not more than one of each kind into any one box, ſome one box, if not more, will have two articles, one of each kind, put into it. The common doctrine has it, that an article of one particular kind muſt be put into every box, and then ſome one or more of another kind into one or more of the boxes, before it may be affirmed that one or more of different kinds are found together. This excluſion is a ſimple miſtake, the mere ſubſtitution of the aſſertion that none but a certain law of inference *can* exiſt, for the determination that no other *ſhall* exiſt. Any one is at liberty to limit the inferences he will uſe, in any manner he pleaſes : but he may err if he declare his own arbitrary boundary to be a natural limit impoſed by the laws of thought.

The other excluſion may involve, on the ſame terms, an error of the ſame kind; or may equally be the expreſſion of arbitrary will: but there is what is more reaſonably matter of opinion about it. Ariſtotle will have no contrary terms: not-man, he ſays, is not the name of anything. He afterwards calls it an indefinite or *aoriſt* name, becauſe, as he aſſerts, it is both the name of exiſting and non-exiſting things. If he had here made the diſtinction between ideal and objective, he would have ſeen that *man* and *not-man* equally belong to both (objectively) exiſting and non-exiſting things : *man*, for example, belongs as a name to Achilles and the ſeven champions of Chriſtendom, whether they ever exiſted in objective reality or not : and *not-man* belongs, in either caſe, to their horſes. I think, however, that the excluſion was probably dictated by the want of a definite notion of the extent of the field of argument, which I have called the *univerſe* of the propoſitions. Adopt ſuch a definite notion, and, as ſufficiently ſhown, there is no more reaſon to attach the mere idea of negation to the contrary, than to the direct term.

The excluſion of contraries throws out the propoſitions E' and I', or x.y and xy, which cannot be expreſſed without either contraries, as in x.y=x)Y=y)X, and xy=x:Y=y:X, or reference to things not named by X and Y, as in ' Every *thing* is either X or Y' and ' Some things are neither Xs nor Ys,' the moſt natural readings of 'No not-Xs are not-Ys, and 'Some not-Xs are not-Ys.' There remain then ſix modes of connexion of X and Y, namely X)Y and Y)X, X:Y and Y:X, and XY(=

YX) and X.Y(=Y.X). Thefe fix are made eight; for in the common fyftem, XY and YX are confidered as diftinct in form, and alfo X.Y and Y.X. But thefe eight are only treated as four: for reference to order is not made in the fimple propofition. Thus X)Y and Y)X are both denoted by A, XY and YX by I, X.Y and Y.X by E, and X:Y and Y:X by O. But the ftandard of order which is neglected as to·the propofition by itfelf, is adopted in the fyllogifm in the following manner.

The *predicate* of the conclufion is called the *major* term, and the fubject of the conclufion the *minor* term. This language is fafhioned upon the idea of an affirmative propofition, in which major and minor have reference to *magnitude*. In ' every X is Z ' Z is a name which entirely contains X and is therefore *at leaft as great* as X, *greater than or equal to* X. Here is, before it was introduced into mathematics, the idea now fo familiar to the mathematician, of allowing his language to include the extreme limit of its meaning. When the fame terms are applied to negative propofitions, the notion of magnitudinal inclufion is loft; and major and minor, being ftill retained, muft be prefumed to refer to real or fuppofed importance. The premifes are called major and minor, according as they contain the major or minor term of the conclufion: and the major premife is always written firft. Accordingly, Z and X being the major and minor terms, there are four poffible arrangements, which are called the four figures. Ariftotle gives three, and tradition has it that Galen fupplied the fourth in number and order.

1. YZ	2. ZY	3. YZ	4. ZY
XY	XY	YX	YX
----	----	----	----
XZ	XZ	XZ	XZ

To me, the moft fimple arrangement is that which takes up what was left off with, as in the fourth figure: and ' X is in Y, Y is in Z, therefore X is in Z ' is more natural than ' Y is in Z, X is in Y, therefore X is in Z.'

It is now plain, that whenever one only of the three propofitions is convertible, there are two diftinct ways in which the fyllogifm may be written: when two only, four: and when all three (if there were fuch a thing), eight.

The system rejects all conclusions which may be made stronger: thus when $X.Z$ follows, it does not allow $X:Z$ to make a distinct form. But when $X)Z$ is the conclusion, it does not reject ZX, for, not considering ZX as identical with XZ, it does not consider $X)Z$ as a strengthened form of ZX. But it does not reject syllogisms in which as strong a conclusion can be deduced from a weaker premise: accordingly, we must search for Aristotelian forms among the strengthened syllogisms of chapter V, as well as among the fundamental ones. Now, taking all the forms which show neither E' or I', let us write down the symbols of them, and the number of cases we may expect from each. Moreover, since transformation of order makes no difference here, I put the syllogisms together as in page 96, into twelve pairs.

Fundamental $A_1A_1A_1$, $A'A'A'$, 1; $O'A_1O'$, $A'O_1O_1$, 1; $A_1O'O'$, $O_1A'O_1$, 1; $E'A_1E'$, $A'E'E'$, rejected; $I_1A_1I_1$, $A'I_1I_1$, 4; $E'O'I_1$, $O_1E'I_1$, rejected; $E_1A'E_1$, $A_1E_1E_1$, 4; $I'A'I'$, $A_1I'I'$, rejected; E_1O_1I', $O'E_1I'$, rejected; $E'E_1A'$, $E_1E'A_1$, rejected; $I_1E_1O_1$, E_1I_1O', 4; $E'I'O_1$, $I'E'O'$, rejected.

Weakened $A_1A_1I_1$, 1.

Strengthened $A'A_1I_1$, 1; $A_1A'I'$, rejected; $A'E_1O_1$, E_1A_1O', 2; $A_1E'O'$, $E'A'O_1$, rejected; $E'E'I_1$, rejected; E_1E_1I', rejected.

There are then fifteen fundamental, one weakened, and three strengthened, forms of syllogism in the received system. I now put them down, with their derivations, forms of expression in full, ordinary symbols, figures into which they fall, and the magic words by which they have been denoted for many centuries, words which I take to be more full of meaning than any that ever were made.

Fundamental.

$A_1A_1A_1$	$A'A'A'$	$Y)Z + X)Y = X)Z$	AAA	I	*Barbara*
$O'A_1O'$	$A'O_1O_1$	$Y:Z + Y)X = X:Z$	OAO	III	*Bokardo*
$A_1O'O'$	$O_1A'O_1$	$Z)Y + X:Y = X:Z$	AOO	II	*Baroko*
$I_1A_1I_1$	$A'I_1I_1$	$Y)Z + XY = XZ$	AII	I	*Darii*
—	—	$Y)Z + YX = XZ$	AII	III	*Datisi*
—	—	$ZY + Y)X = XZ$	IAI	IV	*Dimaris*
—	—	$YZ + Y)X = XZ$	IAI	III	*Disamis*

Fundamental.

$E_1A'E_1$	$A_1E_1E_1$	$Y.Z+X)Y=X.Z$	EAE	I	*Celarent*
—	—	$Z.Y+X)Y=X.Z$	EAE	II	*Cesare*
—	—	$Z)Y+Y.X=X.Z$	AEE	IV	*Camenes*
—	—	$Z)Y+X.Y=X.Z$	AEE	II	*Camestres*
E_1I_1O'	$I_1E_1O_1$	$Y.Z+XY=X:Z$	EIO	I	*Ferio*
—	—	$Z.Y+XY=X:Z$	EIO	II	*Festino*
—	—	$Y.Z+YX=X:Z$	EIO	III	*Ferison*
—	—	$Z.Y+YX=X:Z$	EIO	IV	*Fresison*

Weakened.

$A_1A_1I_1$	$A'A'I_1$	$Z)Y+Y)X=XZ$	AAI	IV	*Bramantip*

Strengthened.

$A'A_1I_1$	$A'A_1I_1$	$Y)Z+Y)X=XZ$	AAI	III.	*Darapti*
$A'E_1O_1$	E_1A_1O'	$Y.Z+Y)X=X:Z$	EAO	III	*Felapton*
—	—	$Z.Y+Y)X=X:Z$	EAO	IV	*Fesapo*

The words which represent the different *moods* (as they are called) are usually collected under their figures in the following lines.

Barbara, Celarent, Darii, Ferioque prioris.
Cesare, Camestres, Festino, Baroko, secundæ.
Tertia Darapti, Disamis, Datisi, Felapton,
Bokardo, Ferison habet. Quarta insuper addit
Bramantip, Camenes, Dimaris, Fesapo, Fresison.

The vowels of the different words give the symbol of the syllogism; thus A,A,A, are seen in *Barbara*. The consonants in the first figure have no special meaning: but in the other figures every consonant except T and N (which are only euphonic) has its meaning as follows;—every mood of every figure can (with two exceptions) in one way or another, be reduced to a mood of the first figure: and the letters show the way of doing it. The initial tells to which mood the reduction brings us: thus Cesare is reduced to Celarent, and also Camestres; Festino is reduced to Ferio, and so on. The two exceptions are denoted by the letter K (as in Baroko and Bokardo); we shall presently notice them further. And S means that the preceding premise is to be simply converted. P, that what was called conversion *per acci-*

dens is to be made, ZX for X)Z, or X)Z for ZX : accordingly, P only occurs in the weakened or strengthened syllogisms. M means that the premises are to be transposed. Thus the meaning of the word *Disamis* is nothing less than what follows. 'There is a syllogism in which the middle term is the subject of both premises, and when reduced to the first figure it becomes *Darii*: the major premise, which must be converted in reduction, is a particular affirmative: the minor premise, which must become the major one in reduction, is a universal affirmative: and the conclusion, which must be converted in reduction, is a particular affirmative.' Thus,

$$\text{YZ} + \text{Y)X} = \text{XZ} \quad Disamis$$
$$\text{becomes} \quad \text{Y)X} + \text{ZY} = \text{ZX} \quad Darii$$

The moods *Baroko* and *Bokardo* do not admit of reduction to the first figure, by any fair use of the phrase : but the logicians were determined they should do so, and they accordingly hit upon the following plan, which they called reduction *per impossibile*. AOO and OAO being the opponent forms (pages 88, and 102) of AAA, the two moods in question were connected with *Barbara* (whence their letter B) by showing that the latter would make the denial of their conclusions force one premise to contradict the other. Thus, *Baroko*, or if Z)Y and X:Y then X:Z was proved in the first figure as follows. If under these premises, X:Z be not true, then X)Z is true; but Z)Y is true : and Z)Y + X)Z, by *Barbara*, gives X)Y. But X:Y : therefore, if *Baroko* be not a legitimate form, X)Y and X:Y are both true at once, which is absurd. Had contraries been used, Z)Y + X:Y = X:Z would have been thrown into the first figure as y)z + Xy = Xz, *Darii*, or y.Z + Xy = X:Z, *Ferio*. And Y:Z + Y)X = X:Z, *Bokardo*, is seen reduced to the first figure in Y)X + zY = zX, *Darii*.

Aristotle did not use the fourth figure, considering it, as is said, to be only an inversion of the first. The introduction of it among the figures is attributed to Galen, and it does not often appear in ordinary works of logic before the beginning of the last century. If the order of the premises be inverted, so as to make the first figure appear, the major and minor terms will appear wrongly placed in the conclusion. The words used for these indirect moods of the first figure were usually the fifth and following ones in

> *Barbara*, *Celarent*, *Darii*, *Ferio*, Baralip-*ton*
> Celantes, Dabitis, Fapesmo, Frisesom-*orum*

the final syllables in Italics being only euphonic (Frisesmo-orum would have been more correct). Some used the words *Faresmo* and *Firesmo*.

In calling the moods of the fourth figure by the name of indirect moods of the first figure, notice was taken of the circumstance that a transposition of the premises would give the arrangement of the first figure, in every thing but the proper arrangement of major and minor terms, which is inverted. A little consideration will show the reader that the earlier Aristotelians were wiser than the later ones in this matter. Consider the fourth and first figures as coincident, and the arbitrary notion of arrangement by major and minor vanishes. It was not till this mere matter of discipline was made an article of faith that the fourth figure had any ground of secession from the first.

It might seem as if the union of the first and fourth figures would demand that of the second and third: the first pair containing all the moods in which the middle term occupies different places in the two premises, the second pair those in which it has the same place in both. If this were done, each of the two main subdivisions must be itself subdivided into two. And this would perhaps have been the more skilful mode of division.

The distinction of figures has been condemned by many, and particularly by Kant. Whether attacked or defended, it is essential that the true grounds of the side taken should be more explicitly stated than is often done. The root of the distinction of figure is undoubtedly the distinction between the two forms XY and YX, X.Y and Y.X. It would be equally absurd, either to deny the identity of XY and YX, considered as material of inference, or to deny their difference in many other points of view. In this work I am concerned only with what can be inferred, and to what extent of quantity, and accordingly the distinction is to me immaterial. But if I had not merely to study the way of using premises, but also that of arriving at them, it might very well happen that the aspects under which the same inference is feen in different figures would give it very different fhades of character. A fimple inftance will fhow that though the comparifon, and its extent, are all that can be attended to in forming the conclufion, thefe points of meaning are not the only ones. A perfon who wifhed to conteft the old ufe of the word *green*, as applied to unripe fruit, would fay that ' fome green fruits are ripe,' if he wanted fpecially to fhow the mifapplication of the word. But if he rather wanted to fhow the badnefs of the method of denying ripenefs, he would fay ' fome ripe fruits are green.' The propofitions are endlefs in which, X and Y being the terms, it is at one time X which is brought to Y for comparifon, and at another Y to X. The fubject of a propofition is always the object of examination; whether the form be X)Y, X.Y, XY, or X:Y, we examine and report upon the Xs.

If we arrange the four figures feparately, we fhall better fee their feveral peculiarities.

Firft Figure.

Barbara	Y)Z + X)Y = X)Z		*Celarent*	Y.Z + X)Y = X.Z
Darii	Y)Z + XY = XZ		*Ferio*	Y.Z + XY = X:Z

What is here declared, is in every cafe the *dictum de omni et nullo* in its fimpleft form, in a manner which juftifies the preference given to this figure. The middle term being completely contained in, or completely excluded from, the major term; fuch inclufion or exclufion then follows of all fuch part of the minor term as is declared in the fecond premife to be in the middle term. The inference then is in this fentence ' What is true of the whole middle term, is true of its part.' And it is obvious that in this figure the major premife muft be univerfal, the minor premife affirmative. The four forms are all found among the conclufions. I think that the inverfion of the premifes which the fyftem of chapter V. employs will be found to give the forms which are moft eafily tranflated into language independent of the middle term. The fentence ' All (or fome) of the Xs are what muft be Zs, therefore all (or fome) of the Xs are Zs' includes *Barbara* and *Darii:* and ' All (or fome) of the Xs are what cannot be Zs, and therefore cannot be Zs,' contains *Celarent* and *Ferio.*

Second Figure.

Cesare	$Z.Y+X)Y=X.Z$	*Camestres*	$Z)Y+X.Y=X.Z$
Festino	$Z.Y+XY=X:Z$	*Baroko*	$Z)Y+X:Y=X:Z$

In this figure (in which only negatives can be proved) the appearance of the dictum is not so direct. The terms of the conclusion are both objects of examination, and one is wholly included, and the whole or part of the other excluded (Cesare, Camestres, and Baroko) or one is wholly excluded, and the whole or part of the other included (Cesare, Camestres, and Festino). Or rather, to justify the distinction, we should say that the whole of the major term is $\frac{\text{included}}{\text{excluded}}$ and the whole of the minor $\frac{\text{excluded}}{\text{included}}$ which gives $\frac{\text{Camestres}}{\text{Cesare}}$ in which the whole of the minor is therefore excluded from the major; or else the whole of the major is $\frac{\text{included}}{\text{excluded}}$ and part of the minor $\frac{\text{excluded}}{\text{included}}$ which gives $\frac{\text{Baroko}}{\text{Festino}}$ in which that part of the minor is excluded from the major. And it is evident enough why the premises must be of different signs.

In the first figure, though all the forms be essentially one, (page 98,) the reduction of either to the form *Barbara* requires either the explicit use of contraries, or invention of a name subidentical to X. Accordingly, no mood of that figure is reducible to any other by the usually admitted reductions. But this cannot be said of any of the other figures. In the one before us, Cesare and Camestres are identical, even without changing the figure. That which is *Cesare* when X is major and Z minor, is *Camestres* when X is minor and Z major. In the first figure, the same attempt made on *Celarent* or *Darii*, removes them into another figure.

Third Figure.

Darapti	$Y)Z+Y)X=XZ$	*Felapton*	$Y.Z+Y)X=X:Z$
Disamis	$YZ+Y)X=XZ$	*Bokardo*	$Y:Z+Y)X=X:Z$
Datisi	$Y)Z+YX=XZ$	*Ferison*	$Y.Z+YX=X:Z$

The first and second figures contain a pair of universals each, with one particular derived from each, by a legitimate weakening of one premise and the conclusion at the same time : but in no instance is the quantity of the middle term weakened. And all the syllogisms in these two figures are fundamental (page 77). In the case now before us, both the leading syllogisms are not fundamental, but strengthened, and capable of being weakened in two different ways. The middle term is here examined in both premises : if it be wholly included in, or excluded from, one of the concluding terms, and wholly or partly included in, or excluded from, the other (but not so that there shall be exclusion from both) we have it that the whole or part mentioned in one case is included in, or excluded from, that which the whole is included in, or excluded from, in the other. There can be none but particular conclusions.

Fourth Figure.

Bramantip $Z)Y + Y)X = XZ$ | Camenes $Z)Y + Y.X = X.Z$
Dimaris $ZY + Y)X = XZ$ | . . . .

Fesapo $Z.Y + Y)X = X:Z$
Fresison $Z.Y + YX = X:Z$

We have now one universal syllogism in a form which does not admit of being weakened *in this figure*, and two strengthened syllogisms, each of which has one weakened form, one of them, *Bramantip*, admitting a stronger conclusion in another figure. Every conclusion except A appears. The mode of inference of the three first syllogisms has been described in the other figures. In Fesapo and Fresison, the perfect exclusion of the major term from the middle, accompanied by the total or partial inclusion of the middle in the minor, secures the exclusion from the major, of as much of the minor as it has in common with the middle.

I shall now proceed to the rules usually given, and to some remarks on the degree in which they apply to the more general system in chapter V. Aldrich gives them as follows—

> Distribuas medium : nec quartus terminus adsit :
> Utraque nec præmissa negans, nec particularis :
> Sectetur partem conclusio deteriorem ;
> Et non distribuat, nisi cum præmissa, negetve.

Thefe rules, I need hardly fay, are perfectly correct, when the contraries of the terms are excluded, and alfo all notion of quantity except all, or the indefinite fome. Taking them in the natural order, which verfification has a little difturbed, we have ;—

1. There are to be but three terms, of which it is underftood two only appear in the conclufion, the excluded or middle term appearing in both of the premiies. This is true in my fyftem, when by terms are underftood alfo contraries of terms. I fhould fuppofe that there can be no objection to the admiffion of contraries, unlefs there be one to the conception of a contrary. Any one may, with Ariftotle, object to the word not-man, as not the name of anything : on the grounds which immediately induced him to call it an aorift, or indefinite, name. But it can hardly be affirmed that any one admitting not-man as a name, fhould thereupon refufe to recognife the identity of ' horfe *is not* man,' with ' horfe *is* not-man.'

2. The middle term is to be *diftributed* in one or the other of the premifes. By diftributed is here meant *univerfally fpoken of* I do not ufe this term in the prefent work, becaufe I do not fee why, in any deducible meaning of the word *diftributed*, it can be applied to univerfal as diftinguifhed from particular. In ufing a name, it feems to me that we always diftribute : that is, fcatter as it were, the general name over the inftances to which it is to apply. When I fay fome horfes are animals, I diftribute certain horfes among the animals ; and when all, all. Leaving the word, the principle is one which clearly muft be true whenever we are reftricted in quantity to all or fome (indefinite), and when contraries are not admitted. In the former cafe we have, in one form or another, to make $m + n$ greater than n (chapter VIII.) when we cannot know what relation either m or n has to n, unlefs one of them, or both, be equal to n. We have no alternative then, but to require that m or n fhall be n. The cafes in which there is apparently no dependence on n will be difcuffed in the next chapter.

But when contraries are introduced, this rule is not univerfally true. The exception is feen in

$$A_1A'I' \text{ or } X)Y + Z)Y = xz.$$

If all the Xs be Ys, and alfo all the Zs, it follows that there are things which are neither Xs nor Zs, namely, all which are not

Ys. It is here, as elsewhere, implied that the middle term is not the universe of the proposition.

When we come, then, to use contraries, the simple rule of the middle term is no longer universally true. What other rule are we to put in its place? We know, of course, that every syllogism can be reduced to an Aristotelian syllogism, and even to one or other of two among them, $A_1A_1A_1$ or $I_1A_1I_1$, or to the first of these, if we be at liberty to use invention of names (page 97). Again, each term, or its contrary, is mentioned universally in every proposition: so that there is certainly one way in which every pair of premises may be made to exhibit a middle term universally used in one of them. The rule to be substituted for the *distribuas medium* is, that all pairs of universals are conclusive, but a universal and a particular require that the middle term should also be a universal and a particular, that is, universal in one and particular in the other. Thus, in $X)Y + Z)Y$, as it stands, the middle is particular in both; transpose into $y)x + y)z$ and the middle is now universal in both, by which we see the Aristotelian conclusion. Again, in $X)Y + ZY$, which is of the same kind, the transposition gives $y)x + Z:y$, which is faulty, because, though there be a particular premise, there is not anywhere a particular middle term. The cases in which the middle is of the same name in both places (universal in four, particular in four), are the strengthened syllogisms only. There is nothing to be surprised at in its thus appearing that the particularity of the middle term is just as much a test of a good syllogism as its universality: of every name and its contrary, one enters universally, and one particularly, in every proposition which contains it; and the system in chapter V is as much concerned with contrary as with direct terms. It is thence visible beforehand, to the mathematician at least, that any test must be defective, unless universal and particular enter into it in the same manner.

The above contains a complete canon of validity, as soon as the law of the three terms is understood, which is only a law of definition. We may state it as follows: Two premises conclude —when both are universal, always; when one only is universal, so often as it happens that the middle term (be it Y or y) is once only universal; when neither is universal, never. By this rule alone the thirty-two conclusive cases can be distinguished from the thirty-two inconclusive ones.

3. When both premifes are negative, there is no Ariftotelian fyllogifm. In the fyftem completed by contraries, there are eight fuch fyllogifms, as many in fact, as there are with premifes both affirmative. But a pair of negative premifes never conclude with both terms of the premifes, but with the contrary of one or both : and this muft be fubftituted, as a rule of conclufion, for the one juft named.

4. Both premifes muft not be particular. This rule, which relates wholly to quantity, muft be preferved in every fyftem which admits no definite ratio, except that of one to one, or *all* (pages 56, 57). I cannot learn that any writer on logic ever propounded even the very fimple cafe of ' Moft Ys are Xs, moft Ys are Zs, therefore fome Xs are Zs,' as a legitimate inference. And this, though it is certain that the quantitative prefix *moft (plurimi)* has before now excited difcuffion as to whether it belonged to a univerfal or a particular.

5. By *fectetur partem conclufio deteriorem* it is underftood that the negative is called weaker or lower *(deterior)* than the affirmative, and the particular than the univerfal ; and that the conclufion is to be as weak as negative, or as particular, if there be a premife which is negative or particular. This rule muft be preferved, when contraries are introduced, fo far as relates to particulars. But fo far as negatives are concerned, the rule muft be that *one* negative premife gives a *negative* conclufion, and *two* an affirmative one.

7. The laft line, *et non diftribuat, nifi cum premiffa, negetve,* fpoils the fymmetry to procure a verfe. The conclufion is not to be negative without a negative premife : that is, affirmative premifes give an affirmative conclufion. Alfo, no term is to be diftributively, (*i. e.* univerfally) taken in the conclufion, unlefs it were fo taken in its premife. A breach of this rule would be equivalent to drawing a conclufion about what was not (or about more than was) introduced into the premifes.

When contraries are introduced, the diftinction between pofitive and negative is made to appear, what it really is, one of language, or rather one of choice of names. But the diftinction of form is not abolifhed, but is exactly what it was before. We cannot lay down any rules for the formation of the conclufion unlefs, in our eight ftandard forms, we preferve the mode of writing which belongs to the fundamental derivation of the forms (page 61). Thus, the order being XY, A' is x)y and not Y)X, and O' is x:y and not Y:X. This method of writing being reftored, when neceffary, in pages 89 and 91, it follows immediately that the rule of accentuation in the notation gives the rule by which we determine whether the conclufion takes the terms from the premifes, or prefers contraries. According as the prepofition of the conclufion agrees with or differs from that of a premife, fo does the conclufion take a term from that premife, or its contrary. Thus, $A_1A_1A_1$ takes both terms from the premifes, but $A_1A'I'$ takes a contrary from the firft premife only. This laft we fee if we write the fyllogifm as $X)Y + y)z = xz$. Accordingly, we have—

Syllogifms taking both concluding terms direct from the premifes. Univerfals which begin with A ; particulars which begin with I . eight in number ; being all which ifolate no accent.

Taking the firft term only from the premife. Univerfals beginning with E ; particulars beginning with O : eight in number ; being all which ifolate the middle accent.

Taking the fecond term only from the premife. Strengthened forms and particulars which begin with A : eight in number, being all which ifolate the firft accent.

Taking neither term from the premifes. Strengthened forms and particulars which begin with E : eight in number, being all which ifolate the third accent.

This is a new mode of ftating the law of accentuation (pages 92-3) which I have preferred to place here, for fear of overloading chapter V. with rules. I have not ftated one half of thofe which fuggefted themfelves. This multiplicity of relations is a prefumption of the completenefs of the fyftem.

In the Ariftotelian fyftem, there is multiplication of the fame modes of inference, under the varieties of figure. In that which I propofe, there is a reduplication of moft of the effential cafes ; for whatever cafe is found, the fame is alfo found with X and Z interchanged, and alfo the order of the premifes. Again, whatever cafe is found, it is found contranominally ; or with all the accents (or prepofitions) altered. There are other ways (and many of them) in which the fyftem is only in one half a duplicate of what it is in the other. If all thefe modes of dividing the fyftem into

two correlative parts divided it into the same two parts, there can
be no question that one alone of those parts should have been
presented as the object of consideration. But this does not happen in any instance : so that it is impossible to dispense with the
whole of the thirty-two cases. The Aristotelian cases do not
form or include any half whatever of this system.

CHAPTER VIII.

On the numerically definite Syllogism.

IN the last chapter I considered no other quantity in names
except all and some : the latter meaning ' one or more, it
may be all.' To this extent of quantity we are limited in most
kinds of reasoning, by want of knowledge of the definite extent
of our propositions : and the few phrases (page 58), as ' most,'
' a good many,' &c. by which we endeavour to establish differences of extent in ordinary conversation, have been hitherto held
inadmissible into logic. In this science it seems to have been
always intended that the bases on which its forms are constructed
shall be nothing but the supposition of the most imperfect and
inaccurate knowledge. Though in geometry we are permitted
to assume as the object of reasoning the ideal straight line, the
' length without breadth' of Euclid, which has no objective pro-
totype, and though we see the advantage of reasoning upon ideas,
and allowing the essential inaccuracies of material application to
produce no effect except in material application,—yet in the consideration of the pure forms of thought, the learner has always
been denied the advantage of studying the more perfect system of
which his inferences are the imperfect imitation.

The ordinary universal propositions are of a certain approach
to definite character, both of them with respect to their subjects,
and the negative one with respect to its predicate also. In X)Y
for example, what is known·is as much known of any one X as
of any other. Perfect definiteness would consist in a more exact
degree of description, and would require a higher degree of knowledge. But in this chapter I speak only of *numerical* definite-

nefs, of the fuppofition that we know *how many* things we are
talking about. We may be well content to examine what we
fhould do if we were a ftep or two higher in the fcale of creation,
if by fo doing we can manage to add fomething to our methods
of inference in the higheft to which we have as yet attained.

A numerically definite propofition is of this kind. Suppofe
the whole number of Xs and Ys to be known: fay there are
100 Xs and 200 Ys in exiftence. Then an affirmative propofition of the fort in queftion is feen in ' 45 Xs (or more*) are
each of them one of 70 Ys' : and a negative propofition in
' 45 Xs (or more) are no one of them to be found among
70 Ys.'

But it muft be particularly noticed that in fpeaking of a number of Xs, as 45 Xs, I do not mean *certain* 45 Xs which can
be diftinguifhed from all the reft, fo that of any X it is poffible
to be known whether it belong to the 45 of the propofition, or to
the remaining 55. This degree of definitenefs is one ftep higher
than that which I here propofe to confider, and which is defcribed
by ' there are 45 Xs which are contained among 70 Ys, it not
being known which Xs are the 45 Xs, nor which Ys are the
70 Ys :' or elfe by ' there are 45 Xs which are not any of them
identical with any one of 70 Ys, the precife Xs and Ys in
queftion being unknown.'

It cannot of courfe be difputed that if any thing fhould neceffarily follow from *any* 45 Xs being found among *any* 70 Ys, it
will not the lefs follow from our knowing which are the Xs and
which are the Ys. But this laft fuppofition only brings us to
really univerfal propofitions. If, there being 100 Xs, 45 of
them can be fpecifically feparated from the reft, fo as to be
known, the procefs of feparation is equivalent to putting them

---

* Thefe words (or more) fhow that the word *definite* has reference only
to the *lower* boundary. Of courfe nothing can be fhown in right of " 45
or more, *perhaps*" except what is true in right of the 45. It is defirable
that as the premifes, fo fhould be the conclufion, of a fyllogifm : this would
not be the cafe if we ufed premifes definite both ways. For example, there
being 100 Ys in exiftence, it will prefently appear that ' *Exactly* 55 Ys are
Xs and *exactly* 60 Ys are Zs,' though it enable us to fay that ' 15 Xs are
Zs' does not allow us to fay ' Exactly 15 Xs are Zs,' but only ' 15 Xs (or
more) are Zs.'

under a feparate name, fubidentical to X, and the reft, which are
equally diftinguifhable, under another name, alfo fubidentical to
X, and contrary of the firft name, when the univerfe is X.
Whether the name be long or fhort, does not matter, nor
whether it carry the feparating diftinction in its etymology or
not. To feparate in any way inftance from inftance by language, is to name.

If then 45 definite Xs were known to be contained among
70 definite Ys, and if thefe Xs were each named M, and thofe
Ys each N, and if the reft of the Xs and Ys were named P
and Q, we fhould have the following propofitions,

M)X, P)X, N)Y, Q)Y, M)N, M.P, N.Q,

and all inferences. Moreover, in each cafe; we fhould have the
total number of inftances which are contained under each name ;
the numbers carrying with them evidence that every X is either
M or P, and every Y either N or Q. Subftitute M.N for
M)N and we have the correfponding negative propofition.

But if 45 unfeparated and infeparable Xs be fuppofed known
each to be among 70 fimilarly fituated Ys, there is no immediate
method of making any other propofition out of the terms X and
Y except its converfe, that 45 of thefe 70 Ys are 45 Xs, and (if
the whole number of Ys be known, fay 200) that there are 45
Xs which are not any one among 200—70, or 130 Ys. This is
then a fimple propofition, which becomes of a highly complex
character, when the Xs and Ys named in it are taken as definitely feparable from the reft. I fhall call it the *fimple numerical*
propofition.

The diftinction may be eafily illuftrated by example. " All
the planets but one' is a particular propofition ; it is ' fome
planets:' there is no one planet of right included in it. But
' all the planets except Neptune' is a univerfal propofition : ' a-
planet-not-Neptune' is a name of Mercury, of Venus, &c. ;
and of every planet it can be ftated whether it be in the name
or not. That which is true inferentially of ' all the planets but
one' left particular, is true of ' all the planets but Neptune:'
but that which is true of the latter is not neceffarily true of the
former.

Taking X, Y, Z as the terms of the fyllogifm, & the number

of Xs in exiftence, η the number of Ys, and ζ the number of Zs, and ν the number of inftances in the univerfe, there are of courfe fixteen poffible cafes of knowledge, more or lefs, of thefe primary quantities, from all unknown to all known. Of thefe fixteen cafes, it will be requifite to confider two only. Firft, when the extent η of the middle term is known, and all the reft unknown; fecondly, when all are known. The *algebraical* formulæ of the latter cafe will enable us to point out how the fuppofition of any lefs degree of knowledge would affect our power of inference.

I propofe the following notation. Let mXY denote either of the equivalent propofitions, that m Xs are to be found among the Ys, or that m Ys are to be found among the Xs. Let mX:nY denote either of the equivalent propofitions, that there are m Xs which are not any one among n Ys, or n Ys which are not any one among m Xs.

The fymbol 10X is the algebraical fymbol for ten equal Xs added together, X being a magnitude: it is then a *collective* fymbol. In this work, X being a name, it implies every one out of ten inftances of that name, *diftributively*, but not *collectively*. This diftinction is very material, not only in this chapter, but throughout every part of logic. ' Every X is Y' is diftributively true, when, by ' Every X' we mean each one X: fo that the propofition is ' The firft X is Y, *and* the fecond X is Y, *and* the third X is Y, &c.' In this cafe the fubject is X, and the word *every* belongs to the quantity of the propofition. But ' every X is Y' is *collectively* true, when we do not mean that any one X is a Y, nor that any number of Xs are Ys, but that all the Xs make a Y. In this cafe the propofition is *fingular*: there is but one inftance of the fubject mentioned, that fubject being, not X, but the collection ' all the Xs.' Thus ' the ten men are members of a committee' is diftributive: ' the ten men are a committee' is collective.

If, in fuch a propofition as 10XY, we were to fuppofe the 10 Xs fpecifically feparated from the reft, being certain affignable ten individuals from among all the Xs, then 10X *becomes a name* for each of the ten, as much as X, and may be confidered as a univerfal term. And now 10XY and {10X})Y mean the fame things.

Let η be known, and η only of the four, υ, ξ, η, ζ. The only collections of premifes which it is neceffary to confider are

$$mXY + nYZ$$
$$mXY + nZ : sY$$
$$mX : rY + nZ : sY$$

Without fome knowledge of the number of ys, of which by fuppofition we have none, it would be ufelefs to attempt to draw an inference from a pair in which Y and y enter together, partially quantified, as in $mXY + nZ : ry$. And nZy merely amounts to $nZ : {}_\eta Y$.

The above three are all we need confider : and even of thefe the third is incapable of inference, fince both premifes are negative, and moreover, not reducible to a pofitive form by ufe of contraries, the only way in which negative premifes really acquire a conclufion in chapter V.

Let us firft confider the premifes $mXY + nYZ$. They tell us that among the η Ys we find m Xs and n Zs : accordingly, neither m nor n exceeds η. If m and n together fall fhort of η, nothing can be inferred : Y is extenfive enough (that is, there are inftances enough of Y) to hold the m Xs and the n Zs without any coincidence of an X with a Z. As to other Xs or Zs, we do not know whether they exift ; or, if they exift, we do not know that any one of them is a Y. But if m and n together exceed η, it is impoffible that m Xs and n Zs can find place among η Ys, except by putting either two Xs or two Zs, or an X and a Z, with *one* of the Ys. Now as by the nature of the fuppofitions, there cannot be two Xs, nor two Zs, to one Y, we muft have the inference 1XZ as often as there are units in the excefs of $m + n$ over η. That is,

$$mXY + nYZ = (m + n - \eta)\, XZ$$

Next, let us take $mXY + nZ : sY$. There may be two inferences, perfectly diftinct from each other, the connexion of which can only be explained in the more general fyftem to which we fhall prefently come. Firft, let m and s together exceed η. Then $m + s - \eta$ of the Ys have the common property of being Xs, and of being clear of the n Zs. Accordingly, we have

$$mXY + nZ : sY = (m + s - \eta)X : nZ$$

Next, let $n+s$ be greater than n.   Take the s Ys among which
no one of the n Zs is found.   Becaufe $n+s$ is greater than n, n
is greater than $n-s$, the number of Ys left.   Accordingly,
$n-(n-s)$ of the n Zs cannot be *any* Ys, and therefore cannot
be any of the m Xs which are Ys.   Hence we have

$$mXY + nZ : sY = mX : (n+s-n)Z$$

In the appendix to this chapter (at the end of the work) will
be feen the manner in which all the Ariftotelian fyllogifms can
be brought under the firft cafe, and the firft* inference of the
fecond cafe.   No Ariftotelian fyllogifm can be deduced from the
fecond inference except when $s=n$, in which cafe it agrees with
the firft.   For, when s is not n, we muft, to make fuch a fyllogifm, have $m=n$, and then, to make $nZ:sY$ Ariftotelian, s not
being n, we muft have all the Zs in n, or $n=\zeta$.   We thus get
$Y)X + sY:Z$, the premifes of *Bokardo*.   But the conclufion is
$nX:(\zeta+s-n)Z$, that of *Bokardo* being $sX:Z$.   And this will
be found to be the only Ariftotelian fyllogifm which has this
fecond and numerically quantified inference, depending upon the
number of Zs exceeding the number of Ys unnamed in the particular premife.

I now proceed to fuppofe that all the quantities are taken into
account.   Some preliminary confiderations will be ufeful, as
follows.

Let two propofitions be called identical, when, either of them
being true, the other muft be true alfo : fo that nothing can be
inferred from the one, which does not equally follow from the
other.   Such propofitions are X.Y and Y.X, fuch are X)Y
and y)x, and fo on.   Again, two propofitions may be identical
relatively to a third : thus, P being true, Q and R may either
follow from the other; accordingly, as long as it is underftood
that P is true, Q and R may, relatively to that fuppofition, be
treated as identical.

The word *identical*, as applied to *propofitions*, is here made to
mean more than ufual, but not with more licenfe than when the
word is applied to *names*.   Thus, *man* and *rational animal* are

---

* I was not in poffeffion of the fecond inference till I had written what is
in page 157.

not identical names, *qua* names, for they neither spell nor sound alike: the identity understood is that of meaning; where one applies, there shall the other apply also. Similarly, as to propositions (of which subject, predicate, and copula are the material parts, just as spelling and sound are those of names), identity does not consist in sameness of parts, nor in reducibility to sameness, but in simultaneous truth or falsehood, so that what either is, be it true or false, the other is also, in every case. Thus two propositions, one of which signifies that an end has been gained, and the other that the sole and sufficient means of gaining it have been used, are identical.

All the theory of *names*, their *application* or *non-application*, may be applied to *propositions*, their *truth* or *falsehood*. To say that a proposition is true in a certain case, is to say that a certain name applies to a certain case: to say that it is false, is to say that a certain name does not apply, but that its *contrary* does. That contrary is what logicians usually call *contradictory*: and the *name* is not simply *true* or *false*, but the adjective attached to the proposition. The conditions under which we are to speak limit us to a number of cases which constitute what we may now call, not the *universe* of the *names in the propositions*, but the universe of the *truth or falsehood* of the propositions. Thus we shall suppose ourselves now to be speaking, not of all instances to which the *name* U applies, but of all in which the *proposition* U is *true*, or in which the name 'true U' applies. A case in which a proposition P is true may be marked P, one in which it is false, p, We may now apply the names subidentical, &c. and the symbols, together with all the syllogisms, complex and simple; but on each a remark may be necessary.

Subidentical, identical, and *superidentical.* If P be a proposition subidentical of Q, that is, if every case in which P is true be one in which Q is true, but so that Q is sometimes true when P is not, the proposition Q is usually mentioned as *essential* to P, and as a *necessary consequence* of it. Whenever P is true, Q is true; Q necessarily follows from P; if Q be false, P cannot be true; Q is essential to P; are all mere synonymes. Accordingly ' *necessary consequent*' and ' *superidentical or identical*' are synonymous terms: that is (page 68), *necessary consequent* and *superaffirmative.* Identity of course consists in *each* proposition being true when the other is true. I think that, according to general notions, it would be held more juſt to ſay that a propoſition *contains* its neceſſary conſequence than that it *is contained* : but a moment's conſideration will ſhow that the latter analogy is at leaſt as ſound. If the ſecond be true whenever the firſt is true, it may be true in other caſes alſo : ſo that we only ſay the ſecond contains the firſt, and it may be more.

Subcontrary, contrary, and *ſupercontrary.* It is uſual to call ' No X is Y.' and ' Every X is Y ' by the name of contraries, and to ſay that ' contraries may be both falſe, but cannot be both true.' This is a technical uſe of the word : in common language we ſhould ſay that either a propoſition or its contrary muſt be true ; ' have you any thing to ſay to the *contrary*' generally means what a logician would expreſs by putting the word *contradictory* in the place of *contrary.* I am compelled to uſe the words contrary and contradictory as ſynonymous : at which compulſion I am well pleaſed, never having ſeen any good reaſon why, in the ſcience which conſiders the relations of *dicta,* the *contraria* ſhould be any thing but the *contra dicta.* The proper word for contrary, commonly uſed to expreſs the relation of X)Y and X.Y, is *ſubcontrary.* Here are two propoſitions P and Q which cannot both be true, but may both be falſe : here is a pair which can never be aſſerted of the ſame inſtance, and of which, in many inſtances, neither can apply. In the ſame manner, the propoſitions XY and X:Y, uſually called *ſubcontrary* (for no reaſon that I can find except that they are written *under* the ſo called *contraries* in a ſcheme or diagram very common in books of logic) ſhould be called *ſupercontrary :* they are never both falſe, and may be both true. This is a complete inverſion of the uſual propoſitions : an inverſion which ſeems to me to be imperatively required, if only my uſe of *ſub* and *ſuper* in Chapter IV. be allowed.

In applying theſe names to propoſitions, it muſt be remembered that we make the ſame ſort of aſcent which we make in paſſing from ſpecific to univerſal arithmetic, in uſing a ſymbol to ſtand for any number at pleaſure. For inſtance ;—Perhaps it may be thought that XY and X:Y may ſometimes be only contraries, and not ſupercontraries, becauſe there may be names which make one only true and not both. But this is not correct :

for we are considering the proposition itself *as an instance among propositions*, not the proposition as subdivisible into instances, in which name is compared with name. In speaking of propositions, it is change from use of one name to use of another, or from use of one number to use of another, which is change of instance: not change from one instance of name to another. And just as in a universe of names, every name introduced is supposed to belong, or not to belong, to every instance in that universe: so in a universe of propositions, I suppose every proposition, or its contrary, to apply (whether it be or be not known which applies) in every instance. We have never considered such a thing as the universe U, in which there are cases in which neither X nor x applies: we suppose there is always a power of declaring that the name X must either belong or not belong to each instance. In like manner, all the propositions in each universe now considered, are supposed to be connected with all the names in question: so that X, Y, being two of them in their order of reference, A_1 or O_1 is true in each case, and A' or O', E_1 or I_1, and E' or I'. We might, if we pleased, enter upon a wider system. For though we cannot imagine of any object of thought, but that it is either X or not X, be X what *name* it may, yet we can imagine of *propositions* that they may be wholly inapplicable, as being neither true nor false. The first assertion is all the more true, that it could hardly be exemplified without exciting laughter: as I should do if I reminded the reader that a book is either a cornfield or not a cornfield. We have never considered names under more predicaments than two; never, for instance, as if we were to suppose three names X_1, X_2, X_3, of which everything must be one or the other, and nothing can be more than one. But we should be led to extend our system if we considered propositions under three points of view, as true, false, or inapplicable. We may confine ourselves to single alternatives either by introducing not-true (including both false and inapplicable) as the recognized contrary of true: or else by confining our results to universes in which there is always applicability, so that true or false holds in every case. The latter hypothesis will best suit my present purpose.

This digression is somewhat out of place here, but I have preferred to retain the matter of it until I had occasion to use it.

I now proceed to assert that the simple numerical proposition has no occasion for a numerically definite predicate. Let us confider first an affirmative proposition, say 'Of 10 Xs, each is to be found among some 15 Ys.' Of course it is supposed there are 15 or more Ys in existence. With this let us compare ' 10 Xs are to be found among the Ys.' These two propositions are identical : if 10 Xs be among 15 Ys, there are 10 Xs among the Ys : and if 10 Xs be among the Ys they are certainly 10 Ys; put on 5 more Ys at pleasure, and they can be said to be among 15 Ys in just as many ways as we can choose 5 more Ys to make up the 15. Note, that if the 10 Xs were among certain specified 15 Ys, then, though the first proposition would give the second, the second would not necessarily give the first. But we are now supposing that numerical selection is only *numerically* definite : definite as to the number, not as to the instances which make up that number. When therefore we say ' 10 Xs are among 15 Ys' we say neither more nor less than when we say 10 Xs are among the Ys.' It is in fact ' 10 of the Xs are 10 of the Ys' and the converse ' 10 of the Ys are 10 of the Xs' is the same proposition.

Now let us take a negative proposition, ' 10 of the Xs are not to be found, *any one of them*, among some 15 Ys;' abbreviated into ' 10 Xs are not in 15 Ys.' If there be 25 Ys in existence this proposition must be true; mean X and Y what they may. It is as true as that the X which is one Y is not any other Y. Say there are 25 or more Ys : take any 10 Xs you choose, and put them down on any 10 Ys you choose. Then certainly there are 15 Ys left, no one of which is any of those 10 Xs. Again, if there be 25 Xs in existence, still the proposition must be true. For if the 15 Ys were all there are, and they were all Xs, there still remain 10 Xs which are not any one in the 15 Ys. Accordingly, the proposition ' *m* Xs are all clear of *n* Ys,' whenever either the whole number of Xs, or the whole number of Ys, exceeds *m* + *n*, says no more than is conveyed in our permanent understanding that no object of thought can be more than one X or one Y. But let it be otherwise ; let neither Xs nor Ys be as many as *m* + *n* in number. Say there are 20 Xs and 23 Ys and let 10 Xs be clear of 15 Ys. There must now be at least 15 + 10 — 20, or 5 Ys which are *no* Xs *at all*, and at least 15 + 10 — 23, or 2 Xs which are *no* Ys *at all*. First, it is

plain that there are no 10 Xs among thofe Ys which are clear of
15 Ys: for there are but 23 Ys in all. Therefore, 2 at leaft
of thefe 10 Xs muft be Xs which are not Ys: which with 8
Xs that may be Ys, will be clear of the remaining 15 Ys.
Therefore 2 Xs at leaft are not Ys. Again, there are no 15
Ys among thofe Xs which are clear of 10 Xs, for there are but
20 Xs in all. Five Ys which are not Xs muft exift, which
with 10 that may be Xs, will be clear of the remaining 10 Xs.
Accordingly, if the whole number of Xs be ξ, and the whole
number of Ys be n, the propofition ' there are m Xs which are
no one to be found among n Ys' is effentially true of every cafe of
that univerfe, whenever $m+n$ is lefs than either ξ or n. But
when $m+n$ is greater than both ξ and n, there are two propofitions, neceffarily involved, which are not effentials of all cafes
of that univerfe: namely, that there are $m+n-\xi$ Ys which are
not any Xs, and $m+n-n$ Xs which are not any Ys.

But, it may be afked, if n fhould be lefs than ξ, and $m+n$
greater than n, but ftill lefs than ξ, may we not affirm that
$m+n-n$ Xs are not Ys? Undoubtedly we may, but then we
do not affirm fo much as already belongs to every cafe of the
univerfe. For if ξ be greater than n, no more than n Xs can be
Ys, and there are left $\xi-n$ Xs which cannot be Ys: and $\xi-n$
is, in the cafe fuppofed, more than $m+n-n$.

Let v be the number of inftances in the univerfe, ξ and n being
the number of Xs and of Ys. The following ufes of the notation
will be readily feen to exprefs preceding refults, or others immediately deducible.

$\qquad \xi$ greater than n $\qquad (\xi-n)X:nY$ $\qquad$ or $(\xi-n)Xy$
$\qquad n$ greater than ξ $\qquad (n-\xi)Y:\xi X$ $\qquad$ or $(n-\xi)Yx$

$m+n$ greater than ξ and than n gives
$$mX:nY=(m+n-n)X:nY=(m+n-\xi)Y:\xi X$$

A_1	$X)Y=\xi XY$	$=(v-n)yx$	$=\xi Y:(v-\xi)x$
O_1	$X:Y=mX:nY$	$=mXy$	$=my:(v-\xi)x$
A'	$Y)X=nXY$	$=(v-\xi)xy$	$=nX:(v-n)y$
O'	$Y:X=mY:\xi X$	$=mYx$	$=mx:(v-n)y$
E_1	$Y.X=\xi Xy$	$=nYx$	$=\xi y:(v-\xi)x =nx:(v-n)y$
I_1	$XY=mXY$	$=mX:(v-n)y=mY:(v-\xi)x$
E'	$x.y=(v-\xi)xY=(v-n)yX=(v-\xi)Y:\xi X=(v-n)X:nY$
I'	$xy=mxy$	$=mx:nY$	$=my:\xi X$

I now examine the modes of contradicting mXY and mX:nY. As to the firſt, it is obvious that (m always meaning that m are, but that more may be) either m or more Xs are Ys, or elſe $\xi - m + 1$ or more Xs are not Ys. The contradiction then is either of the equivalents

$$(\xi - m + 1)\text{X}:{}_n\text{Y} \text{ and } (n - m + 1)\text{Y}:\xi\text{X}$$

It will be ſatisfactory to evolve the contradiction of mX:nY by a method which will again demonſtrate the caſes in which no contradiction exiſts; or in which the propoſition is always true. Let us put the two names in the leaſt favourable poſition for making mX:nY true. Let p then be the number of Xs which are not Ys, all the reſt being Ys. Take the p Xs which are not Ys (p muſt not be ſo great as m, for then the propoſition is made good by the Xs which are not any Ys) and $m - p$ from thoſe which are Ys. All the m Xs thus obtained are clear of $n - (m - p)$ or $n - m + p$ Ys. Let this juſt be n: that is, let $p = m + n - n$. Then $\xi - p$, the number of Xs which *are* Ys, is $\xi - (m + n - n)$ or $\xi + n - m - n$. Let but one more X be Y, and the propoſition begins to be contradicted: for now $m + n - n - 1$ Xs are not Ys, we muſt take up $n + 1 - n$ of thoſe which are Ys to make m Xs, and there only remain $n - (n + 1 - n)$ or $n - 1$ Ys clear of the m Xs. And it is plain that if we cannot do it by uſing firſt all the Xs which are not Ys at all, ſtill leſs can it be done by uſing thoſe which are. Accordingly the contradiction of mX:nY is

$$(\xi + n - m - n + 1)\text{XY}$$

Then, in order to have a propoſition which can be contradicted, $m + n$ muſt be greater than ξ, or equal to $\xi + 1$ at leaſt, for otherwiſe $\xi + n - m - n + 1$ would be greater than n, or more Ys than n muſt be Xs, which is abſurd: and ſimilarly $m + n$ muſt be greater than n. Otherwiſe, all contradiction is abſurd, or mX:nY is always true.

Aſſuming theſe laſt conditions, however, the contradiction of mX:nY is made eaſier. To be capable of contradiction, it muſt amount to $(m + n - n)$X:nY. Now when $m + n - n$ Xs are not Ys, and no more, $\xi + n - m - n$ Xs *are* Ys. One or more above this, or let $(\xi + n - m - n + 1)XY$, and $mX:nY$ cannot be true.

Thus much for contradictory or contrary propositions. I shall presently consider the *contranominal* propositions.

We must guard ourselves from prescribing the use of any premise which necessarily belongs to all cases in the universe (of propositions). Let P be a proposition which may or may not be true, laid down as a premise, and Q a proposition which is true in every case. Let R be their necessary consequence, or legitimate inference: then it is not ' whenever P and Q are true, R is true,' but ' whenever P is true, R is true.' So far as R is a consequence of Q, so far it is a consequence of every thing which necessarily gives Q; and thus it is a consequence of the supposed constitution of the universe from which the propositions are taken. Now this constitution is always understood; it may be a convenience that R should be deduced by first deducing Q, but it cannot be a necessity. And R is a consequence of P and this constitution, not of P and Q

For example, let the universe of propositions be all that can be formed out of the suppositions of the existence of 20 Xs, and 30 Ys, and 40 Zs, in one universe of names. Let us join together 15XY and 10Z : 20Y. Our rules of inference will presently show us, that $5X : 10Z$ is the necessary consequence of these premises : but this result is not only true when 15XY is true, without anything else, but even without that ; because $5 + 10$ falls short of 40.

Again, we must guard ourselves from adopting the conclusion which follows from premises, when that conclusion is true in all cases by the constitution of the universe : it is then a sort of *spurious** conclusion, legitimate enough as an inference, but of a perfectly distinct character from inferences which would bear

---

* To this word, as here used, I have heard much objection ; and when I first took it, it was unwillingly, and for want of a better. But on further consideration I am well satisfied with it. The objection arises from the idea of false or worthless being generally attached to the word. But, though it may be usual for spurious things to be worthless, it is not necessary. If a London maker of razors should put the name of a great Sheffield house upon them, those razors would be spurious. Suppose them as good as those of the Sheffield maker, or better, they are still spurious : though it may be true enough

doubt but for the premises, or would bear contradiction under other premises. Say that in the above universe we join the propositions 15XY and. 30Z : 20Y. Both these propositions are capable of contradiction : the second is 20Z : ηY (η means 30, but the symbol reminds the reader that 30 is *all*) or 10Y : ζZ (ζ being 40). Now, by laws of inference, 15XY + 30Z : 20Y yields 5X : 30Z, which is always true in that universe.

Here is a case in which premises capable of contradiction give a conclusion which is not.

The rule of inference is obviously as follows. We cannot show that Xs are Zs by comparison of both with a third name, unless we can assign a number of instances of that third name, *more than filled up* by Xs and Zs : that is to say, such that the very least number of Xs and Zs which it can contain are together more in number than there are separate places to put them in. If our premises, for example, separate some 30 Ys, and dictate that among those 30 Ys there must be 20 Xs and 15 Zs, it is clear that there must be at least 5 Zs which are Xs. For if we put down the 20 Xs which are to go in, and try to put the Zs into separate places, we are stopped as soon as we have filled up the 10 remaining out of the 30 Ys, and must put the other 5 Zs among the Ys which have been made Xs. Accordingly, so many Xs at least must be Zs as there are units in the number by which the Xs and Zs to be placed, together exceed the number of places for them. All the other rules of inference are modifications of this. For example, to prove that 10 Xs are not Zs, we must show some number of instances (be they Ys or ys, or part one and part the other) *overfull* (in the above sense) of Xs and zs, to the amount of 10 at least ; so that 10 Xs are zs, or are not Zs. To prove that some xs are ys, we must show a number of instances in which the *least* numbers of xs and zs

that the chances are rather in favour of their resembling the ware of Peter Pindar's hero. In this work, a spurious inference is that which passes for the consequence of certain premises, but does not in reality follow from those premises any more than from an infinity of others: being true by the constitution of the universe. It is made to have the mark of those premises, when in truth we cannot know whether those premises be possible or not, until we have first examined a constitution which virtually contains our conclusion.

which it can contain, *overfill* it, or in which the *greateſt* number
of Xs and Zs which it can contain *underfill* it, or do not fill it,
though made completely ſeparate.

In examining the fundamental laws of ſyllogiſtic inference, it
is not neceſſary to conſider any thing but the poſitive forms.
For $mX:nY$, when not ſpurious (and we ſhall ſee that the
ſpurious caſes may be rejected) is $(m+n-n)X:nY$, which is
$(m+n-n)Xy$ or $(m+n-\xi)xY$. There are, then, but two
fundamental caſes: one in which the predicates are the ſame,
one in which they are contraries. We ſhall accordingly have to
conſider

$$mXY+nZY \qquad \text{and} \qquad mXY+nZy:$$

and it will preſently appear that not more than one, even of
theſe, is abſolutely neceſſary. In each caſe we muſt aſk, what
collective inſtances of Y or of y, or partly of one and partly of
the other, receive any dictation as to how they are to be filled
with Xs, with xs, with Zs, or with zs: and what is the leaſt
number of each which can be allowed to every ſuch collection.
But there is yet ſomething to do, ſuggeſted by the preceding
remarks. Let us take one propoſition, a type of all we ſhall
have to conſider, ſay mXY. This means that XY is true to at
leaſt m inſtances. Now, this propoſition may involve Xy, or
xY, or xy. Firſt, as to Xy. To get the leaſt number of Xs
among the ys, we muſt put the greateſt number among the Ys.
If all the Xs will go among the Ys (or if n be greater than or
equal to ξ) there need be no Xs among the ys: but if not (or if
n be leſs than ξ) then $\xi-n$ Xs muſt be among the ys, in every
caſe. Accordingly

$$mXY \text{ gives } (\xi-n)Xy$$

where by $\xi-n$ underſtand o, not only when ξ is equal to n, but
when it is leſs. This reſult is ſpurious, ſince it is true or falſe,
by the mere conſtitution of the univerſe, independently of mXY.
Secondly, as to xY. Since mXY is equally mYX, the ſame
reaſoning ſhows that

$$mXY \text{ gives } (n-\xi)xY$$

where $n-\xi$ is to be underſtood in the ſame way. This reſult is
alſo ſpurious for a like reaſon.

Thirdly, as to xy. Since there muſt be *m* Xs among the Ys, the greateſt poſſible number of xs is *n*—*m*. If this be as great as *v*—*ξ*, the whole number of xs, there need be no xs among the ys: but if *n*—*m* be leſs than *v*—*ξ*, there muſt then be at leaſt *v*—*ξ*—(*n*—*m*) xs among the ys, or *v*+*m*—*ξ*—*n*. Conſequently

$$mXY=(v+m-n-\xi)\,xy.$$

I here put the ſign = becauſe theſe propoſitions are really equivalents. Treat the ſecond in the ſame way as that which deduced it from the firſt, and we have

$$(v+m-n-\xi)xy=(v+\overline{v+m-n-\xi}-\overline{v-n}-\overline{v-\xi})XY$$
$$=mXY$$

If *v*+*m* be not greater than *n*+*ξ*, the equivalent does not exiſt. We are already well acquainted with one caſe of this propoſition. Let *m*=*ξ*: then *m*XY is X)Y and the equivalent becomes (*v*—*n*)xy, which, as *v*—*n* is the whole number of ys, is y)x.

The rule is, if two names have a certain number of inſtances at leaſt in common, to the whole number in the univerſe add that number of inſtances, and ſee if the ſum exceed the whole number of inſtances of both names together. If it do ſo, the exceſs ſhows the leaſt number of inſtances which the contraries of theſe two names muſt have in common. Follow this rule, and we have

$$mXY=(v+m-n-\xi)\,xy$$
$$mxY=(\xi+m-n)Xy$$
$$mXy=(n+m-\xi)xY$$
$$mxy=(\xi+n+m-v)XY$$

Here are exhibited the equivalent *contranominal* forms. The following reſults may now be deduced.

Firſt, theſe contranominals being formed in the ſame way, each from the other, in any one pair, whatever we prove of the firſt from the ſecond, we alſo prove of the ſecond from the firſt. The mathematician would call them *conjugate* pairs. Next, ſince all the four pairs are but verſions of the firſt, with difference of names, whatever we prove univerſally of the firſt pair, we prove of all. Now, taking the firſt of any pair and making it poſſible, which is done by allowing *m* not to exceed the number of either of the names mentioned, the ſecond may be poſſible or impoſſible, according as the fubtraction indicated can be done or not. *But whenever the fecond is impoffible, the firft is fpurious.* Take mXY, and let $(v+m-\xi-n)$xy be impoffible, or $v+m$ (and ftill more v) lefs than $\xi+n$. Now as all the ξXs and nYs muft find place in the v inftances of the univerfe, and $\xi+n$ exceeds v, we muft, in every cafe of the univerfe of propofitions, have at leaft $(\xi+n-v)$XY. But $v+m$ is lefs than $\xi+n$ or $\xi+n-v$ greater than m: confequently, mXY is fpurious, a larger propofition being always true.

As we are not to admit fpurious propofitions among our premifes, we had better write all premifes double, putting down each of the forms, and making double forms of inference. The prefence of the fymbols of all neceffary fubtractions will remind the reader of the fuppofitions which muft be made, to infure a legitimate fyllogifm. I now take the feveral forms.

$$\begin{array}{ccc} m & n & (m+n-n)\text{XZ} \\ \dfrac{\text{XY}}{(v+m-\xi-n)\ \text{xy}} & + \dfrac{\text{ZY}}{(v+n-\zeta-n)\ \text{zy}} = \dfrac{}{(v+m+n-n-\xi-\zeta)\text{xz}} \end{array}$$

The law of inference here tells us (page 154,) that $m+n$ being greater than n, $(m+n-n)$XZ, be it fpurious or not, follows from the upper premifes. The lower premifes alfo give their inference if

$$(v+m-\xi-n)+(v+n-\zeta-n) \text{ be greater than } v-n.$$

$v-n$ being the number of the ys. This laft is equivalent to faying that $v+m+n$ is greater than $\xi+n+\zeta$. Firft, remark that one fpurious premife neceffarily gives a fpurious conclufion. Say that $v+m$ is lefs than $\xi+n$, or that mXY is fpurious. Then, fince $v+m$ is lefs than $\xi+n$, and n does not exceed ζ, it follows that $v+m+n$ is lefs than $\xi+n+\zeta$; whence the contranominal of the conclufion does not exift, or the conclufion is fpurious, as afferted.

Next, obferve that the conclufion may be fpurious, though neither of the premifes be fo. For though $v+m$ be greater than $\xi+n$, and $v+n$ than $\zeta+n$, and therefore $2v+m+n$ greater than $\xi+\zeta+2n$, or $v+m+n+(v-n)$ greater than $n+\xi+\zeta$, it by no means follows that $v+m+n$ alone is greater than $n+\xi+\zeta$. It is alfo vifible in the mode of formation of the fecond inference, that to fay $v+m$ exceeds $\xi+n$, and $v+n$ exceeds $\zeta+n$, only gives *exiſtence* to the premiſes: to give them concluſion, the ſum of the two exceſſes muſt itſelf exceed $v-n$, the whole number of ys.

Thirdly, we muſt not omit to examine the poſſible caſe in which a premiſe is *partially ſpurious*. For example, there are 10 Xs and 20 Ys in a univerſe of 25 inſtances; accordingly, $10+20-25$, or 5, of the Xs *muſt* be Ys. Let one of the premiſes be 8XY: this is not then all contingent, and capable of contra-diction; we only learn ſomething about 3 out of the 8 Xs. And I call this propoſition partially ſpurious. But it will give no trouble: for we muſt deal with the premiſes and their contra-nominal equivalents before we can pronounce for a concluſion; and of two propoſitions which are contranominal equivalents of each other, one *muſt* be partially ſpurious. To ſhow this, obſerve that if mXY be *not* partially ſpurious, it is becauſe v is greater than $\xi+n$; or $2v$ than $\xi+n+v$; or $(v-\xi)+(v-n)$ than v. But then the numbers of xs and ys together exceed the whole num-ber of inſtances in the univerſe; whence ſome xs muſt be ys, or the contranominal equivalent of mXY *is* partially ſpurious.

Now, to write down the various forms of inference. There are ſixteen ways of trying for an inference: we may combine a propoſition in XY, or xy, or xY, or Xy, with one in XZ, or xz, or xZ, or Xz. But theſe ſixteen caſes really combine four and four into only four diſtinct caſes. Thus the one we have been conſidering, really contains the combinations of XY and YZ, XY and yz, xy and YZ, and xy and yz. It is in our power to make either pair the principal pair, and to give the other pair as contranominals of the firſt pair.

Thus, we may write the caſe of inference we have been con-ſidering, as in the firſt of the following liſt, the others being ob-tained from the firſt, by changing X into x, or Z into z, or both. The ſign $+$ placed in the middle implies the coexiſtence of the four propoſitions: and independent numeral letters are introduced as ſeen, which will preſently be connected with the others by equations, inſtead of being expreſſed in terms of them.

$$\left.\begin{array}{l} m\text{XY} \\ m'\text{xy} \end{array}\right\} + \left.\begin{array}{l} n\text{YZ} \\ n'\text{yz} \end{array}\right\} = \left\{\begin{array}{l} p\text{XZ} \\ p'\text{xz} \end{array}\right.$$

The equations preſently given for this caſe apply with certain changes to the other caſes.

$$\left.\begin{array}{l} m\text{xY} \\ m'\text{Xy} \end{array} +\begin{array}{l} n\text{YZ} \\ n'\text{yz} \end{array}\right\} = \left\{\begin{array}{l} p\text{xZ} \\ p'\text{Xz} \end{array}\right.$$

Here X and x are made to change their former places: in the equations, ξ and ξ' muſt change places.

$$\left.\begin{array}{l} m\text{XY} \\ m'\text{xy} \end{array} +\begin{array}{l} n\text{Yz} \\ n'\text{yZ} \end{array}\right\} = \left\{\begin{array}{l} p\text{Xz} \\ p'\text{xZ} \end{array}\right.$$

Here Z and z change places: as muſt ζ and ζ' in the equations.

$$\left.\begin{array}{l} m\text{xY} \\ m'\text{Xy} \end{array} +\begin{array}{l} n\text{Yz} \\ n'\text{yZ} \end{array}\right\} = \left\{\begin{array}{l} p\text{xz} \\ p'\text{XZ} \end{array}\right.$$

Here X and x, and alſo Z and z, change places; as muſt ξ and ξ', and ζ and ζ', in the equations.

In the new manner of writing the form we have already conſidered, being the firſt of the four, we have juſt written

$$m' \text{ for } \upsilon+m-\xi-\eta \quad . \quad p \text{ for } m+n-\eta$$
$$n' \text{ for } \upsilon+n-\zeta-\eta \quad . \quad p' \text{ for } \upsilon+m+n-\eta-\xi-\zeta$$

Let us write ξ', η', ζ', for $\upsilon-\xi$, $\upsilon-\eta$, $\upsilon-\zeta$, the numbers of xs, ys, and zs : and then, $\xi+\xi'$, $\eta+\eta'$, $\zeta+\zeta'$, being all the ſame, (for each is υ) we may write $\eta-\xi'$ for $\xi-\eta'$, $\zeta'-\xi$ for $\xi'-\zeta$, and ſo on. That is, in the difference of two, one of which is accented, we may interchange the letters if we pleaſe. The equations of connection for the firſt or ſtandard caſe, are then

$$m'=m+\xi'-\eta=m+\eta'-\xi \quad \big| \quad m=m'+\xi-\eta'=m'+\eta-\xi'$$
$$n'=n+\zeta'-\eta=n+\eta'-\zeta \quad \big| \quad n=n'+\zeta-\eta'=n'+\eta-\zeta'$$

$$\left.\begin{array}{l} p=m+n-\eta=m'+n'+\eta-\zeta'-\xi' \\ \text{or } m'+n'+\xi-\eta'-\zeta' \\ \text{or } m'+n'+\zeta-\xi'-\eta' \end{array}\right\} \begin{array}{l} =m+n'-\zeta' \\ =n+m'-\xi' \end{array}$$

$$\left.\begin{array}{l} p'=m'+n'-\eta'=m+n+\eta'-\zeta-\xi \\ \text{or } m+n+\xi'-\eta-\zeta \\ \text{or } m+n+\zeta'-\xi-\eta \end{array}\right\} \begin{array}{l} =m'+n-\zeta \\ =n'+m-\xi \end{array}$$

For the ſecond caſe we muſt write $m'=m+\xi-\eta=m+\eta'-\xi'$, and ſo on. I now proceed to the ſeveral diviſions into which our uſual modes of thinking make it convenient to ſeparate the caſes of this moſt general form.

Firſt, when every thing is numerically definite. In this caſe, as ſeen, every form requires an examination of the premiſes and concluſion, as to whether they are or are not ſpuriouſ.

Secondly, when v, the number of inſtances in the whole univerſe of names, is wholly unknown. In this caſe ξ' is indefinite when ξ is definite, and *vice verſâ*; and ſimilarly one at leaſt of each two, η or η', ζ or ζ', is indefinite. There are then no ſpurious concluſions; or, which is the ſame thing, nóne which are known to be ſuch: for the ſpuriouſneſs of a premiſe or concluſion conſiſts in our *knowing* that it muſt be true of its two terms, independently of all compariſon of thoſe terms with a third.

Thirdly, when ξ, η, ζ, are all indefinite, as well as v. In this caſe, as here ſtated, there is no poſſibility of inference. We cannot tell whether $m+n$ be or be not greater than η, if we do not know what η is, in any manner, or to any extent.

But here we introduce that degree of definiteneſs by which we diſtinguiſh the *univerſal* from the *particular* (or *poſſible* particular, ſee page 56) propoſition. If we can know that either of the two, m and n, is the ſame as η (greater neither can be) then we know that $m+n$ is greater than η. And at the ſame time we make Y univerſal, in one or the other of the premiſes. And the ſame if we can know that either m' or n' is η'.

The following are the forms which may all be derived from the firſt, by uſing all the varieties of contrary names and contranominal equivalents. If we want, for inſtance, to ſhow the connection of the fourteenth with the firſt, we throw the firſt into the form

$$(m+\eta'-\xi)xy+n\mathrm{YZ}=(m+n+\eta'-\xi-\zeta)xz$$

We then change x into X, and Z into z, changing at the ſame time ξ into ξ' and ζ into ζ': and thus we get

$$(m+\eta'-\xi')\mathrm{X}y+n\mathrm{Y}z=(m+n+\eta'-\xi'-\zeta')\mathrm{XZ}$$

Now, for $m+\eta'-\xi'$ write m', that is, for m write $m'-\eta'+\xi'$ and we have

$$m'\mathrm{X}y+n\mathrm{Y}z=(m'+n-\zeta')\mathrm{XZ}$$

which is one of the forms of the fourteenth. And $(n+m'-\xi)xz$ is only the contranominal of $(m'+n-\zeta')\mathrm{XZ}$.

1. $mXY + nYZ = (m+n-n)XZ = (m+n+n'-\xi-\zeta)xz$
2. $m'xy + nYZ = (n+m'-\xi')XZ = (m'+n-\zeta)xz$
3. $mXY + n'yz = (m+n'-\zeta')XZ = (n'+m-\xi)xz$
4. $m'xy + n'yz = (m'+n'+n-\zeta'-\xi')XZ = (m'+n'-n')xz$
5. $mxY + nYZ = (m+n-n)xZ = (m+n+n'-\xi'-\zeta)Xz$
6. $m'Xy + nYZ = (n+m'-\xi)xZ = (m'+n-\zeta)Xz$
7. $mxY + n'yz = (m+n'-\zeta')xZ = (n'+m-\xi')Xz$
8. $m'Xy + n'yz = (m'+n'+n-\zeta'-\xi)xZ = (m'+n'-n')Xz$
9. $mXY + nYz' = (m+n-n)Xz = (m+n+n'-\xi-\zeta')xZ$
10. $m'xy + nYz = (n+m'-\xi)Xz = (m'+n-\zeta')xZ$
11. $mXY + n'yZ = (m+n'-\zeta)Xz = (n'+m-\xi)xZ$
12. $m'xy + n'yZ = (m'+n'+n-\zeta-\xi')Xz = (m'+n'-n')xZ$
13. $mxY + nYz = (m+n-n)xz = (m+n+n'-\xi'-\zeta')XZ$
14. $m'Xy + nYz = (n+m'-\xi)xz = (m^r+n-\zeta')XZ$
15. $mxY + n'yZ = (m+n'-\zeta)xz = (n'+m-\xi')XZ$
16. $m'Xy + n'yZ = (m'+n'+n-\zeta-\xi)xz = (m'+n'-n')XZ$

The syllogisms of chapter V are all particular cases of the above list, obtained as follows:—

1.	$m = n$	A'I,I,	9.	$m = n$	A'O,O,
	$m = n,\ n = \zeta$	A'A'A'		$m = n,\ n = \zeta'$	A'E'E'
	$n = n$	I,A,I,		$n = n$	I,E,O,
	$n = n,\ m = \xi$	A,A,A,		$n = n,\ m = \xi$	A,E,E,
	$m = n,\ n = n$	A'A,I,		$m = n,\ n = n$	A'E,O,
	$m = \xi,\ n = \zeta$	A,A'I'		$m = \xi\ n = \zeta'$	A,E'O'
2.	$m' = \xi'$	A'I,I,	10.	$m' = \xi'$	A'O,O,
	$n = \zeta$	I'A'I'		$n = \zeta'$	I'E'O'
	$m' = \xi',\ n = \zeta$	A'A'A'		$m' = \xi',\ n = \zeta'$	A'E'E'
3.	$m = \xi$	A,I'I'	11.	$m = \xi$	A,O'O'
	$n' = \zeta'$	I,A,I,		$n' = \zeta$	I,E,O,
	$m = \xi,\ n' = \zeta'$	A,A,A,		$m = \xi,\ n' = \zeta$	A,E,E,
4.	$m' = n'$	A,I'I'	12.	$m' = n'$	A,O'O'
	$m' = n',\ n' = \zeta'$	A,A,A,		$m' = n',\ n' = \zeta$	A,E,E,
	$n' = n'$	I'A'I'		$n' = n'$	I'E'O'
	$n^L = n',\ m' = \xi'$	A'A'A'		$n' = n',\ m' = \xi'$	A'E'E'
	$m' = n',\ n' = n'$	A,A'I'		$m' = n',\ n' = n'$	A,E'O'
	$m' = \xi'\ n' = \zeta'$	A'A,I,		$m' = \xi'\ n' = \zeta$	A'E,O,

5.	$m = \eta$	$E_\iota I_\iota O'$	13.	$m = \eta$	$E_\iota O_\iota I'$
	$m = \eta,\ n = \zeta$	$E_\iota A'E_\iota$		$m = \eta,\ n = \zeta'$	$E_\iota E'A_\iota$
	$n = \eta$	$O'A_\iota O'$		$n = \eta$	$O'E_\iota I'$
	$n = \eta,\ m = \xi'$	$E'A_\iota E'$		$n = \eta \quad m = \xi'$	$E'E_\iota A'$
	$m = \eta,\ n = \eta$	$E_\iota A_\iota O'$		$m = \eta \quad n = \eta$	$E_\iota E_\iota I'$
	$m = \xi',\ n = \zeta$	$E'A'O_\iota$		$m = \xi' \quad n = \zeta'$	$E'E'I_\iota$
6.	$m' = \xi$	$E_\iota I_\iota O'$	14.	$m' = \xi$	$E_\iota O_\iota I'$
	$n = \zeta$	$O_\iota A'O_\iota$		$n = \zeta'$	$O_\iota E'I_\iota$
	$m' = \xi,\ n = \zeta$	$E_\iota A'E_\iota$		$m' = \xi,\ n = \zeta'$	$E_\iota E'A_\iota$
7.	$m = \xi'$	$E'I'O_\iota$	15.	$m = \xi'$	$E'O'I_\iota$
	$n' = \zeta'$	$O'A_\iota O'$		$n' = \zeta$	$O'E_\iota I'$
	$m = \xi',\ n' = \zeta'$	$E'A_\iota E'$		$m = \xi',\ n = \zeta$	$E'E_\iota A'$
8.	$m' = n'$	$E'I'O_\iota$	16.	$m' = n'$	$E'O'I_\iota$
	$m' = n',\ n' = \zeta'$	$E'A_\iota E'$		$m' = n',\ n' = \zeta$	$E'E_\iota A'$
	$n' = n'$	$O_\iota A'O_\iota$		$n' = n'$	$O_\iota E'I_\iota$
	$n' = n',\ m' = \xi$	$E_\iota A'E_\iota$		$n' = n' \quad m' = \xi$	$E_\iota E'A_\iota$
	$m' = n',\ n' = n'$	$E'A'O_\iota$		$m' = n',\ n' = n'$	$E'E'I_\iota$
	$m' = \xi \quad n' = \zeta'$	$E_\iota A_\iota O'$		$m' = \xi \quad n' = \zeta$	$E_\iota E_\iota I'$

We have thus another mode of eſtabliſhing the completeneſs of the ſyſtem of ſyllogiſm, laid down in the laſt chapter : that is, of the ſyſtem in which there is only the common univerſal and particular quantity. Theſe ſyllogiſms of numerical quantity, in which conditions of inference belonging to every imaginable caſe are repreſented by the general forms which numerical ſymbols take in algebra, muſt of neceſſity be the moſt general of their kind. And examination makes it clear that, except the preceding, there can be no ſyllogiſm exiſting between X, Y, Z, and their contraries. Many ſubordinate laws of connexion might be noticed between the general forms and their particular caſes. Thus, each univerſal occurs three times, each fundamental particular twice, and each ſtrengthened particular twice. The firſt form in pages 158, 159, gives only affirmative, the fourth only negative, premiſes: the ſecond and third one of each kind, commencing with a negative in the ſecond, and with an affirmative in the third.

There are two remarkable ſpecies of ſyllogiſm (or rather, which ought to have been remarkable): which I ſhall now proceed to notice.

The diſtinction of larger and ſmaller part, when diviſion into two parts is made, is as much received into the common idiom of language as the diftinction of whole and part itfelf. 'Moft of the Xs are Ys,' is nearly as common as 'All the Xs are Ys:' though 'feweft of the Xs are Ys,' is only feen as 'moft of the Xs are not Ys.' The fyllogifms which can be made legitimate by the ufe of this language will do equally well for any fraction, provided we couple with it the fraction complemental to unity (which in the cafe of one half is one half itfelf). Let α and β ftand for two fractions which have unity for their fum, as $\frac{3}{7}$ and $\frac{4}{7}$. Let $_{\alpha}XY$ and $_{\alpha}X:Y$ indicate that lefs than the fraction α of the Xs are or are not Ys. Let $^{\alpha}XY$ and $^{\alpha}X:Y$ indicate that more than the fraction α of the Xs are or are not Ys.

Then the following fyllogifms arife from the cafes with the numbers prefixed.

1. $^{\alpha}YX +^{\beta}YX =XZ$	9.	$^{\alpha}YX +^{\beta}Y:Z=X:Z$
4. $^{\alpha}y:X +^{\beta}y:Z =xz$	12.	$^{\alpha}y:X +^{\beta}yZ =Z:X$
5. $^{\alpha}Y:X+^{\beta}YZ =Z:X$	13.	$^{\alpha}Y:X+^{\beta}Y:Z=xz$
8. $^{\alpha}yX +^{\beta}y:Z =X:Z$	16.	$^{\alpha}yX +^{\beta}yZ =XZ$

It will be feen that here are but three really diftinct forms; of which the fimpleft examples are as follows,

Moft Ys are Xs; Moft Ys are Zs; therefore fome Xs are Zs.
Moft Ys are Xs; Moft Ys are not Zs; therefore fome Xs are not Zs.
Moft Ys are not Xs; Moft Ys are not Zs; therefore fome things are neither Zs nor Xs.

It is hardly neceffary to obferve that in one of the premifes 'more than' may be reduced to 'as much as:' but not in both. Thus, if two-fevenths exactly of the Ys be Xs, and more than five-fevenths of the Ys be Zs, it follows that fome Xs are Zs.

The above fyllogifms admit a change of premife, as follows: If we fay that more than $\frac{3}{7}$ths of the Ys are Xs, we thereby fay that lefs than $\frac{4}{7}$ths of the Ys are xs: or $^{\alpha}YX$ and $_{\beta}Y:X$ are the fame propofitions. Thus, 'moft are' is equivalent to 'a minority (none included) are not.' Hence we have

$$_{\beta}Y:X + _{\alpha}Y:Z=XZ$$

and fo on. Or we may combine the two forms, as in

$$^{\alpha}YX + _{\alpha}Y:Z=XZ$$

The above are the only fyllogifms in which indefinite particulars give conclufions, by reafon of that approach to definitenefs which confifts in defcribing what fractions of *the middle term* are fpoken of, at leaft, or at moft. But they are not the only fyllogifms of the fame general fpecies. In every cafe inference follows when there is a certain preponderance; and the largenefs of the inference depends upon the extent of that preponderance. Thus in (12) there is an Xz inference when $m' + n' + n$ exceeds $\zeta + \xi'$: fo many units as there are in this excefs, fo many Xs (at leaft) are zs. Now in every cafe, a pair of univerfal premifes give inference: and in every cafe there muft be a degree of approach to univerfality at which inference begins. The ordinary fyllogifms, I fufpect, are, and are meant to be, not fuch as 'Every X is Y, *every* Y is Z, therefore *every* X is Z,' but 'generally fpeaking X is Y, and generally fpeaking Y is Z, therefore generally fpeaking X is Z.' And by 'generally fpeaking' is meant the affertion that an enormous majority of inftances make the affertion true. A fyllogifm of this fort is the oppofite of the *à fortiori* fyllogifm; and might be faid to be true *ab infirmiori*. If we have X)Y with p exceptions, and Y)Z with q exceptions; then, in form (1.) we have $m = \xi - p, n = n - q$, and $m + n - n = \xi - p - q$. As long, then, as the number of exceptions altogether fall fhort of the number of Xs, there is inference: if the total number of exceptions be very fmall, compared with the number of Xs, there is the 'generally fpeaking' kind of inference. Examine all the univerfal cafes, and it will be found that the fame law prevails; namely, that there is inference when the numbers of exceptional inftances in both premifes together do not amount to the number of inftances in the univerfal term of the conclufion; and that there is *exceptional* univerfality (as we may call it) in the conclufion, whenever the whole amount of exception is very fmall, compared with that number of inftances.

This leads us to what I will call the theory of *exceptional* particular fyllogifms. We have feen that the eight complex affirmatory fyllogifms, which are all *à fortiori* in their conclufions, afford each two particular fyllogifms. We have denoted coexiftence by $+$; and the coexiftence of two propofitions gives more than either. Let us denote exceptive coexiftence by $-$: thus, P$-$Q means that the propofition P is true except in the

inftances contained in Q. Thus, X)Y—X:Y means that every
X (with fome exceptions) is Y. This is, of courfe, A_1—O_1, and
only differs from I_1 in the mode of expreffion not being 'fome
more than none at all' but 'fome lefs than all.' In the expreffion

$$(A_1—O_1)(A_1—O_1)(A_1—O_1)$$

we have the fymbol of the *ab infirmiori* fyllogifm ftated above,
fubject to the poffibility of nonexiftence if the number of exceptions in the two premifes fhould exceed the number of inftances
in the univerfal term of the conclufion. If we look at A_1O_1, as
a fymbol defcriptive of premifes, we fee one of the inconclufive
forms; that is, a form from which we cannot draw an inference.
But this is only becaufe our inferences are all pofitive, and imply
affertion of *fufficiency* in the premifes. There is no ufe (except
to fhow the manner in which the parts of a fyftem hang together)
in declarations of *infufficiency :* for we know that all collections
of premifes, whatever they may be fufficient for, will be infufficient for an infinite number of different things. And it is
important to remember that while fufficiency is accompanied by
muft be, infufficiency only allows *may be*. From A_1A_1 the conclufion A_1 *muft* be true : from A_1O_1 (and as far as thefe are concerned) it may be falfe. Accordingly $A_1O_1O_1$ and $O_1A_1O_1$ may
ferve to exprefs the two defects of $(A_1—O_1)(A_1—O_1)(A_1—O_1)$
from $A_1A_1A_1$, exifting in the *ab infirmiori* fyllogifm, and poffibly
preventing conclufion altogether : juft as $A_1O'O'$ and $O'A_1O'$
fhow the additional conditions by the fulfilment of which $A_1A_1A_1$
is elevated into the *à fortiori* fyllogifm $D_1D_1D_1$. It is worth
while to dwell upon the varieties of this cafe. The *ab infirmiori*
fyllogifms of the ftrengthened particulars were previoufly confidered.

In all the cafes yet treated, we have had, *more or lefs*, the
power of giving inftances in common language, without recourfe
to numerical relation expreffed in unufual terms. This of courfe,
is always the cafe in the fyllogifms óf chapter V.; and we
have given one *common inftance* (though never met with in books
of logic) from each fet of *ab infirmiori* fyllogifms. But there are
ftill cafes of the fame fort to be confidered. Though in our definite relation (page 56) of *all*, we ufually (in books of logic at

leaft) make the relation exift, for each propofition, between the terms of the propofition itfelf, yet it may be afked whether we cannot fometimes infer fuch a fpecies of univerfal as this, 'for every Z there is an X which is Y;' Z being one of the names of the fecond premife. If we examine the firft two cafes, which will be guide enough, we fhall find the following refults from the new fuppofitions now made.

1. $m=\zeta$, $n=\xi$, gives $\zeta XY + \xi YZ = \eta'xz$: or if for every Z there be an X which is Y, and for every X a Z which is Y, then, fo many ys as there are, fo many things which are neither X nor Z. This fyllogifm has little new meaning, and no new application : it requires $\xi=\zeta$, and therefore X)Y and Z)Y.

2. $m'=\zeta$, gives $\zeta xy + n \, YZ = nxz$, or if for every Z there be that which is neither X nor Y, and if fome Ys be Zs, there are as many inftances which are neither X nor Z. This is a new and effective form.

2. $n=\xi'$, gives $m'xy + \xi'YZ = m'XZ$, a new form.

Thefe two cafes will be prefently further confidered. Now, obferve that if $m+n$ in the firft form, or $m'+n$ in the fecond, be v, that is, if the pair m and n be ξ and ξ', or η and η', or ζ and ζ', we have inference of the kind required. The firft form gives no new fyllogifm: fince v is more than η, Ys which are Xs, and Ys which are Zs, to the number of v, give the form (1.) by the main law of inference (page 154). In the fecond form, if $m'+n=v$, we diftribute among the Ys and ys, Zs and xs to the full number of both, fo that wherever there are not xs (that is, wherever there are Xs) there are Zs : or X)Z as obtained from the form.

But every way of conftructing $m'xy + nYZ = (m'+n-\xi')XZ$ which gives $m'+n=v$, is only a cafe of $A_1A_1A_1$. For m' cannot exceed η', and n cannot exceed η : and $m'+n$ being v or $\eta+\eta'$, we muft have $m'=\eta'$ and $u=\eta$; whence the affertion made. The forms we are now in fearch of, fo far as quite new, are all contained in the two new ones above noted ; and of thefe, the fecond is but a transformation of the firft. The eight varieties derived from ufe of contraries, or from the forms in page 161, beginning with the fimpleft, are

$$\zeta XY + nZ:\eta Y = nX:\zeta Z \qquad \zeta'XY + nyz = nXZ$$
$$\zeta X:\eta Y + nZY = nX:\zeta Z \qquad \zeta'X:\eta Y + nY:\zeta Z = nXZ$$

$$\zeta Y : \xi X + nZ : nY = nxz \quad | \quad \zeta' Y : \xi X + nyz \quad = nZ : \xi X$$
$$\zeta xy \quad + nZY \quad = nxz \quad | \quad \zeta' xy \quad + nY : \zeta Z = nZ : \xi X$$

These are syllogisms, which exhibit a curious kind of antagonism to the particular syllogisms. Take the syllogism $A_1O'O'$, the terms being M, Y, Z; we have then M)Y + Z : Y = Z : M. Of course the conclusion M : Z is not legitimate from these premises alone: but if M have as many instances as Z, then M : Z *is* legitimate. For if Ms, as many as there are Zs, be among the Ys, and some of the Zs be not among the Ys, though all the rest were, there would not be enough to match all the Ms, or some Ms are not Zs. Now, let M be a name given to an X which is Y, and let such Xs have as many instances as Z, and the above becomes the first of the syllogisms in the last list. Thus, $I_1O'O_1$ is legitimate, if the quantity of the subject mentioned in I_1 be taken from the Zs. The second syllogism is E_1I_1O', altered into $O_1I_1O_1$ in the same manner.

The reader may find all the results of the above case in the following rule, in which it is understood that all the super-propositions are to be written either way: thus, A' is written x)y, or Y)X, and O' is $nx : ny$, or $nY : \xi X$ (page 62). Write down any pair of particulars, followed by I if the pair be of the same sign, and O if the pair be of different signs : as in OOI or IOO. Accent the pair in contradiction to either the direct rule (page 62) as far as the words affirmative and negative are concerned : that is, let a *negative* beginning isolate nothing, and an *affirmative* beginning isolate the middle proposition : or else, accent the pair according to the inverse rule. Thus, $O_1O_1I_1$ and $O'O'I'$ contradict the direct rule, and $O'O'I_1$ and O_1O_1I' preserve the inverse rule. To make these syllogisms good (in the particular way in question) proceed thus :—When the *direct* rule is contradicted, take the quantity of the *first* concluding term from the total of the *second*, if the second premise be affirmative, and from its contrary, if negative. When the inverse rule is preserved, take the quantity of the *second* from the total of the *first*. Thus, in $O'O'I'$ the direct rule is contradicted : and it stands $m'x : n'y + n'y : \zeta'z = p'xz$. The second premise is negative, the total of its predicate ζ' instances, that of the contrary ζ. Accordingly, $\zeta x : n'y + n'y : \zeta'z = n'xz$, or $\zeta Y : \xi X +$

$n'Z : _{n}Y = n'xz$, which is one of the forms already obtained. Again, O'O'I, preferves the inverfe rule, and is $m'x : _{n}'y + nZ : _{n}Y = pXZ$. The total of the firft term is ξ' inftances. Hence, $m'x : _{n}'y + \xi'Z : _{n}Y = m'XZ$, or $m'Y : \xi X + \xi'Z : _{n}Y = m'XZ$, which is derived from one of the forms given, by interchanging X and Z.

This clafs of fyllogifms *with tranfpofed quantity* naturally leads to the queftion, Is it ufed? Do fuch fyllogifms occur in ordinary or in literary life? If not, there is no reafon for felecting them from the infinite number of cafes which the numerically definite fyftem affords. To try this, fuppofe a perfon, on reviewing his purchafes for the day, finds, by his countercheques, that he has certainly drawn as many cheques on his banker (and may be more) as he has made purchafes. But he knows that he paid fome of his purchafes in money, or otherwife than by cheques. He infers then that he has drawn cheques for fomething elfe except that day's purchafes. He infers rightly enough; but his inference cannot be reduced to a common fyllogifm, with the names in queftion for terms. It is really a fyllogifm of tranfpofed quantity, as follows :—

For every 'memorandum of a purchafe' a 'countercheque' is a 'tranfaction involving the drawing of a cheque.'

Some 'purchafes' are not 'tranfactions involving, &c.'

Therefore fome 'countercheques' are not 'memoranda of purchafes.'

It may be worth while to give one inftance of the verification of the contradictory form. By page 152 it appears that the contradiction of mXY is $(\xi - m + 1)Xy$, or $(_{n} - m + 1)xY$, and that of $m'Xy$ is $(\xi - m' + 1)XY$, or $(_{n}' - m' + 1)xy$.

To mXY join the contrary of $(m + n - _{n})XZ$, or $(\xi + _{n} - m - n + 1)Xz$: we have then

$$mYX + (\xi + _{n} - m - n + 1)zX ;$$

the inference of which is $(m + \xi + _{n} - m - n + 1 - \xi)Yz$, that is, $(_{n} - n + 1)Yz$, the contrary of nYZ.

Returning to the forms in page 161, it will be obferved that we have no double inferences. In every cafe we have made ufe of one form of inference: if v be known, the other is a real equivalent; or elfe it is impoffible, and as we have feen, then the firſt is ſpurious. If v be not known, then the ſecond is either perfectly indefinite, or elſe identical with the one choſen. Examination will ſhow that in every one of the caſes cited in page 161, the neglected form of inference is only ſaved from perfect indefiniteneſs when we are able to apply the word *all* to one or other of the terms: the number being as indefinite as before; the *relation* thus obtained being definite. Take the firſt form, and make $n = \eta$; by the firſt inference we then get the ſyllogiſm $I_1A_1I_1$: by the ſecond, we get $(m + v - \xi - \zeta)xz$, indefinite both in number and relation. We do not know what v, ξ, and ζ are. If we knew as much as that $m + v$ is leſs than $\xi + \zeta$, we ſhould know our inference to be ſpurious,* it being not the leſs an inference. Now, add the condition $m = \xi$: the firſt inference gives the ſyllogiſm $A_1A_1A_1$, the ſecond inference now becomes $(v - \zeta)xz$: definite relation enters, and we have z)x, or X)Z, or A_1, as before. And the ſame of the other forms.

The reader may perhaps ſuppoſe that I ought to have commenced this chapter with the complex numerical ſyllogiſm, in imitation of the method which I followed in treating the ordinary ſyllogiſm. But in truth there is no ſyſtem of complex ſyllogiſm of perfect numerical definiteneſs both in premiſes and concluſion. To ſhow this, let m,XY with the comma, mean that there are exactly m Xs which are Ys, neither more nor fewer. Accordingly m,XY is a ſynonyme for $mXY + (\eta - m)xY$. Now combine m,XY and n,ZY, or

$$(mXY + \overline{\eta - m}\, xY)(nZY + \overline{\eta - n}\, zY)$$

We then have
$$mXY + nZY = (m + n - \eta)XZ$$
$$(\eta - m)xY + (\eta - n)zY = (\eta - m - n)xz$$
$$mXY + (\eta - n)zY = (m - n)Xz$$
$$(\eta - m)xY + nZY = (n - m)xZ$$

---

* I muſt again remind the reader, of the diſtinction between *ſpurious* and *illegitimate*, which exiſts in my language. The ſpurious inference follows from the premiſes, and is perfectly good and true: but from the conſtitution of the univerſe, it will always be true, whatever premiſes in that univerſe are taken. The illegitimate inference is that which does not follow from the premiſes. A concluſion not *known* to be ſpurious, that is, there not being the means of knowledge, *is not* ſpurious: but an illegitimate concluſion cannot be made legitimate, that is, following from the premiſes, by any further knowledge.

Two only of thefe have meaning: let them be the two upper ones. We can affign then Z or z to $(m+n-\eta)+(m-n)$, or to $2m-\eta$ of the Xs. But there are not all of the Xs here: for m is less than η, and than ξ, whence $2m$ is less than $\eta+\xi$, or $2m-\eta$ less than ξ. The reft of the Xs, $\xi+\eta-2m$ in number, may, for aught thefe premifes declare, be either Zs or zs.

CHAPTER IX.

On Probability.

THE moft difficult inquiry which any one can propofe to himfelf is to find out what any thing *is* : in all probability we do not know what we are talking about when we afk fuch a queftion. The philofophers of the middle ages were much concerned with the *is*, or *effence*, of things : they argued to their own minds, with great juftice, that if they could only find out what a thing is, they fhould find out all about it : they tried, and failed. Their fucceffors, taking warning by their example, have inverted the propofition; and have fatisfied themfelves that the only way of finding what a thing is, lies in finding what we can about it ; that modes of relation and connexion are all we can know of the effence of any thing; in fhort, that the proverb 'tell me who you are with, and I will tell you what you are,' applies as much to the nature of things as to the characters of men. We are apt to think that we know more of the effence of objects than of ideas ; or rather, of ideas which have an objective fource, than of thofe which are the confequence of the mind's action upon them. I doubt whether the reverfe be not the cafe : at any rate, when we content ourfelves with inquiry into properties and relations, we have certain knowledge upon our moft abftract ideas. The object of this chapter is the confideration of the degrees of knowledge itfelf. That which we know, of which we are certain, of which we are well affured nothing could perfuade us to the contrary, is the exiftence of our own minds, thoughts, and perceptions, the two laft when actually prefent. This higheft knowledge, this abfolute certainty, admits of no imagination of the poffibility of falfehood. We cannot, by ftopping to confider, make ourfelves more fure than we are already, that we exift, think, fee, &c. Next to this, come the things of which we cannot but fay *at laft* we are as certain of them as of our own exiftence ; but of which, neverthelefs, we are obliged to fay that we arrive at them by procefs, by reflection. Thefe we call *neceffary truths* (page 33). The *neceffity* of admitting thefe things caufes fome to imagine that they are merely identities, that they amount to faying that when a thing is, it is : but this is not correct. To fay that two and two make four (which muft be), and that a certain man wears a black coat (when he does fo) both involve the pure identity that whatever is, is ; and not one more than the other. Nor is two and two identically four, though neceffarily fo. Our definitions of number arife in the procefs of fimple counting. Throw a pebble into a bafket, and we fay *one :* throw in another, and we fay *two;* yet one more, and we fay *three*, and fo on. · The full definitions of the fucceffive numbers are feen in

$$1 \quad (1+1) \quad \{(1+1)+1\} \quad [\{(1+1)+1\}+1], \&c.$$

That three and one are four is *definition :* it is our pleafure to give the name *four* to $3+1$. But that $3+1$ is $2+2$ is neither definition nor pure identity. It is not even true that ' two and two' *is* four; that

$$[\{(1+1)+1\}+1] \ is \ (1+1)+(1+1)$$

It is true, no doubt, that ' two and two' is four, in amount, value, &c. but not in form, conftruction, definition, &c.

There is no further ufe in drawing diftinction between the knowledge which we have of our own exiftence, and that of two and two amounting to four. This abfolute and inaffailable feeling we fhall call *certainty*. We have lower grades of knowledge, which we ufually call *degrees of belief*, but they are really *degrees of knowledge*. A man knows at this moment that two and two make four : did he know it yefterday ? He feels perfectly certain that he knew it yefterday. But he may have been feized with a fit yefterday, which kept him in unconfcioufnefs all day : and thofe about him may have been warned by the medical man not to give him the leaft hint of what has taken place. He could fwear, as oaths are ufually underftood, that it was not fo : if he

could not fwear to this, no man could fwear to anything except
neceffary truths. But he could not regard the affertion that it
was not fo, as incapable of contradiction : he knows it well, but,
as long as it may poffibly be contradicted, he cannot but fay that
he might know it better.

It may feem a ftrange thing to treat *knowledge* as a magnitude,
in the fame manner as length, or weight, or furface. This is
what all writers do who treat of probability, and what all their
readers have done, long before they ever faw a book on the fubject.
But it is not cuftomary to make the ftatement fo openly as I
now do : and I confider that fome juftification of it is neceffary.

By degree of probability we really mean, or ought to mean,
degree of belief. It is true that we may, if we like, divide probability into ideal and objective, and that we muft do fo, in order
to reprefent common language. It is perfectly correct to fay ' It
is much more likely than not, *whether you know it or not*, that
rain will foon follow the fall of the barometer.' We mean that
rain does foon follow much more often than not, and that there
do exift the means of arriving at this knowledge. The thing is fo,
every one will fay, and can be known. It is not remembered,
perhaps, that there is an *ideal probability*, a pure ftate of the mind,
involved in this affertion : namely, that the things which have been
are correct reprefentatives of the things which are to be. That
up to this 21ft of June, 1847, the above ftatement has been true,
ever fince the barometer was ufed as a weather-glafs, is not denied by any who have examined it : that the connexion of
natural phenomena will, for fome time to come, be what it has
been, cannot be fettled by examination : we all have ftrong reafon to believe it, but our knowledge is *ideal*, as diftinguifhed
from *objective*. And it will be found that, frame what circumftances we may, we cannot invent a cafe of purely objective probability. I put ten white balls and ten black ones into an urn,
and lock the door of the room. I may feel well affured that,
when I unlock the room again, and draw a ball, I am juftified
in faying it is an even chance that it will be a white one. If all
the metaphyficians who ever wrote on probability were to witnefs
the trial, they would, each in his own fenfe and manner, hold me
right in my affertion. But how many things there are to be
taken for granted ! Do my eyes ftill diftinguifh colours as be-

fore? Some perfons never do, and eyes alter with age. Has the black paint melted, and blackened the white balls? Has any one elfe poffeffed a key of the room, or got in at the window, and changed the balls? We may be *very fure*, as thofe words are commonly ufed, that none of thefe things have happened, and it may turn out (and I have no doubt will do fo, if the reader try the circumftances) that the ten white and ten black balls will be found, as diftinguifhable as ever, and unchanged. But for all that, there is much to be affumed in reckoning upon fuch a refult, which is not fo objective (in the fenfe in which I have ufed the word) as the knowledge of what the balls were when they were put into the urn. We have to affume all that is re-quifite to make our experience of the paft the means of judging the future.

Having made this illuftration to draw a diftinction, I now pre-mife that I throw away objective *probability* altogether, and con-fider the word as meaning the ftate of the mind with refpect to an affertion, a coming event, or any other matter on which ab-folute knowledge does not exift. ' It is more probable than im-probable' means in this chapter ' I believe that it will happen more than I believe that it will not happen.' Or rather ' I *ought* to believe, &c. :' for it may happen that the ftate of mind which *is*, is not the ftate of mind which fhould be. D'Alembert be-lieved that it was *two* to *one* that the firft head which the throw of a halfpenny was to give would occur before the third throw : a jufter view of the mode of applying the theory would have taught him it was *three* to *one*. But he *believed* it, and thought he could fhow reafon for his belief: to him the probability *was* two to one. But I fhall fay, for all that, that the probability *is* three to one : meaning, that in the univerfal opinion of thofe who examine the fubject, the ftate of mind to which a perfon *ought* to be able to bring himfelf is to look three times as confidently upon the arrival as upon the non-arrival.

Probability then, refers to and implies belief, more or lefs, and belief is but another name for imperfect knowledge, or it may be, expreffes the mind in a ftate of imperfect knowledge. There is accurate meaning in the phrafe ' to the beft of his *knowledge* and *belief*,' the firft word applying to the ftate of his circumftances with refpect to external objects, the fecond to the ftate of his mind with refpect to the circumftances. But we cannot make any ufe of the diftinction here: what we know is to regulate what we believe; nor can we make any effective ufe of what we know, except in obtaining and defcribing what we believe, or ought to believe. According to common idiom, belief is often a lower degree of knowledge: but it is imperative upon us to drop all the *quantitative* diftinctions of common life, or rather to remodel them, when we come to the conftruction of a fcience of quantity.

I have faid that we treat knowledge and belief as magnitudes: I will now put a broad illuftration of what I mean. We know, (fuppofe it *known*) that an urn contains nothing but two balls, one white and one black, undiftinguifhable by feeling: and we know (fuppofe this alfo) that a ball is to be drawn. Disjunctively then we know 'white will be drawn: black will be drawn,' one or the other muft be. How do we ftand as to 'white will be drawn,' and 'black will be drawn,' feparately? Clearly in no preponderance with refpect to either. May we then properly and reafonably fay that we divide our knowledge and belief of the event 'one or the other' into two halves, and give one half to each. I can conceive much objection to this fuppofition: but, whether they formally make it or not, I am fure writers on probability act upon it, and are accepted by their readers.

Let us confider what magnitude is, that is to fay, how we know we are talking about a magnitude. We know that whenever we can attach a diftinct conception of more and lefs to different inftances, fo as to fay this has more than that, we are talking of comparable magnitudes. We fpeak of a quantity of talent, or of prudence: we fay one man has more talent than another, and one man more prudence than another: but we never fay that one man has more talent than another has prudence. If we occafionally fay he (the fame one man) has more talent than prudence, it is only as an abbreviation: we mean that he has not prudence enough to guide his talent. Juft as we might fay (though we do not) that there is more cart than horfe, when the horfe cannot draw the cart: juft as, fpeaking very loofely, we *do* fay, the *preffure* of the atmofphere is not fifty *inches*; meaning that it is not enough to balance the preffure of fifty inches of mercury in the barometer. And thus, both up to, and beyond our means of meafurement, we form to ourfelves diftinct notions of comparable magnitudes, and incomparable magnitudes, as well as of the meaning of the fomewhat incorrect, but eafily amended, figures of fpeech by which we fometimes talk of comparing the latter.

But the object of all quantitative fcience is not merely magnitude, but the *meafurement* of magnitude. And when are we entitled to fay that we can meafure magnitude? As foon as we know how, from the greater, to take off a part equal to the lefs: a procefs which neceffarily involves the teft of which is the greater, and which is the lefs, and, in certain cafes, as it may happen, of neither being the greater nor the lefs. As to fome magnitudes, the clear idea of meafurement comes foon: in the cafe of length, for example. But let us take a more difficult one, and trace the fteps by which we acquire and fix the idea: fay *weight*. What weight is, we need not know: the Newtonian, who makes it depend on the earth's attraction, and the Ariftotelian, who referred it to an impulfe which all bodies poffefs to feek their *natural places*, are quite at one on their notions of the meafurable magnitude which their feveral philofophies difcufs. We know it as a magnitude before we give it a name: any child can difcover the *more* that there is in a bullet, and the *lefs* that there is in a cork of twice its fize. Had it not been for the fimple contrivance of the balance, which we are well affured (how, it matters not here) enables us to poife equal weights againft one another, that is, to detect equality and inequality, and thence to afcertain how many times the greater contains the lefs, we might not to this day have had much clearer ideas on the fubject of weight, as a magnitude, than we have on thofe of talent, prudence, or felf-denial, looked at in the fame light. All who are ever fo little of geometers will remember the time when their notions of an angle, as a magnitude, were as vague as, perhaps more fo than, thofe of a moral quality: and they will alfo remember the fteps by which this vaguenefs became clearnefs and precifion.

Now a very little confideration will fhow us that, the moment we begin to talk of our belief (the mind's meafure of our knowledge) of propofitions fet before us, we recognize the relations called more and lefs. Does the child feel that the bullet has

more fomething than the cork one bit better than an educated
man feels that his belief in the ftory of the death of Cæfar is
more than his belief in that of the death of Remus. Let any
one try whether he have not in his mind the means of arranging
the following fet in order of magnitude of belief, including within
that term all the range which comes between certain knowledge
of the falfehood, and certain knowledge of the truth, of an affertion. Let them be 1. Cæfar invaded Britain with the fole view
of benefiting the natives. 2. Two and two make five. 3. Two
and two make four. 4. Cæfar invaded Britain. 5. Romulus
founded Rome. He will probably difcover the gradations of
neceffary truth, moral certainty, reafonable prefumption, utter
incredibility, and neceffary falfehood. Thefe are but names given
to different ftates of the mind with refpect to knowledge of propofitions afferted; and I fay they exprefs different ftates of
quantity.

The only difficulty, and a ferious one it can be made, may be
ftated in the following queftion;—Are we to confider the fort
of belief which we have of a neceffary propofition (as two and
two make four), that is, abfolute knowledge, to which contradiction is glaring abfurdity—as only a ftrengthened or augmented
fpecimen of the fort of knowledge which we have of any contingent propofition (fuch as Cæfar invaded Britain) which may
have been, or might have been, falfe, and can be contradicted
without abfurdity? I anfwer, we can eafily fhow that the difference of the two cafes is connected with the difference between finite and infinite, not between two magnitudes of different kinds. The mathematician will eafily apprehend this,
and will look upon the various difficulties which furround even
the explanation as upon things to which he is well accuftomed,
and which he underftands by many parallel inftances. We can
invent circumftances under which a contingent propofition fhall
make any degree of approach to neceffity which we pleafe, but
fo that no actual attainment fhall be arrived at. If an urn contain balls, and if one ball muft be drawn, then, the balls being
all white, it is neceffary that a white ball muft be drawn, as
neceffary as that two and two being in any place, there are four
in that place : for there are no degrees of neceffity. But let it
be that there are black balls alfo, at the rate of one to a thoufand

white ones: the drawing of a white ball is no longer neceſſary ; but there is ſtill a ſtrong degree of aſſurance that a white ball will be drawn. We do not readily ſee how much : becauſe the urn has no viſible relation to our uſual caſes of judgment. But let it be made to repreſent the life of a youth of twenty : and let the drawing of a white ball repreſent his living to come of age, and of a black one his death in the interval. There ought to be *ſeven* black balls to the thouſand white ones to make the caſes parallel. And yet we know that our aſſurance of his ſurvival is generally very ſtrong : be it wiſe aſſurance or not, it exiſts, and we act upon it. Now ſuppoſe the rate to be one black to a million of white : the aſſurance is much increaſed, but ſtill there is no neceſſity ; the black ball may be drawn. Take one black to a million of million of white, or a million of million of million, &c.: long before we have arrived at ſuch a point, we have loſt all conception of the quantitative difference between our belief in drawing a white ball, and our belief that two and two are four. We ſay it is *almoſt impoſſible* that one trial ſhould give a black ball : and this very phraſe is a recognition of the ſameneſs for which I am contending. Except on the ſuppoſition of ſuch ſameneſs, there is no *almoſt impoſſible*, nor *nearly certain*. Between the impoſſible and the poſſible, the certain and the not certain, there muſt be every imaginable difference, if we do not admit unlimited approach. For it will clearly not be contended that, repreſenting certainty, ſay by 100, we can make an approach to it by an uncertainty counting as, ſay 90, but nothing higher. Repreſenting the ſtate of abſolute knowledge by 100, any one, with a little conſideration, will ſay that the laws of thought fix no numerical limit to our approach towards this ſtate : but that things ſhort of certainty are capable of being brought within any degree of nearneſs to certainty. On ſuch conſiderations, I ſhall aſſume that neceſſity on the one hand, a certainty for, and impoſſibility on the other, a certainty againſt, are extreme limits, which being repreſented by quantities, may allow our knowledge of all contingent propoſitions to be repreſented by intermediate quantities.

It muſt be fully allowed, nay, imperatively inſiſted on, that nothing in the numerical view, tending to connect neceſſary and contingent propoſitions, can at all leſſen the diſtinction between

N

them : nor give the latter any resemblance to the former, except only in the quantities by which they are indicated. Though there be only one black ball to as many white ones as would fill the visible universe, yet between that case and the one of no black balls must always exist the essential difference, that in the former a black ball *may* be drawn, and in the latter it *cannot.* But this very great distinction between the necessarily certain and the contingent, is it compatible with their being represented by numerical quantities as near to one another as we please? I answer that all who are acquainted with the relations of quantity are aware that nearness of value is no bar to any amount of difference of properties. A common fraction, for instance, may be made as near as we please in value to an integer : but there do not exist, even among propositions, more essential, or more striking, differences, than those which exist between the properties of integers and of fractions. There are crowds of theorems (I should rather say unlimited crowds of classes of theorems) which are always true when integers are used, and never true when fractions are used. Let any quantities be named, integer or fractional, and it is easy to make classes of theorems which are true for those quantities, and not for any others, however near to them. The reader who is not a mathematician must rely upon the knowledge of the one who is, that the difference between two quantities, no matter how nearly equal, may be connected with other differences as complete, and by practice as easily recognized, as the difference between necessary and contingent truth.

I will take it then that all the grades of knowledge, from knowledge of impossibility to knowledge of necessity, are capable of being quantitatively conceived. The next question is, are these quantities capable, *in any case*, of measurement, or of comparison with one another. At present, we stand as the child stands with respect to the bullet and the cork : perceptive of more and less, but without a balance by which to make comparisons. To show the postulate on which our balance depends, let us suppose an urn, which, to our knowledge, contains white, black, red, green, and blue balls, one of each colour. It is within our knowledge that a ball must be drawn : accordingly we have full knowledge (and of course *entire belief*) that the result 'no ball' is impossible, and that ' white, or black, or red, or green, or blue' is neceſſary. To the reſult 'white' we accord a certain probability, that is, a certain amount of belief. If a man tell us that white will be drawn, we may hold him raſh, but we do not pronounce his communication incredible : let another tell us that 'black, or red, or green, or blue' will be drawn, and we hold him not ſo raſh, and his communication more credible. We may hold with either, if he will deſcribe his knowledge and belief as partial, and give them their proper amounts. Now, whether we ſhall proceed, or ſtop ſhort at this point, depends upon our acceptance or non-acceptance of the following POSTULATE :—

When any number of events are disjunctively poſſible, ſo that one of them may happen, but not more than one, the meaſure of our belief that one out of any ſome of them will happen, ought to be the amount of the meaſures of our ſeparate beliefs in each one of thoſe ſome.

I mean that any one ſhould ſay, A, B, C, being things of which not more than one can happen, 'my belief that one of the three will happen is the ſum of my ſeparate beliefs in A, and in B, and in C.' This is the poſtulate on which the balance depends ; and there is a ſimilar poſtulate before we can uſe the phyſical balance. The only difference (and that but apparent) is that we are to ſpeak of weights collectively, and of events disjunctively. The weight of the (conjunctive) maſs is the ſum of the weights of its parts : the credibility of the (disjunctive) event is the ſum of the credibilities of its components. There are ſeveral *may-bes*, any one of which may become a *has-been :* when we ſpeak *disjunctively*, it is of the *will-be*, which cannot be ſaid of more than one : the *may-be* of an event deſcribed as contained in 'A, B, C,' is to be repreſented as in quantity the ſum of thoſe in 'A,' in 'B,' and in 'C.'

Is it matter of mere neceſſity that, talking of phyſical weight, the weight of the whole is equal to the ſum of the weights of the parts ? We have learnt to admit this poſtulate, of which no man ever doubted : but no one can ſay that it was neceſſary. The laws of matter and mind being both what they are, the connexion between phyſical collection and mental ſummation is, I grant, neceſſary : the ſimpleſt of manual, and the ſimpleſt of mental, operations, are and, with us, muſt be, concomitants.

But, in the firſt place, it is *not true* that the weight of the whole is equal to the fum of the weights of the parts, in the manner in which the reader probably imagines it to be true. Let the firft part we hang on the balance be the weight which is correctly meafured by W. Then if we hang under it another weight, as correctly reprefented by V, we think we are quite fure when we fay that the collective mafs muft have a weight W + V becaufe its parts have the weights W and V. But its parts have

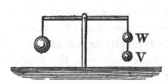

not the weights W and V. The weight of V is diminifhed by the upward attraction of W, and is, fay, V — M : the weight of W is as much increafed by. the downward attraction of V, and is W + M. And though V — M and W + M added together do give V + W, yet it was not in this way that the reader made out his neceffary truth. The univerfal equality of action and reaction did not exift in the thoughts of the firft perfon who formed a diftinct conception of the weight of the whole as compofed of the fum of that of the parts : and he was only right by the (fo far as he was concerned) accidental circumftance, that two things of which he knew nothing, counterbalanced each other's effects. Nor do we know at this moment, as of neceffity, that the propofition is correct. We have much reafon to think that the law of equality of action and reaction is mathematically true : but, let it fail to the amount of only one grain in a thoufand million of tons, and the propofition is not true, but only nearly true.

Again, the co-exiftence of thofe laws of mind and matter which beft, fo to fpeak, fit each other, and which make the phenomena of the external world, after due confideration, appear to be almoft what they *muft* have been, is not, to our apprehenfions, a neceffary coexiftence. We can imagine the following refult, though we cannot trace what the full confequences of it would be on the expreffion of the laws of thought. Conceive fentient beings, to whom the fimpleft mode of arithmetical fucceffion is not 0, 1, 2, 3, &c. but 1, 10, 100, 1000, &c. *their* powers of numeration being fo conftructed that the fecond of thefe fucceffions has that character of fundamental fimplicity which we attach to the firft. Of courfe, their primary fymbols would be

significative of 1, 10, 100, &c. It would be impoffible for us to
conceive any mode by which *ten* or any other number could be
thus fundamentally attached to unity, in a manner fhared by no
third number : but, I am not faying, 'Imagine how this could
be,' but, 'Imagine that it is.' There is no contradiction in the
fuppofition, either to itfelf, or, till we know much more of the
mind than we now do, to anything elfe. Beings fo conftituted
would have *logarithmic* brains ; and if, thus conftituted, they
were placed among our material laws of exiftence, the manner
in which the weight of the whole is to be inferred from thofe of
the parts, would be a profound myftery for ages, only to be folved
in an advanced ftage of mathematical fcience. A recent mode
of conftructing mathematical tables, which generally carries with
it the name of its eminent inventor, Gaufs, would conftitute
one of their principal neceffities : they. would have to ufe it as
their only mode (except actual experiment) of finding out that
what we reprefent by 156 and 200, together make (and this
making would be a complicated procefs) 356.

Inftead, then, of trying to eftablifh it as perfectly natural and
neceffary to fay that our belief of 'one of the two A or B, when
both cannot happen,' is, quantitatively fpeaking, the fum of our
belief in A, and our belief in B, I have rather endeavoured to
fhow that the analogous cafes with which we firft think of comparing this propofition, other kinds of compofition, are not fo
natural and neceffary as is fuppofed. There are two ways of
levelling ; by bringing up the lower, or bringing down the higher
And I particularly wifh in this chapter to prevent the reader
from accepting the arithmetical doctrine of probability quite fo
rapidly as is ufually done. In furtherance of this object, I proceed to the following poffible objection.

It may be faid, you have, by thus formally identifying probability with belief, and ftating a poftulate which, in exprefs terms,
has not the moft axiomatic degree of evidence, rendered fomewhat difficult that which in the ordinary view of fimple chances,
is very eafy. This charge, I hope, is true : fuch was my intention, at leaft. And my reafon is, that in the ordinary view of
the fubject, one of two things occurs : either probability is feparated by definition from ftate of belief, though it be known that
the two words will afterwards be confounded without any per-

miffion; or elfe the poftulate is tacitly affumed, and the difficulty which I fuppofe myfelf charged with introducing, is flurred over. Take a common queftion;—An urn has two white balls and five black ones: there are feven equally likely drawings, two white; therefore the chance or probability of drawing a white ball is called two-fevenths. But the chance of either particular white ball is one-feventh. Now firft, if any one fhould fay that this is mere definition, I can, of courfe, allow it : but it then remains to fhow what connexion this defined probability has with any ordinary acceptation of the word. But if, probability mean-ing belief, or fentiment of probability actually exifting in the mind, or index of the proper degree of belief, &c. &c.—the above ftatement be made as fundamentally evident, I fhould then afk how it is known that the probability of 'one or the other white ball being drawn' is properly fet down as the *fum* of the probabilities of the feparate white balls. And I cannot conceive any anfwer except that it is by an affumption of the poftulate. That fuch affumption will finally be knowingly made, on the fulleft conviction, by every one who ftudies the theory, I have no doubt whatever : nor that it has been made, no matter in what words, nor with what clearnefs of avowal, by every one who has ftudied that theory. And therefore I hold it defirable that the beginner fhould know what I have here told him.

It is indifferent, as far as the theory is concerned, what nu-merical fcale of belief we take. We might, if we pleafed, copy Fahrenheit's thermometer, fet down knowledge of impoffi-bility as 32°, perfect certainty as 212°, and other ftates of mind accordingly. Thus, 122° would reprefent perfect indecifion, belief inclining neither way, an even chance. But this would complicate our formulæ : the ufual and preferable plan is to af-fume o as the index of knowledge of impoffibility, 1 as that of certainty, and intermediate fractions for the intermediate ftates. This mode of eftimation makes formulæ and proceffes fo much more eafy than any other, that it muft be adopted ; but there is a ftrong objection to it in one point of view: as follows.

When we fpeak of belief in common life, we always mean that we confider the object of belief more likely than not: the ftate of mind in which we rather reject than admit, we call *un*belief. When the mind is quite unbalanced either way, we

have no word to exprefs it, becaufe the ftate is not a popular*
one. The quantitative theory calls by the name of belief every
admiffion of poffibility. When there is only one black ball to a
million of white ones, there is fome belief that a black ball will
be drawn; a much larger belief in a white one. It would be
advantageous in fome refpects that o fhould reprefent the ftate
of indifference, + 1, that of knowledge of certainty, and − 1,
that of knowledge of impoffibility. But this would complicate
formulæ too much. I confider it therefore defirable to ufe the
common meafures and formulæ, but to affociate them with the
one juft propofed, in the following manner.

When a perfon tells us that his belief in an affertion is, fay $\frac{3}{10}$,
meaning that he confiders it 3 for and 7 againft, or 7 to 3 againft,
we fhould fay in common talk that he difbelieves, but not very
ftrongly. In the language of this theory, we fay that he both
believes and difbelieves, the latter more ftrongly than the former.
Let us add that his *authority* is againft the conclufion. If he fay
that it is in his mind an even chance, or that he has no opinion
one way or the other, let us fay that he gives no authority either
way. If we adapt this definition to the fuppofition that + 1 and
− 1 reprefent the extremes of authority for and againft, we have
the following rules. The meafure of authority is twice the meafure of belief diminifhed by unity, for, when pofitive, againft,
when negative: the meafure of belief is half of unity increafed
(algebraically) by the meafure of authority. If *a* reprefent the
meafure of belief, and A that of authority, then

$$A = 2a - 1 \quad , \quad a = \tfrac{1}{2}(1 + A)$$

It is alfo advifable to have a term to reprefent what are ufually
called the *odds*. Some might think it defirable to rid the fubject
as much as poffible of words derived from gambling: aftrono-
mers have done the fame thing with the phrafes of aftrology, and
chemifts with thofe of alchemy. When it is 7 for and 3 againft,

---

* Many minds, and almoft all uneducated ones, can hardly retain an
intermediate ftate. Put it to the firft comer, what he thinks on the queftion
whether there be volcanoes on the unfeen fide of the moon larger than
thofe on our fide. The odds are, that though he has never thought of the
queftion, he has a pretty ftiff opinion in three feconds.

it might be faid that the *relative teſtimony* for, is $\frac{4}{7}$, and that againſt, $\frac{3}{7}$. But the brevity of the firſt phraſe will inſure its continuance, let who will try to change it.

The ordinary rule is a conſequence of the notions hereinbefore laid down, and of the particular mode of meaſurement adopted. It is as follows ;—When all the things that can happen can be reſolved into a number of equally probable (or credible) caſes, ſome favourable and ſome unfavourable to the event under conſideration, then the fraction which the favourable caſes are of all the caſes, meaſures the probability (or credibility) of the arrival of the event : and the fraction which the unfavourable caſes are of all the caſes, meaſures the probability (or credibility) of the non-arrival. There are, for inſtance, in an urn, 5 white, 4 black, and 3 red balls, 12 in all. It is aſſumed that we know them to be equally likely to be drawn ; which here means no more than that we know nothing to the contrary. That one ball muſt be drawn, is ſuppoſed certainly known. Accordingly, our belief in ‘ one or another’ is repreſented by 1 : which is, by the poſtulate, the ſum of the ſeveral credibilities of the balls ; which laſt are all equal. Therefore each ball has $\frac{1}{12}$: and by the ſame poſtulate, the event ‘ one or other of the white balls’ or the drawing of a white ball, has $\frac{5}{12}$; of a black ball $\frac{4}{12}$; of a red ball, $\frac{3}{12}$.

Inſtances like the above, in which we invent all the caſes and have arbitrary power over their number, are the only ones on which we can employ *à priori* numerical reaſoning. They are alſo the only ones on which we can try experiments. It is important to know whether, as a matter of fact, our belief, numerically formed, will be approximately juſtified by the reſults of trial. And this juſtification is found to exiſt, in the following way. It is a remote, but certain, concluſion from the theory, requiring mathematical reaſoning too complicated to introduce here, that events will, in the long run, happen in numbers proportional to the objective probabilities under which the trials are made. For inſtance ;—if a die be correctly formed, ſo that no one face has more tendency than another to fall upwards, the probability of throwing an ace is $\frac{1}{6}$; that of not throwing an ace is $\frac{5}{6}$. The theory tells us its own worthleſſneſs, if in the long run, *not-ace* do not occur five times as often as *ace*. If 60,000 trials were made, the theory would tell us to expect *about* 10,000 aces and about 50,000 *not-aces.* Practice confirms the theory: not, that I know of, in the actual cafe juft cited, but in fimilar ones. I will ftate an inftance.

Throw a half-penny up, and if it give *tail*, repeat the throw, and fo on, till *head* arrives : and let this fucceffion be called a *fet.* The probability that a fet fhall confift of one throw, is fhewn by the theory to be $\frac{1}{2}$; that it fhall have two throws, $\frac{1}{4}$; three throws, $\frac{1}{8}$; and fo on. If a very large number of fets be tried, we are to expect that about half will be of one throw, about a quarter of two throws, about an eighth of three throws ; and fo on, as long as the number is large enough to give any profpect of fomething like an average. This experiment has been tried twice : once by the celebrated Buffon, and once by a young pupil of mine, for his own fatisfaction ; both in 2,048 fets. The refults were as follows ; the third column fhowing the number of each kind which the theory afferts to be moft probable.

	B	H	
Head at the firft throw	1061	1048	1024
No head till the 2nd throw	494	507	512
3rd —	232	248	256
4th —	137	99	128
5th —	56	71	64
6th —	29	38	32
7th —	25	17	16
8th —	8	9	8
9th —	6	5	4
10th —	0	3	2
11th —	0	1	1
12th —	0	0	⎫
13th —	0	0	⎪
14th —	0	1	⎬ 1
15th —	0	0	⎪
16th —	0	1	⎪
&c —	0	0	⎭
	2048	2048	2048

In Buffon's trials, there were altogether 1992 tails to 2048 heads, and in Mr. H's there were 2044 tails to 2048 heads.

Inftances in which we can command all the cafes are to the

mind, in this theory, what acceffible lengths are to the eye. We can meafure the latter by a rule, and fo train the organ to judge of lengths which cannot be approached, or cafes in which the rule is not at hand.

I fhall now refer the reader to other works on the fubject, for further details on the operative part, and proceed to juft as much as is neceffary for the particular purpofe of the next chapter, namely, the application of the hypothefis of meafure of belief to queftions of argument and teftimony. Two theorems will be enough : the firft relating to independent events, the fecond to the probability of events which are neither wholly independent, nor wholly confequent, either upon the other. The word *event* is ufed in the wideft poffible fenfe : it does not even neceffarily mean future event. Unlefs our knowledge, either of the cir-cumftances, or of the event itfelf, thereby undergo fome altera-tion, it is nothing to us now whether it has happened, or is to happen.

Let there be two events, P and Q, of which the probabilities are the fractions *a* and *b*; and let them be wholly independent of one another, the arrival or non-arrival of either being perfectly independent of that of the other. The probability that both fhall happen is the product of *a* and *b* : and fimilarly for more events than two. Suppofe, to take an inftance, that *a* is $\frac{4}{7}$ and *b* is $\frac{3}{5}$. We muft then confider P as an event which has 3 ways of failing to 4 of happening : if we would have an urn from which the credibility of drawing a white ball fhould be that of the happening of P, we muft put in 4 white balls and 3 not white (fay black) balls. Similarly to reprefent Q, we muft have an urn of 3 white and 2 black balls. Now to afcertain the profpect of drawing white from both urns, we muft count all the cafes. A ball from the urn of 7 may be combined with one from the urn of 5, in 7×5 or 35 ways. But a *white* ball from the firft urn may be combined with a *white* ball from the fecond, in 4×3 or 12 different ways. There are then 35 cafes in all, 12 of which are favourable : hence the probability in favour of white from both (which is that of the two events both happen-ing) is

$$\frac{12}{35} \quad \text{or} \quad \frac{4 \times 3}{7 \times 5} \quad \text{or} \quad \frac{4}{7} \times \frac{3}{5} \quad \text{or} \quad ab.$$

Similar reafoning may be applied to more events than two. This theorem has a large number of confequences, fome of which we may notice.

When a is the probability for, $1-a$ is the probability againft. This I fhall always denote by a': fimilarly b' will ftand for $1-b$; and fo on.

Required the probability that of a number of independent events, P,Q,R, &c one or more fhall happen. Let a,b,c, &c. be the feveral probabilities, then that of their all failing is the product $a'b'c'$ .... and that of their *not* all failing (or of one or more happening) is $1-a'b'c'$ .... Accordingly, if there be only two events, for 'one or both' we have $1-(1-a)(1-b)$ which is $a+b-ab$. If the number of events be n, and all equally probable (fo that $a=b=c$, &c.) for 'one or more' we have $1-a'^n$ or $1-(1-a)^n$.

It is a confequence of this laft that, however unlikely an event may be, it is fure (in the common fenfe of the word) to happen, if the trial can be repeated as often as we pleafe. However fmall a may be, or however near to unity $1-a$, n may be taken fo great that $(1-a)^n$ fhall be as fmall as we pleafe, or $1-(1-a)^n$ as near to unity as we pleafe, or the probability that the unlikely event will happen once or more in n times, as great as we pleafe. Let $a=1:(k+1)$, which means that the odds are k to 1 againft the event on any one trial: the following rough deductions will fhow what kind of refults the formula gives, true within an inftance or two when k is confiderable. In $\frac{7}{10}k$ inftances it is an even chance that the event happens once or more; in $2\cdot3k$, it is 9 to 1; in $4\cdot6k$, 99 to 1; $6\cdot9k$, 999 to 1; $9\cdot2k$, 9999 to 1: and in $23k$, it is ten thoufand millions to 1. Thus, fuppofe at each trial it is a hundred to one againft fuccefs. Then of thofe who try 70 efforts, as many will fucceed once or more as will altogether fail, in the long run. Of thofe who try 6900 times, only one of a thoufand will always fail. A perfon who will not examine an affertion that comes to him with ten to one againft it, muft count it an even chance that he throws away one or more truths, if he follow his plan feven times.

Let us now fuppofe that there are reafons why the feveral inftances which can arrive are not equally credible. Suppofe the urn to contain a white, a black, and a red ball, and ourfelves to have reasons to think the balls not equally probable or credible, but that 6, 5, and 2 are the proportions of the degrees of belief we should accord to them severally. If then 6*x* represent the probability of a black ball, 5*x* and 2*x* will represent those of the other two severally. By the postulate, 13*x* represents that of one or the other. But this is certainty; whence *x* must be $\frac{1}{13}$, and $\frac{6}{13}$, $\frac{5}{13}$, and $\frac{2}{13}$ are the probabilities of the white, black, and red balls. That is to say, when the several instances are unequally probable, we must count each instance as though it occurred a number of times proportioned to its probability, and then proceed as in the case of equally probable instances. Thus, in the above, instead of saying (as we should do if the balls were equally probable) that the probability of the white ball is

$$\frac{1}{1+1+1}, \quad \text{we say it is} \quad \frac{6}{6+5+2}; \quad \text{or} \quad \frac{6m}{6m+5m+2m}$$

would do, *m* being any number or fraction whatsoever.

Now suppose two urns, one of all white balls, and the other of all black ones. If we actually draw a ball, and find it white, we know that the urn chosen to draw from must have been the first: the second *could not* have given that drawing. But suppose the first urn to have 99 white balls to one black, and the second one white to 1000 black. If we now draw again, and draw a white one, not knowing from which we drew, we feel almost certain, from the drawing, that we have chosen the first urn. We still feel *almost* certain that the second urn would have given a black ball. This inversion of circumstances, this conclusion that the circumstances under which the event did happen, are most probably those which would have been most likely to bring about the event, is of the utmost evidence to our minds: but the question now before us is, are we to call it a second postulate, or is it deducible from the other one? It is so deducible, and is not a second postulate; but it has not been usual to give a very distinct account of the deduction.* If it could not be made,

---

* So well established is this species of inversion in the mind, that both Laplace and Poisson, the two most eminent mathematical writers on the subject, of the present century, have in a certain case assumed that an equation which gives the most probable value of *x* in terms of *y*, is therefore the one which gives the most probable value of *y* in terms of *x*. This is carrying the principle too far.

the following process would, no doubt, be sufficient: it has often been held so. Let the urns have 6 white balls to 1 black, and 2 white balls to 9 black. Then the probabilities of drawing a white ball from the two are $\frac{6}{7}$ and $\frac{2}{11}$, which are in the proportion of 33 to 7. If, because when we choose the first urn, we have nearly five times as much chance of a white ball as the second one would give, we conclude that a known white ball from an unknown urn is in that proportion more likely to have come from the first urn ; we shall have $\frac{33}{40}$ and $\frac{7}{40}$ for the proper degrees of belief in the two urns. For if $33x$ be that for the first urn, then $7x$ must, by the assumption, be that for the second : and for one or the other, we have $40x$. But this is certainty ; whence x must be $\frac{1}{40}$.

To reduce this result to dependence upon the first postulate, proceed as follows. The probability that two events are *connected*, our belief, that is, in the *connexion*, must be the same whether the two events, or either of them, have happened, or whether they be yet to happen : unless there be something in the happening which alters our knowledge, and puts us in a different state for forming a judgment. Suppose I make up my mind, rightly or wrongly, as to how far I will believe that a white ball, *if* drawn, *will have been* drawn from the first urn. An instant after, I am told that the trial I anticipated has been made, and the contingency which I supposed has occurred ; a white ball has been drawn. I know no more than I took myself to know in my hypothesis ; and cannot therefore have any means of altering my opinion. Now, without altering the proportions in the urns, change the numbers of the balls, so that there may be the same total number in each : let them be

{66 white, 11 black}   {14 white, 63 black}

Now put each ball in an urn by itself, 154 urns in all. This gives $\frac{1}{154}$ to any one ball, if I choose an urn at hazard. But it was so before : as to the first of the two urns for instance, $\frac{1}{2}$ was the probability of choosing that urn, and $\frac{1}{77}$ that of choosing one particular ball from it : and $\frac{1}{2} \times \frac{1}{77}$ is $\frac{1}{154}$. If we then remove all the urns with black balls, so that a white ball must be drawn, the chance of its being one of the 66 is $\frac{66}{80}$ or $\frac{33}{40}$. If without removing the black balls, we think of the probability of a white ball, if drawn, being of the 66, or of the 14, the credibilities of those suppositions are as 66 to 14. If, having chosen an urn, we find it contains a white ball, the same probabilities are still in that proportion.

The rules derived from similar reasoning, whether for judging of the probabilities of precedents from an observed consequent, or for judging of the probabilities of events which restrict each other, are precisely the same, as follows. If the probability of the observed event, supposed still future, from the several possible precedents, severally supposed actually to exist, be a,b,c, &c : then, when the event is known to have happened, the probabilities that it happened from the several precedents are

$$\frac{a}{a+b+c+\ldots} \quad \text{for the first,} \quad \frac{b}{a+b+c \ldots} \quad \text{for the second, &c.}$$

Again, if there be several events, which are not all that could have happened; and if, by a new arrangement (or by additional knowledge of old ones) we find that these several events are now made all that can happen, without alteration of their relative credibilities: their probabilities are found by the same rule. If a, b, c, &c. be the probabilities of the several events, when not restricted to be the only ones : then, after the restriction, the probability of the first is $a \div (a+b+\ldots)$, of the second, $b \div (a+b+\ldots)$ and so on.

We may obtain a very distinct notion of this last theorem, as follows. Suppose two events, which are *among* those that can happen, and let one, say, be twice as probable as the other. This means, that among all the independent, and equally likely, cases, there are twice as many favourable to the first as to the second. Now, suppose by some alteration of suppositions, the introduction of new knowledge, for instance, it is found, all the cases remaining as before, that all are prevented from happening except these two events. This new state of things does not alter the cases in number: accordingly, the *proportion* of the probabilities of the two events is as before, two to one. But now one of them must happen : or the sum of these probabilities must be unity. It follows then that one of them is $\frac{2}{3}$, and the other $\frac{1}{3}$. The same reasoning may be applied to more complicated cases.

It frequently happens, when different problems are solved by

the fame formula, that they may be confidered as the fame problem in two different points of view : and alfo that one·and the
fame problem may be confidered as belonging to either clafs.
For inftance ;—Let there be two witneffes, whofe credibilities
(or the probabilities that in any given inftance they are correct)
are *a* and *b*. As long as we do not know that they are talking
about the fame thing, the probability that both will tell truth is
ab. But the moment we know that they both affert the fame
thing, the problem is changed : they muft now be either both
right or both wrong ; before, one might have been right and
the other wrong. To take the firft view of the problem, we
have now an obferved event, both ftate that the circumftance
did happen. There are two precedents ; the event did, or did
not, happen. If it did, the probability of the obferved event
(which is then that both are right) would be *ab* ; if it did not, it
would then be $(1-a)(1-b)$. Accordingly, the probability that
the obferved event did happen, will be, by the rule above, *ab*
divided by $ab+(1-a)(1-b)$.

If we take the fecond view, we have, before the reftriction,
four poffible cafes, the probabilities of which are ab, $a(1-b)$,
$b(1-a)$ and $(1-a)(1-b)$. After the reftriction, only the firft
and fourth are poffible : whence the conclufion is as juft given.
Full exemplifications of thefe methods will be found in the next
chapter.

CHAPTER X.

On probable Inference.

THERE are two fources of conviction, *argument* and *teftimony*, reafon why the thing fhould be, ftatement that the
thing is. When the argument is neceffarily good, we call it
demonftration when the ftatement can be abfolutely relied on,
we call it *authority*. Both words are ufed in lower than their
abfolute fenfes ; thus, very cogent arguments are often called
demonftration, and very good evidence, authority.

I fhall fuppofe all the arguments I fpeak of to be logically valid ; that is, having conclusions which certainly follow from the premises. If then the premises be all true, the conclusion is certainly true. If *a*, *b*, *c*, &c. be the probabilities of the independent premises, or the independent propositions from which premises are deduced, then the product *abc* . . . is the probability that the argument is every way good.

Argument being offer of proof, its failure is only failure of proof : and the conclusion may yet be true. But testimony is an affertion of the truth of the conclusion ; and its failure can only be failure of truth. If a proposition of Euclid turn out to be badly demonftrated, the enunciation need not therefore be false. An argument may prove, disprove, or neither prove nor disprove : a teftimony cannot be true, false, or neither true nor false. This diftinction generally gains no more than a one-sided admiffion : perfons begin to see it when some over-zealous brother writes weakly on their own side of a queftion ; but they are very apt to think, with respect to the other side, that anfwering the arguments is disproving the conclusion.

Teftimony is, for the above reafon, more eafily underftood than argument. It is the moft effective mode of conveying knowledge to the uneducated. But it muft not be suppofed that, in any ftage of reafon, argument can be the only vehicle of information, even on fubjects called argumentative. This point is one of great importance.

When argument is demonftration, it eftablishes its conclusion againft all teftimony. The idea of an infallible witnefs bearing evidence againft a demonftrated conclusion, is a contradiction. That *n* confecutive numbers have a sum which is divifible by *n*, whenever *n* is odd is demonftrated. If a thoufand of the beft qualified witneffes that ever lived, both for honefty and arithmetic, were to fwear that they had difcovered 101 very high confecutive numbers, the sum of which is not divifible by 101, any mere beginner in mathematics would be more fure that a thoufand good witneffes had loft their wits or their characters, than any one elfe can be of anything not admitting of demonftration.

But when argument does not amount to demonftration, not only is the truth or falfehood of the conclusion matter of credibility, but the iffue of the argument is not that mere truth or falfehood. It does not ftand thus : ' According as this argument is good or bad, fo is the conclufion true or falfe,' but ' According as this argument is good or bad, fo is the conclufion *true in this. way*, or *not true in this way*, (that is, either falfe, or true in fome other way).' If we were to fay ' men are trees, and trees have reafon, therefore men have reafon,' we have a perfectly logical argument, falfe in the matter of both premifes : but we cannot deny the conclufion.

Suppofe now that an argument is prefented to us of which we are fatisfied that the like will prove their conclufions to be true in the particular modes afferted, in nine cafes out of ten. What are we to fay of the truth or falfehood of the conclufion ? We have $\frac{9}{10}$ of belief to its being true in one particular way : how much fhall we add for other poffible ways? Are we to reft in the conclufion as having 9 to 1 for it, or are we to allow more? We cannot fay, let us confine ourfelves to the grounds we have got, and believe or difbelieve, not in the conclufion, but in the conclufion as obtained in that one way.

I take it for granted that the mind muft have a ftate with refpect to every affertion prefented to it, with reafon, or without reafon. Every propofition, the terms of which convey any meaning, at once, when brought forward, puts the hearer into fome degree of belief, or, if we ufe the common phrafe, of belief or unbelief: including, of courfe, the intermediate ftate, which is as clearly marked upon our fcale as any other. Men who are accuftomed to fufpend their opinion, as it is called, that is, to throw themfelves into the intermediate ftate when they have no definite reafon to chink either way, are interefted in this queftion as much as any others. If there be fome ftate, though not numerically appreciable, in which their belief muft be, there is fome ftate, which they would rather know numerically than not, in which it ought to be. In the preceding cafe, fuppofe it known that 9 to 1, or $\frac{9}{10}$, is granted to the conclufion from the argument alone, and any one wifhes to fufpend his opinion as to the remaining $\frac{1}{10}$. Is he to grant half of that $\frac{1}{10}$, and fay that $\frac{9}{10} + \frac{1}{20}$ or $\frac{19}{20}$ is what he would wifh to make the meafure of his belief, if he knew how ? The confideration of this queftion will enter among others.

The manner in which he deals with the refult of the argument muſt depend upon *teſtimony*, uſing the word in its wideſt ſenſe. Firſt, every man has, as juſt noticed, a teſtimony in his own mind as to every propoſition. He may ſet out with the intermediate ſtate : he may have no reaſon to lean either way, *and may know it ;* that is to ſay, he may have to apply an argument of $\frac{9}{10}$ to an exiſting probability of $\frac{1}{2}$. Or he may have previous good reaſon, or bad reaſon, which makes him lean to the aſſertion or denial ; and the meaſure of this leaning muſt then be combined with $\frac{9}{10}$. Or he may have other teſtimony to combine with that of his own previous ſtate. Any way, he cannot have a definite opinion on the bare truth or falſehood of the concluſion of the argument, without appeal to the previous ſtate of his own mind at leaſt, if not to that of others.

It is generally ſaid that we are to throw away authority, and judge by argument alone ; that our reaſon is to be convinced, and not biaſſed by the opinion of others ; that no concluſions are worth anything, except thoſe which a man forms for himſelf. All the forms in which this frequent caution is expreſſed, I take to be diſtortions of the very needful warning not to allow authority more weight than is properly due to it : a warning, by the way, which is juſt as much wanted with reſpect to argument as to authority: For every miſtake which has been made by taking authorities *on truſt* (that is, taking bad witneſſes to prove the goodneſs of aſſerted good ones), one miſtake at leaſt has been made by taking arguments *on preponderance :* that is, treating them as proving their concluſion, as ſoon as they ſhow it to be more likely than its contradiction.

To form the habit of allowing authority *no more* weight than is due to it, and the ſame of argument, is undoubtedly one great object of mental cultivation : but it ought not to be forgotten that it is another and juſt as great an object to form the habit of allowing them *no leſs.* Suppoſe an argument of value $\frac{9}{10}$ is preſented, and that at the ſame time we have the teſtimony of a witneſs againſt the concluſion, of whom we *know* that he leads us right 1000 times for each once that he miſleads us. Is there any ſenſe in reducing this witneſs to one of no authority, or of an even chance, upon the principle of depending on argument only? Except the argument be demonſtration, we muſt be prepared to admit that a witneſs may be as good as an argument, or better.

I shall now proceed to the several problems which this subject requires, considering first testimony alone, next argument alone, and then the two in combination.

Problem 1. There are independent testimonies to the truth of an assertion, of the value μ, ν, ρ, &c. (one of them being the initial testimony of the mind itself which is to form the judgment): required the value of the united testimony.

Let μ be $1-\mu$, &c. as in page 187. Here is a problem of the same class as in page 190; the restrictions are, that all the testimonies are right, or all wrong, the independent chances of which are $\mu\nu\rho\ldots$ and $\mu'\nu'\rho'\ldots$ Hence the probabilities are

$$\frac{\mu\nu\rho\ldots}{\mu\nu\rho\ldots+\mu'\nu'\rho'\ldots} \text{ for ;} \qquad \frac{\mu'\nu'\rho'\ldots}{\mu\nu\rho\ldots+\mu'\nu'\rho'\ldots} \text{ against.}$$

Observe, first, that any numbers proportional to μ, μ' &c. will do as well: and if the products have a common denominator, (as generally they have) the numerators only need be used. Secondly, the easiest way of expressing the result is by saying that it is $\mu\nu\rho\ldots$ to $\mu'\nu'\rho'\ldots$ for, or $\mu'\nu'\rho'\ldots$ to $\mu\nu\rho\ldots$ against.

For instance, let it be in my mind 99 to one against an assertion, that is, I bear only the testimony $\frac{1}{100}$ *in favour* of it. Let four witnesses, for whose accuracy it is 2 to 1, 3 to 1, 4 to 1, 5 to 1, depose in favour of it: I want to know how it ought to stand in my mind. The testimonies for and against, are

$$\frac{1}{100}, \frac{2}{3}, \frac{3}{4}, \frac{4}{5}, \frac{5}{6}; \text{ and } \frac{99}{100}, \frac{1}{3}, \frac{1}{4}, \frac{1}{5}, \frac{1}{6};$$

Hence, neglecting the common denominator, it ought to be $1\times2\times3\times4\times5$ to $99\times1\times1\times1\times1$, or 120 to 99, or 40 to 33, for the assertion.

Observe that in saying the witness gives testimony, say $\frac{2}{3}$, it is of no consequence whether it be a question of judgment, or of veracity, or of both together. I mean that, come how it may, I am satisfied that when he says anything, it is 2 to 1 he says what is correct.

An easy rule for the more common modes of expression presents itself thus. The combined relative testimony is the product of the separate relative testimonies. Thus, two witnesses of 6 truths to one error, and of 7 truths to one error, are equivalent

to one witnefs of 42 (or 6 × 7) truths to one error. Three witneffes of 8, 6, 5 truths to 7, 3, 11 errors are equivalent to one
witnefs of 80 truths to 77 errors.

A jury of twelve equally truftworthy perfons, after conferring
together, agree to an affertion on which previoufly I had no
leaning. Suppofing me fully fatisfied that fuch agreement gives
100 to 1 for their refult, what am I to think of the *deliberate*
opinion of any one among them, that is, of his opinion after he
has had the advantage of difcuffion with others.

Let μ be the value of fuch teftimony from any one; then by
the queftion

$$\mu^{12} : (1-\mu)^{12} :: 100 : 1, \quad \text{or } \mu : 1-\mu :: \sqrt[12]{100} : 1$$

fay as 1·468 to 1. That is, I think inconfiftently if I rely on
the united verdict as upon 100 to 1, unlefs I am prepared to
think it 1468 to 1000, or about 3 to 2, for each juror alone.

Of $m+n$ equally truftworthy jurors, a majority m are for, and
n againft, a conclufion. If μ be the value of the teftimony of
each, then the odds are to be taken as being $\mu^{m}(1-\mu)^{n}$ for, and
$\mu^{n}(1-\mu)^{m}$ againft. But

$$\mu^{m}(1-\mu)^{n} : \mu^{n}(1-\mu)^{m} :: \mu^{m-n} : (1-\mu)^{m-n}$$

which are exactly as if the majority $m-n$ had been all, and
unanimous. From the original formula it will appear that two
equally good teftimonies on oppofite fides produce no effect on
the refult.

If then, the unanimity of the jury box in this country could
be confidered as that of deliberate conviction, we might fay that
a larger jury, with the condition that the majority fhould exceed
the minority by 12 at leaft, would be always as good, and often
better. But there are various confiderations which prevent the
above refult from being applicable. The neceffity of being unanimous, as our law ftands, may lower the value of the verdict. On
the other hand, a jury of 30, required to find by a majority of 12,
would generally proceed to a vote before they had put the matter
to each other with the real defire to gain opinion which the prefent practice produces : confequently, the value of their verdict
would perhaps be lower than that of the majority only, required
to be unanimous.

The theory thus appears to confirm the notion on which we often act, that a given excefs of majority over minority, is of the fame value whatever the numbers in the two may be. And this might be the cafe, if the thing called deliberation in a large body, were as well adapted to the difcovery of truth as the fame thing in a fmaller one. The reader muft remember that this teft does not compare the one witnefs on his own judgment with a number after common deliberation; but the firft, after common deliberation with others, is compared with the whole.

But in this, and all the problems of this chapter, the diftinction muft be carefully drawn between the credibility of a circumftance at one time and at another. For example, a witnefs enters with 10 to 1 in his favour, and owing to combination with others, the refult comes out that it is 100 to 1 he is in error in the particular matter on which he gives evidence. We cannot believe both that it is 10 to 1 he is right, and 100 to 1 that he is wrong. What we believe is the latter, for the cafe in queftion.

As another inftance, fuppofe m independent witneffes of equal goodnefs (μ) unite in affirming that a certain ball was drawn from a lottery of n balls: collufion being fuppofed impoffible. My knowledge of the circumftances of the affirmation here alters the problem. If n be confiderable, it is almoft impoffible that the witneffes, by independent falfehood or error, fhould all pitch on the fame wrong ball. To find the bias this ought to give me to the conclufion that they have told the truth, I muft obferve that there being $n - 1$ balls not drawn, whichever of thefe any one choofes, by error, the chance of any one of the reft choofing the fame is $1 \div (n - 1)$, the probability that all the $m - 1$ fhall choofe the fame is $1 \div (n - 1)^{m-1}$. Hence, the odds are as μ^m to $(1 - \mu)^m$ multiplied by the laft-named expreffion, or as $(n - 1)^{m-1}\mu^m$ to $(1 - \mu)^m$. If n be very great, the odds may be enormous for the affertion, even though μ, the credibility of each witnefs, may be fmall. In cafes of ordinary evidence, the thing afferted is ufually one out of almoft an infinite number of equally poffible affertions, and the agreement of even two witneffes (for when m is *two* or upwards, n appears in the formula) is certain conviction, if, as affumed, we know the two witneffes to be not in collufion. If $\mu = 1 \div n$, which is as much as to fay that the evidence of each witnefs makes a ball no more likely to have been the one drawn, becaufe he fays it, that it was on our mere knowledge that a ball *had* been drawn, it turns out 1 to $n-1$ for the truth of the affertion, juft as it was before the evidence. But let $\mu = (1+a) \div n$, a being any fraction, however fmall, that is, let each witnefs make the affertion more probable than at firft, however little: then the odds for its truth become

$$(1+a)^m \quad \text{t o} \quad \left(1 - \frac{a}{n-1}\right)^{m-1}(n-1-a)$$

which odds may be made as great as we pleafe, by fufficiently increafing m. That is to fay, however little each witnefs may be good for, in real fupport of the affertion, or in making it more probable than it is of itfelf, a fufficient number of witneffes, certainly independent, will give it any degree of credibility whatever.

The ftudent of this fubject is always ftruck by the frequency of the problems in which the fcience confirms an ordinary notion of common life, or is confirmed by it, according to his ftate of mind with refpect to the whole doctrine. It is impoffible to fay that we have a theory *made to explain* common phenomena, and hence affording no reafon for furprife that it does explain them. The firft principles are too few and two fimple, the train of deduction ends in conclufions far too remote. I believe hundreds of cafes might be cited in which the refults of this theory are found already eftablifhed by the common fenfe of mankind: in many of them, the mathematical fciences were not powerful enough to give the modes of calculation, when the principles of the theory were firft digefted.

There are problems, however, in which we cannot eafily come into poffeffion of data on which many will agree. The fimple queftion of independent witneffes is not one of them: but the queftion of collufion is. One of the difficulties is as follows. We cannot inftitute independent hypothefes upon the goodnefs of the witneffes and the probability of their having conferred upon their evidence. They declare, expreffly or by implication, that they have not done fo: if they have, there is falfehood in one part of their evidence; or, which makes the difficulty ftill greater, there may have been general, but (as they affert or imply) not particular conference: they may have been biaffed by each other, without knowing how or to what extent. The firſt ſtep in one view of the problem is eaſily made, as follows.

Let μ be the value of the evidence of each witneſs, m their number, n the number of aſſertions they have power to chooſe from, all as before. Let λ be the probability that there has been particular conference between them. There are then four caſes to which the problem is reſtricted :—(1) they have conferred and agreed to ſpeak truth ; (2) they have not conferred and all ſpeak truth ; (3) they have conferred and agreed on a falſehood ; (4) they have not conferred and have all lighted upon the ſame falſehood. The a priori probabilities of theſe four caſes are

$$\lambda\mu^{m}, \quad (1-\lambda)\mu^{m}, \quad \lambda(1-\mu)^{m}, \quad \frac{(1-\lambda)(1-\mu)^{m}}{(n-1)^{m-1}}$$

and the odds that they ſpeak the truth (ſuppoſing n ſo great that we may reject the fourth caſe) are μ^{m} to $\lambda(1-\mu)^{m}$. Now comes the practical difficulty of this queſtion ;—How are λ and μ to be connected? Every caſe which is worth examining ſuppoſes that the greater the chance of there having been particular conference, the leſs is the witneſs worth from that very circumſtance. For it is to be remembered that we are not generally able to give the witneſs a character wholly independent of his evidence in the caſe before us ; in hiſtorical queſtions, for inſtance, it frequently happens that we have nothing but the witneſſes to try* the caſe by, and nothing but the caſe to try the witneſſes by. A very common occurrence is this ;—that a caſe is one in which no one would throw any doubt upon the witneſſes, except for ſuſpicion of conference, and juſt as much doubt as there is ſuſpicion of conference. This makes $\mu = 1 - \lambda$, and gives $(1 - \lambda) : \lambda^{m+1}$ for the odds in favour of the aſſertion. On this ſuppoſition, it follows that whenever the chances are againſt all the witneſſes having conferred particularly, their number, if great enough, ought to give any degree of credibility to the aſſertion.

---

* This gives riſe to two great tendencies, which very nearly divide the world among them. Some ſettle the caſe in their own minds, and then try the witneſſes : ſome ſettle the witneſſes and then try the caſe : not a few bring their ſecond reſult back again to juſtify their firſt aſſumption. When there are two unknown quantities with only one equation, it is eaſy for thoſe who will aſſume either to find the other. But the difficulty is to find the moſt probable value of both.

Problem 2. Let there be any number of different affertions, of which one muſt be true, and only one : or of which one may be true, and not more than one : or of which any given number may be true, but not more : required the probability of any one poſſible caſe.

The ſolution of all theſe varieties depends on one principle, explained in page 190 ; requiring the previous probabilities of all the conſiſtent caſes to be compared. As an inſtance, ſuppoſe four affertions, A,B,C,D, and ſuppoſe μ,ν,ρ,σ, to be the probabilities from teſtimony, for each of them. If either of them have ſeveral teſtimonies, their united force muſt be aſcertained by the laſt problem. Firſt, let it be that one of them muſt be true, and one only. The probabilities in favour of A,B,C,D, are in the proportion of $\mu\nu'\rho'\sigma'$, $\nu\mu'\rho'\sigma'$, $\rho\mu'\nu'\sigma'$, and $\sigma\mu'\nu'\rho'$. Either of theſe, divided by the ſum of all, repreſents the probability of its caſe. Secondly, let it be that one of them only can be true, and all may be falſe. Put on the fifth quantity $\mu'\nu'\rho'\sigma'$, for the caſe in which all are falſe. For example, there are four diſtinct affertions, not more than one of which can be true. The ſeparate evidences for theſe four affertions give them the probabilities $\frac{2}{7}$, $\frac{3}{11}$, $\frac{1}{8}$ and $\frac{4}{5}$. There is a certain affertion which is true if either of the firſt three be true : required the probability of that affertion. Here, neglecting the common denominator, which is $7 \times 11 \times 8 \times 5$ in every caſe, the probabilities of the ſeveral affertions, and that of all being falſe, are as 2.8.7.1, 3.5.7.1, 1.5.8.1, 4.5.8.7, and 5.8.7.1, or as 112, 105, 40, 1120, and 280. The odds for one of the firſt three caſes againſt one of the other two are $112 + 105 + 40$ to $1120 + 280$ or as 257 to 1400; or it is 1400 to 257 againſt the truth of the affertion.

Suppoſe the condition were that two of the affertions, but not more, *may* be true, and that one *muſt* be true. Then the poſſible caſes, meaning by an accent that the affertion is not true, are AB'C'D', BA'C'D', CA'B'D', DA'B'C', ABC'D', ACB'D', ADB'C', BCA'D', BDA'C', CDA'B'. Conſequently, the probabilities of theſe caſes are in the proportion of $\mu\nu'\rho'\sigma'$, $\nu\mu'\rho'\sigma'$, $\rho\mu'\nu'\sigma'$, &c. And the odds in favour of, ſay A, being true, are as the ſum of all the terms which contain μ, to the ſum of thoſe which contain μ'.

When we wiſh to ſignify that no evidence is offered either for or againſt one of the aſſertions, we muſt put it down as having the teſtimony $\frac{1}{2}$. To put down o in the place of $\frac{1}{2}$ would be to make an infallible witneſs declare that it is not true. Suppoſe there are four aſſertions, one of which muſt be true and one only : evidence of goodneſs $\frac{4}{9}$ is offered for the firſt, and none either way for the others. Required the probability of the firſt. The probabilities of the four aſſertions are in the proportion of 4.1.1.1, 1.3.1.1, 1.3.1.1, and 1.3.1:1, and it is 4 to 9 for the firſt, or 9 to 4 againſt it.

Problem 3. Arguments being ſuppoſed logically good, and the probabilities of their proving their concluſions (that is, of all their premiſes being *true*) being called their validities, let there be a concluſion for which a number of arguments are preſented, of validities a, b, c, &c. Required the probability that the concluſion is proved.

This problem differs from thoſe which precede in a material point. Teſtimonies are all true together or all falſe together : but one of the arguments may be perfectly ſound, though all the reſt be prepoſterous. The queſtion then is, what is the chance that one or more of the arguments proves its concluſion. That all ſhall fail, the probability is $a'b'c'\ldots$ that all ſhall not fail, the probability is $1-a'b'c'\ldots$ Accordingly, if we ſuppoſe n equal arguments, each of validity a, the probability that the concluſion is proved is $1-(1-a)^n$. And, as in page 187, if the odds againſt each argument be k to 1, then, the number of ſuch arguments being as much as 7 k, the concluſion is rendered as likely as not.

But are we really to believe, having arguments againſt the validity of each of which it is 10 to 1, that ſeven ſuch arguments make the concluſion about as likely to be true as not. If ſuch be the caſe, the theory, uſually ſo accordant with common notions, is ſtrangely at variance with them. This point will require ſome further conſideration.

In this problem I conſider only argument, and not teſtimony, which, nevertheleſs, cannot be finally excluded (ſee page 194). If the concluſion be one on which our minds are wholly un-biaſſed to begin with, it may ſeem that we have no eſcape from the preceding reſult. And to it we muſt oppoſe, for conſidera-tion at leaſt, the common opinion of mankind that ſtrong argu-ments are the preſumption of truth, weak arguments of falſehood.

If a controverfialift were to bring forward a hundred arguments, and if his opponent were fo far to anfwer them as to make it ten to one againft each, there can be no doubt that the latter would be confidered as having fairly contradicted the former.

We muft not forget that argument, in a great many cafes, involves and produces the effect of teftimony, and this in an eafily explicable and perfectly juftifiable manner. If I were to pick up a bit of paper in the ftreets, on which an argument is written, for a conclufion on which I have no previous opinion, and by an unknown writer, and if I could fay that that argument left on my mind the impreffion of ten to one againft its validity, I might be prepared to allow it to ftand as giving $\frac{1}{11}$ of probability, and upon that fuppofition to combine it with my previous opinion, $\frac{1}{2}$, as in the next problem. But fuppofe it is on a queftion of phyfics, and Newton is the propofer of it, and that it is his only argument, and therefore, I conclude, his beft. The cafe is now entirely altered: poffibly the conclufion is one on which the following argument would have great probability : ' If this conclufion were true, it could be proved ; if it could be proved, Newton could have proved it ; therefore if it were true, Newton could have proved it : but Newton cannot prove it ; therefore it is not true.' If the cafe be fuch that the two premifes of this laft argument have each 9 to 1 for it, or $\frac{9}{10}$; then, though the original argument give $\frac{1}{11}$ for the conclufion, the mere circumftance of Newton bringing this argument as his beft is $\frac{81}{100}$ againft it. If Newton at the fame time declare his belief in the conclufion, we have on one fide his argument and his authority, on the other fide the argument arifing from his being reduced to fuch an argument.

That fuch. confiderations have weight, we know : and that they ought to have weight, we may eafily fee. It is of courfe, dependent upon the particular conclufion what weight fhall be attached to the affertion, ' if this conclufion were true it could be proved.' The courts of law conftantly act upon this principle. They confider (very juftly I think) that evidence, however good it may be, is much lowered by not being the beft evidence that could be brought forward. If a man be alive, and capable of being produced with fufficient eafe, they will not take any number of good witneffes to the fact of his having been very recently alive. In enumerating the arguments, then, for or against a propofition, thofe muft be included, if any, which arife out of the nature, mode of production, or producers, of any among them. And until this has been properly done, we are not in a condition to apply the methods of the prefent chapter.

Problem 4. A conclufion and its contradiction being produced, one or the other of which muft be true, and arguments being produced on both fides, required the probability that the conclufion is proved, difproved (*i. e.* the contradiction proved), or left neither proved nor difproved.

Collect all the arguments for the conclufion, as in the laft problem, and let a be the probability that one or more of them prove the conclufion. Similarly, let b be the probability that one or more of the oppofite arguments prove the contradiction. Both thefe cafes cannot be true, though both may be falfe. The probabilities of the different cafes are thus derived. Either the conclufion is proved, and the contradiction not proved, or the conclufion not proved and the contradiction proved, or both are left unproved. The probabilities for thefe cafes are as $a(1-b)$, $b(1-a)$ and $(1-a)(1-b)$, and the probability that the conclufion is *proved* is $a(1-b)$ divided by the fum of the three, and fo on. The fraction $(1-a)(1-b)$ divided by this fum may be called the *inconclufivenefs* of the combined arguments. The manner in which this inconclufivenefs is to be diftributed between the hypo-thefis of the truth and falfehood of the conclufion *muft* depend upon teftimony, in the complete fenfe of the word.

The predominance of one fide or the other, as far as arguments only are concerned, depends on which is the greateft, $a(1-b)$ or $b(1-a)$, or fimply on which is the greateft, a or b. If the arguments on both fides be very ftrong, or a and b both very near to unity, then, though $a(1-b)$ and $b(1-a)$ are both fmall, yet $(1-a)(1-b)$ is very fmall compared with either. The ratio of $a(1-b)$ to $b(1-a)$ on which the degree of predominance de-pends, may, confiftently with this fuppofition, be anything what-ever. But we cannot pretend that, when oppofite fides are thus both nearly demonftrated, the mind can take cognizance of the predominance which depends upon the ratio of the fmall and imperceptible defects from abfolute certainty. The neceffary confequence is, that the arguments are evenly balanced, and are as if they were equal : there is no senfible notion of predominance. This is the ftate. to which moft well conducted oppofitions of argument bring a good many of their followers. They are fairly outwitted by both fides, and unable to anfwer either, and the conclufion to which they come is determined by their own pre- vious impreffions, and by the. authorities to which they attach moft weight; and thefe are, of courfe, thofe which favour their own previoufly adopted fide of the queftion.

When no argument is produced on one fide of the queftion, the cafe is very different from the cafe of the preceding problems, in which no teftimony is produced. Here the queftion is, ' Has the conclufion been proved or not proved ;' and when no argu- ment is produced, we are certain it has not been proved. Ac- cordingly, if no argument were urged for the contradiction, we fhould have $1-b=1$, or $b=0$.

If, in the preceding problem, the two fides of the queftion be not contradictions, but fubcontradictions, of which neither need be true, but both cannot be, the problem is folved in the fame way, for the cafes are juft the fame. But we may introduce a diftinction which the former cafe would not admit. When one muft be true, every argument againft one is of. equal force for the other; which is not the cafe when neither need be true. Let there, then, be arguments for the firft conclufion and againft it, and let a and p be the probabilities that one or more of the arguments for, prove it, or againft, difprove it. Let b and q be the fimilar probabilities for the fecond conclufion. Then, there are thefe cafes :—1. The arguments (or fome of them) for the firft are valid, againft it invalid, and thofe *for* the fecond are invalid (it matters nothing whether thofe *againft* the fecond be valid or invalid). 2. The arguments for the firft are invalid, thofe for the fecond valid, and againft it invalid. 3. The argu- ments againft the firft are valid, and thofe for it invalid. 4. The arguments againft the fecond are valid, and thofe for it invalid. 5. All the arguments are invalid. Accordingly, the probabilities that the firft is proved, that it is difproved, that the fecond is proved, that it is difproved, and that neither of the two is proved nor difproved, are in the proportion of $a(1-p)(1-b)$, $(1-a)p$, $b(1-q)(1-a)$, $(1-b)q$, and $(1-a)(1-b)(1-p)(1-q)$.

Problem 5. Given both teftimony and argument to both fides of a contradiction, one side of which must be true, required the probability of the truth of each side.

This is the most important of our cases, as representing all. ordinary controversy. Collect all the testimonies, and let their united force for the first side be μ, and, from the nature of this case, $1-\mu$ for the other side. Let a and b be the probabilities that the first side and the second side are proved by one or more of the arguments in their favour. Now, observe that, for the *truth* of either side, it is not essential that the argument for it should be valid, but only that the argument against it should be invalid. Accordingly, the probabilities of the two sides are in the proportion of $\mu(1-b)$ and $(1-\mu)(1-a)$, and the probabilities of the two sides are represented by

$$\frac{\mu(1-b)}{\mu(1.-b)+(1-\mu)(1-a)} \qquad \frac{(1-\mu)(1-a)}{\mu(1-b)+(1-\mu)(1-a)}$$

First, let there be no testimony either way: we must then have $\mu=\frac{1}{2}=1-\mu$; consequently, these probabilities are as $1-b$ to $1-a$. Let no argument have been offered for the second side, or let $b=0$. Then we have 1 to $1-a$, for the odds, or $1\div(2-a)$ for the probability of the first side being true. It has been usual to say that if an argument be presented of which the probability is a, the truth of the conclusion has also the probability a. Probably the above was the case intended as to testimony, &c., and the probability should then have been

$$\frac{1}{2-a} \qquad \text{or} \qquad a+\frac{(1-a)^2}{2-a}$$

which is always greater than a. Or, as we might expect, the possibility of the conclusion being true, though the argument should be invalid, always adds something to the probability of its being true. Moreover, $1\div(2-a)$ is always greater than $\frac{1}{2}$: or any argument, however weak, adds something to the force of the previous probability. The same thing is true in every case. Suppose a new argument to be produced for the first side, of the force k. The effect upon the formula is to change $1-a$ into $(1-a)(1-k)$, and the odds in favour of the conclusion are increased in the proportion of 1 to $1-k$. But this is to be under-

ſtood ſtrictly in the ſenſe deſcribed in page 202, namely, we·are
to ſuppoſe that the newly produced argument *is* ſingle, that is,
does not by the circumſtances of its production cauſe itſelf to be
accompanied by an argument for the ſecond ſide, or againſt the
firſt. If this laſt ſhould happen, and the argument thus created
for the ſecond ſide have the force *l*, the odds are altered in the
proportion of $1-l$ to $1-k$.

From the above it appears that oppoſite arguments of the
force *a* and *b* are exactly equivalent to a teſtimony the odds for
the truth of which are as $1-b$ to $1-a$. Thus, ſuppoſe we have
for a concluſion witneſſes whoſe teſtimonies are worth $\frac{2}{3}, \frac{2}{3}, \frac{4}{7}$,
$\frac{9}{10}$; arguments for of the ſeveral forces, $\frac{1}{7}, \frac{11}{12}, \frac{5}{8}$; and arguments
againſt of the forces $\frac{2}{9}, \frac{2}{11}, \frac{6}{7}$. Writing numerators only, we put
down

> For,   2, 2, 4, 9 ; 7, 9, 1 :
> Againſt, 1, 1, 3, 1 ; 4, 1, 3.

> Hence it is, 2. 2. 4. 9. 7. 9. 1 to 1. 1. 3. 1. 4. 1. 3, or
>   252 to 1 for the concluſion.

An argument, we ſhould infer beforehand, is better than a
teſtimony of the ſame force ; for the failure of the argument is
nothing againſt the concluſion, but the failure of the teſtimony is
its overthrow. So ſays the formula alſo : the introduction of a
teſtimony of the value *k*, not before received, alters the exiſting
odds in the proportion of *k* to $1-k$: but the introduction of an
argument of the ſame force alters them in the greater proportion
of 1 to $1-k$. Thus, the introduction of the teſtimony of a
perſon who is as often wrong as right ($\frac{1}{2}$) alters the odds in the
proportion of 1 to 1, or does not alter them at all : but the introduction of an argument which is as likely as not to prove the
concluſion, alters them in the proportion of 1 to $1-\frac{1}{2}$, or of
2 to 1.

Are we not in the habit, unconſciouſly, of recognizing ſome
ſuch diſtinction ?   Do we not give much more weight to argument than to teſtimony ?   I ſuſpect the anſwer ſhould be in the
affirmative : that an argument·of 3 to 1 does convince us much
more than a teſtimony of 3 to 1.   I ſuſpect we ſhow it, not in
numerical appreciation, of courſe, but in liſtening to and allow-

ing weight to arguments, when we fhould refufe teftimony of the fame charaćter.

It may be. doubted, however, whether we have much fcope for experiment on the lower degrees either of teftimony or argument. Perhaps it is not often we meet a witnefs, whether as bearing teftimony of veracity to a faćt, or of judgment to a conclufion, whofe evidence is as low as $\frac{1}{4}$; and the fame perhaps of an argument.

I have fpoken, in the previous part of this chapter, of the rejećtion of authority, that is, of teftimony, authority being only high teftimony. Let us now examine by the formula and fee what it amounts to. Let a be the probability that the argument proves its conclufion: and let us therefore perfift in faying that a is the probability for the truth of the conclufion. In the formula, b being $=0$, let μ be made $a \div (1 + a)$, it will be found that the probability for the conclufion, μ divided by $\mu + (1 - \mu)(1 - a)$, comes out a, as required. Confequently, in the cafe of a fingle argument, the total rejećtion, as it would be thought, of all teftimony, is really equivalent to accompanying every argument by a teftimony lefs than $\frac{1}{2}$, depending upon its own force. It is to declare that, by the laws of thought, an argument of $\frac{7}{10}$ is of its own nature accompanied by a witnefs of $\frac{7}{17}$, one of $\frac{1}{4}$ by a witnefs of $\frac{1}{5}$, and fo on; this is clearly not what was meant. Nor, I fuppofe, can it be meant that we are arbitrarily to ftart with the teftimony $\frac{1}{2}$, and to reduce our own evidence, and that of all others, to the fame. If there be any fenfe in which the rejećtion of authority is defenfible, it muft be when we are required to proceed as if we were in perfećt ignorance what the value of the authority is. We cannot fuppofe it to be as likely to have one value as another. Suppofe, for inftance, that the arguments have unknown propofers: we cannot treat their authorities as if they were juft as likely to be excelfively high or low as to be very near to none at all. The more rational fuppofition is that the authority fhould be more likely to be fmall than great, as likely to be againft as for, and very unlikely to be excelfively great either for or againft. I cannot here enter into the mode in which fuch an hypothefis can be expreffed or ufed: but the refult of the fimpleft formula which fatisfies the above conditions, is as follows:— Let $r = (1 - b) \div (1 - a)$, b and a meaning as above; then the probability that the conclufion is true, which has *a* for. the validity of its argument, &c. is

$$r(r^3 - 6r^2 + 3r + 6r\log r + 2) \div (r-1)^4$$

where log*r* means the *Naperian* logarithm (99-43rds of the common logarithm will be near enough for the prefent purpofe) If, for inftance, *r*=2, which, on the fuppofition of no previous balance of teftimony, would give 2 to 1 for the conclufion, the formula juft written gives ·636, or 636 to 364, fomething lefs than 2 to 1.

In the cafe firft difcuffed in page 202, it may be thought that the weaknefs of a propofed argument, from one who fhould have brought a better, if there had been one, may be confidered as a *teftimony* againft the conclufion rather than an *argument*. Suppofe his argument, for inftance, to have only the probability $\frac{1}{10}$. He tells us then, that after he has done his beft, it is 9 to 1 againft the propofition being proved. If we are very confident that it could be proved, if true, and that he could do it, if any one, he comes before us as a teftimony of 9 to 1 againft the *truth* of the conclufion, or very nearly fo. If we take, then, all that his argument wants of demonftration, as fo much evidence from him againft the conclufion, this amounts to fuppofing that, *a* being the validity of his argument, *a* is alfo his teftimony for the conclufion (and 1—*a* that againft it). If there be only argument for, and none againft, and if our minds be previoufly unbiaffed, we reprefent this cafe by putting *a* for *μ* in the formula, and the odds for the conclufion are then as *a* to $(1-a)^2$. On this fuppofition, which I incline to think well worthy of attention, we fhould not confider an unoppofed argument from an acute reafoner as giving the conclufion to be as likely as not, unlefs $a = (1-a)^2$ or *a*= 382, a little more than $\frac{1}{3}$. Were it not for our peculiar introduction of teftimony, then, the conclufion being as likely as not to begin with, an argument which has any probability of proving it, would have made it more likely than not, as before feen.

But that the introduced teftimony fhould be exactly as above, is a mere fuppofition. If it were a mathematical propofition, for inftance, and Euler were to declare himfelf unable to give more than a probability of proof, I, for one, fhould confider him as giving a much higher rate of testimony against the truth of the assertion than is supposed in the preceding. But all this has reference to the question how to measure testimonies and validities in particular cases, which is quite a distinct thing from the investigation of the way to use them when measured.

In cases in which the number of arguments is multiplied, it generally happens that they stand or fall together, in parcels: namely, that the same failure which makes one invalid, necessarily makes others invalid. In this case, independent arguments must be selected, and the probabilities for them alone employed.

We see in this problem an illustration of the commonly observed result, that the same argument produces very different final conclusions in two different minds; and this when, so far as can be judged, both are disposed to give the same probabilities to the several premises of the argument. The initial odds, come how they may, or μ to $1-\mu$, should be altered by the arguments in the proportion of $1-b$ to $1-a$. Accordingly, b and a being the same to both parties, their belief in the conclusion may have any kind of difference, if μ be not the same thing to both.

Problem 6. Given an assertion, A, which has the probability a; what does that probability become, when it is made known that there is the probability m that B is a necessary consequence of A, B having the probability b? And what does the probability of B then become?

First, let A and B not be inconsistent. The cases are now as follows, with respect to A. Either A is true, and it is not true that both the connexion exists and B is false: or A is false. This is much too concise a statement for the beginner, except when it is supposed left to him to verify it by collecting all the cases. The odds for the truth of A, either as above or by the collection, are $a\{1-m(1-b)\}$ to $1-a$. As to B, either B is true, or B is false and it is not true that A and the connexion are both true. Accordingly, the odds for B are as b to $(1-b)(1-ma)$.

The reader must remember that when B necessarily follows from A, B must be true when A is true, but may be true when A is false; while A must be false when B is false. And now we see that a proposition is not necessarily unlikely, because it is very likely to lead to an incredibility, or even to an absolute impossibility. Let $b=0$, or let B be impossible: then the odds for A are as $a(1-m)$ to $1-a$. Say that it is 9 to 1 that the connection exists; then these odds are as a to $10(1-a)$. If a be greater than $\frac{10}{11}$, still A remains more likely than not, even when it is 9 to 1 that it leads to the absurdity B.

Secondly, let A and B be inconsistent, so that both cannot be true. Either then A is true, B false, and the connexion does not exist; or A is false. The odds for A are then as $a(1-b)$ $(1-m)$ to $1-a$. With respect to B, either B is true and A is false, or B is false, and A and the connexion are not both true. The odds for B are then as $b(1-a)$ to $(1-b)(1-ma)$.

Among the early sophisms with which the Greeks tried the power of logic, as a formal mode of detecting fallacies, was the construction of what we may call *suicidal* propositions, assertions the truth of which would be their own falsehood. If a man should say ' I lie,' he speaks neither truth nor falsehood; for if he say true, he lies, and if he lie, he speaks truth. Such a speech cannot be interpreted. Again, the Cretan, Epimenides, said that all the Cretans were incredible liars; is he to be believed or not? If we believe him, we must, he being a Cretan, disbelieve him. Some stated it thus;—' If we believe him, then the Cretans are liars, and we should not believe him; then there is no evidence against the Cretans, or we may believe him, so that the evidence against the Cretans revives, &c. &c. &c. Refer such a proposition to the theory of probabilities, and the difficulty immediately disappears. Whatever the credit of Epimenides as a witness may be, that is, whatever, upon his word, the odds may be for his proposition, the same odds are there against him from the proposition itself. These equal conflicting testimonies balance one another (problem 1) and leave the effect of other testimonies to the same point unaltered. The sophism of Epimenides, as stated, is but an extreme case of the second of the problems before us. The proposition B is inconsistent with A, and the connexion is certain $(m=1)$: the odds for B must then be as $b(1-a)$ to $(1-b)(1-a)$, or as b to $1-b$, exactly what they are independently of the previous assertion.

CHAPTER XI.

On Induction.

THE theory of what is now called *induction* muſt occupy a large ſpace in every work which profeſſes to treat of the matter of arguments ; but there is not much to ſay upon the genuine meaning of the word, in any ſyſtem of formal logic. And that little would be leſs, if it were not for the miſtaken oppoſition which it has long been cuſtomary to conſider as exiſting between the inductive proceſs and the reſt of our ſubject.

By induction (ἐπαγωγη) is meant the inference of a univerſal propoſition by the ſeparate inference of all the particulars of which it is compoſed : whether theſe particulars deſcend ſo low as ſingle inſtances or not. Thus if X be a name which includes P,Q,R, ſo that every thing which is X muſt be one of the three : then if it be ſhown ſeparately that every P is Y, and that every Q is Y, and that every R is Y ; it follows that every X is Y. And this laſt is ſaid to be proved by induction. Thus (Chapter VI).

$$X)P,Q,R + P)Y + Q)Y + R)Y = X)Y$$

is an inductive proceſs. In form, it may be reduced as ɯ page 123, to one ordinary ſyllogiſm.

Complete induction is demonſtration, and ſtrictly ſyllogiſtic in its character. In the preceding proceſs we have y)p, y)q, y)r, which give y)pqr : and X)P,Q,R is pqr)x ; whence y)x, or X)Y. It is a queſtion of names, that is, it depends upon the exiſtence or nonexiſtence of names, whether a complete induction· ſhall preſerve that form, or loſe it in the appearance of a *Barbara* ſyllogiſm, formed by help of the conjunctive poſtulate of Chapter VI.

But when the number of ſpecies or inſtances contained under a name X is above enumeration, and it is therefore practically impoſſible to collect and examine all the caſes, the final induction, that is, the ſtatement of a univerſal from its particulars, becomes impoſſible, except as a *probable* ſtatement : unleſs it ſhould happen that we can detect ſome law connecting the ſpecies or inſtances, by which the reſult, when obtained as to a certain number, may be inferred as to the reſt.

This laſt named kind of *induction by connexion*, is common enough in mathematics, but can hardly occur in any other kind of knowledge. In an innumerable ſeries of propoſitions, repreſented by P_1, P_2, P_3, P_4, &c, it may and does happen that means will exiſt of ſhowing that when any confecutive number, ſuppoſe three, of them are true, the next muſt be true. When this happens, a formal induction may be made, as ſoon as the three firſt are eſtabliſhed. For by the law of connexion, P_1, P_2, and P_3, eſtabliſh P_4; but P_2, P_3, and P_4, eſtabliſh P_5; and then P_3, P_4, and P_5, eſtabliſh P_6; and ſo on *ad infinitum*. It is to be obſerved that this is really *induction*: there is no way, in this proceſs, of compelling an opponent to admit the truth of P_{100} without forcing him, if he decline to admit it otherwiſe, through all the previous caſes.

As an eaſy inſtance, obſerve the proof that the *ſquare* of any number is equal to the ſum of as many confecutive odd numbers, beginning with unity, as there are units in that number: as ſeen in

$$6 \times 6 = 1 + 3 + 5 + 7 + 9 + 11$$

Take any number, n; and write n ns (repreſenting a unit by a dot) in rank and file. To enlarge this figure into $(n+1)(n+1)$s, we muſt place n more dots at each of two adjacent ſides, and one more at the corner. So that the ſquare of n is turned into the ſquare of $n+1$ by adding $2n+1$, which is the $(n+1)$th odd number. Thus 100×100 is turned into 101×101 by adding the 101ſt odd number, or 201. If then the theorem alleged be true of $n \times n$, it is therefore true of $(n+1) \times (n+1)$. But it is true of the firſt number, 1×1 being 1; therefore it is true of the ſecond, or $2 \times 2 = 1 + 3$; therefore it is true of the third, or $3 \times 3 = 1 + 3 + 5$; and ſo on.

But when we can neither examine every caſe, nor frame a method of connecting one caſe with another, no abſolutely demonſtrative induction can exiſt. That which is uſually called by the name is the declaration of a univerſal truth from the enumeration of ſome particulars, being the aſſumption that the unexamined particulars will agree with thoſe which have been examined, in every point in which thoſe which have been examined agree with one another. The reſult thus obtained is one of probability; and though a moral certainty, or an unimpeachably high degree of probability, can eafily be obtained, and actually is obtained, and though moft of our conclufions with refpect to the external world are really thus obtained, yet it is an error to put the refult of fuch an induction in the fame class with that of a demonftration. There is no objection whatever to any one faying that the former refults are *to his mind* more certain than thofe of the latter: the fact may be that they are fo. The difference between neceffary and contingent propofitions lies in the qualities from which they receive thofe adjectives, more than in difference of credibility. I know that a ftone *will* fall to the ground, when let go: and I know that a fquare number *muft* be equal to the fum of the odd numbers, as above: and though, when I ftop to think, I do become fenfible of more affurance for the fecond than for the firft, yet it is only on reflection that I can diftinguifh the certainty from that which is fo near to it.

The rule of probability of a *pure induction* is eafily given. Suppofing the fimple queftion to be whether X *is* or *is not* Y, there being no previous circumftances whatfoever to make us think that any one X is more likely than not to be Y, or lefs likely than not. Thefe are the circumftances of what I call a *pure* induction. To begin with, it is 1 to 1 that the firft X examined fhall be a Y: if this be done, and X_1 be a Y, then it is 2 to 1 that X_2 fhall be a Y; fhould it fo happen, then it is 3 to 1 that X_3 fhall be a Y. Generally, when the firft *m* Xs have all been examined, and all turn out to be Ys, it is $m + 1$ to 1 that the $(m + 1)$th X fhall be a Y.

The fimplicity of this rule muft not lead the ftudent to fuppofe he can find a fimple reafon for it. Let 10 Xs have been examined and found to be Ys: what do we affert when we fay it is 11 to 1 that the 11th X fhall be a Y? We affert that if an *infinite* number of urns were collected, each having white balls and black balls in *infinite number* but in a definite ratio, and fo that every poffible ratio of white balls to black ones occurs once; and if every poffible way of drawing eleven balls, the firft ten of which are white, were felected and put afide: then, of thofe put afide, there are eleven in which the eleventh ball is white, for one in which the eleventh ball is black. The reader will find fome difficulty in forming a diftinct conception of this, and of courfe will find it impoffible to have any axiomatic perception of the truth or falfehood of the refult.

It may be worth while to fhow that a fuppofition making fome degree of approach to the preceding circumftances will give fome approach to the refult. Firft, in lieu of an infinite number of balls in each box, which is fuppofed only that withdrawal of a definite number may not alter the ratio, let each ball drawn be put back again, which will anfwer the fame purpofe. Let there be only ten urns with ten balls in each, of which let the firft have one white, the fecond two white, &c. and the laft all white. The number of ways of drawing eleven white balls fucceffively out of any one urn is the eleventh power of the number of white balls in the urn: that of drawing ten white balls followed by one black one is the tenth power of the number of white balls multiplied by the number of black ones. If we were to put together all the firft, and then all the fecond, we fhould find about 21 times as many ways of arriving at the firft refult (ten white, followed by a white) as the fecond (ten white followed by a black). But if we now increafed the number of urns, and took a hundred, having one, two, &c. white balls, we fhould find inftead of 21, a number much nearer to 11; and fo on.

Accordingly, when without any previoufly formed bias, we find that m Xs, fucceffively examined, are each of them a Y, we ought then to believe it to be $m+1$ to 1 that the next, or $(m+1)$th X, will be a Y. And further, a being a fraction lefs than unity, we have a right to fay there is the probability $1-a^{m+1}$ that the Xs make up the fraction a or more, of the Ys. Or thus;—if the fraction a be, fay $\frac{6}{7}$, and if m be 10: then if the 10 firft Xs be all Ys, the probability that $\frac{6}{7}$ or more of the Xs are Ys is juft that of drawing one or more black balls in 11 drawings, from an urn in which $\frac{6}{7}$ of the balls are always white.

If, for example, the firft 100 Xs were all Ys, it would be found to be 1000 to 1 that $93\frac{4}{10}$ per cent, at leaft, of all the Xs are Ys.

If as before, the firft m Xs obferved have all been Ys, and we afk what probability thence, and thence only, arifes that the next n Xs examined fhall all be Ys, the anfwer is that the odds in favour of it are $m+1$ to n, and againft it n to $m+1$. No induction then, however extenfive, can by itfelf, afford much probability

to a univerfal conclufion, if the number of inftances to be examined be very great compared with thofe which have been examined. If 100 inftances have been examined, and 1000 remain, it
is 1000 to 101 againft all the thoufand being as the hundred.

This refult is at variance with all our notions; and yet it is
demonftrably as rational as any other refult of the theory. The
truth is, that our notions are not wholly formed on what I have
called the *pure induction*. In this it is fuppofed that we know no
reafon to judge, except the mere mode of occurrence of the induced inftances. Accordingly, the probabilities fhown by the
above rules are merely *minima*, which may be augmented by
other fources of knowledge. For inftance, the ftrong belief,
founded upon the moft extenfive previous induction, that phenomena are regulated by uniform laws, makes the firft inftance *of a
new cafe*, by itfelf, furnifh as ftrong a prefumption as many inftances would do, independently of fuch belief and reafon for it.

With this however I have nothing farther to do, except to
obferve that, in the language of many, induction is ufed in a fenfe
very different from its original and logical one. It is made to
mean, not the collection of a univerfal from particulars, but the
mode of arrival at a common caufe for varied, but fimilar, phenomena. A great part of what is thus called induction confifts
in difcovery of *differences*, not *refemblances*. Under this confufed
ufe of language, the ufual theory is introduced, namely, that
Ariftotle was oppofed to all induction, that Bacon was oppofed
to every thing elfe, that the whole world up to the time of Bacon
followed Ariftotle, that the former was the firft who fhowed the
way to oppofe the latter, that each had a logic of his own, &c.
&c. The whole of this account abounds with miftatements.
The admitted and fufficiently ftriking difference between the
philofophy of modern and ancient times, in all natural and material branches of inquiry, is not fo eafily explained as by choofing
two men, one to bear all the blame, the other all the credit: nor
are Copernicus, Gilbert, Tycho Brahé, Galileo, and the other
predeceffors of the *Novum Organum*, deftined to be always deprived of their proper rank.

What is now called induction, meaning the difcovery of laws
from inftances, and higher laws from lower ones, is beyond the
province of formal logic. Its inftruments are induction properly

so called, separation of apparently related, but really distinct particulars (the neglect of which was far more hurtful to the old philosophy than a neglect of induction proper would have been, even had it existed) mathematical deduction, ordinary logic, &c. &c. &c. It is the use of the whole box of tools : and it would be as absurd to attempt it here, as to append a chapter on carpentry to a description of the mode of cutting the teeth of a saw.

The processes of Aristotle and of Bacon are equally those which we are in the habit of performing every day of our lives. But some perform them well, and some ill. It is extraordinary that there should be such division of opinion on the question whether a careful analysis of them, and study of the parts into which they decompose, is of any use towards performing them well. On this point, and on the character of Bacon's office in philosophy, a living writer, to whom I should think it likely that many yet unborn would owe their first notions of Bacon's writings, expresses himself in a manner which I quote, and comment on at length, as the best exposition I can find, of a class of opinions which is very prevalent, and, I fully believe, to the prejudice of sober thought and accurate knowledge.

The vulgar notion about Bacon we take to be this, that he invented a new method of arriving at truth, which method is called Induction, and that he detected some fallacy in the syllogistic reasoning which had been in vogue before his time. This notion is about as well founded as that of the people who, in the middle ages, imagined that Virgil was a great conjuror. Many who are far too well informed to talk such extravagant nonsense, entertain what we think incorrect notions as to what Bacon really effected in this matter.

The inductive method has been practised ever since the beginning of the world, by every human being. It is constantly practised by the most ignorant clown, by the most thoughtless schoolboy, by the very child at the breast. That method leads the clown to the conclusion that if he sows barley, he shall not reap wheat. By that method a schoolboy learns that a cloudy day is the best for catching trout. The very infant, we imagine, is led by induction to expect milk from his mother or nurse, and none from his father.

Not only is it not true that Bacon invented the inductive method ; but it is not true that he was the first person who correctly analysed that method and explained its uses. Aristotle had long before pointed out the absurdity of supposing that syllogistic reasoning could ever conduct men to the discovery of any new principle, had shown that such discoveries must be made by induction, and by induction alone, and had given the hiftory of the inductive procefs, concifely indeed, but with great perfpicuity and precifion.

Again, we are not inclined to afcribe much practical value to that analy- fis of the inductive method which Bacon has given in the fecond book of the *Novum Organum*. It is indeed an elaborate and correct analyfis. But it is an analyfis of that which we are all doing from morning to night, and which we continue to do even in our dreams. A plain man finds his fto- mach out of order. He never heard Lord Bacon's name. But he proceeds in the ftricteft conformity with the rules laid down in the fecond book of the *Novum Organum*, and fatisfies himfelf that minced pies have done the mifchief. " I eat minced pies on Monday and Wednefday, and I was kept awake by indigeftion all night." This is the *comparentia ad intellectum in- ftantiarum convenientium*. " I did not eat any on Tuefday and Friday, and I was quite well." This is the *comparentia inftantiarum in proximo quæ natura data privantur*. "I ate very fparingly of them on Sunday, and was very flightly indifpofed in the evening. But on Chriftmas-day I almoft dined on them, and was fo ill that I was in great danger." This is the *comparentia inftantiarum fecundum magis et minus*. " It cannot have been the brandy which I took with them; for I have drunk brandy daily for years without being the worfe for it." This is the *rejectio naturarum*. Our invalid then proceeds to what is termed by Bacon the *Vindemiatio*, and pro- nounces that minced pies do not agree with him.

We repeat that we difpute neither the ingenuity nor the accuracy of the theory contained in the fecond book of the *Novum Organum*; but we think that Bacon greatly overrated its utility. We conceive that the inductive procefs, like many other proceffes, is not likely to be better performed merely becaufe men know how they perform it. William Tell would not have been one whit more likely to cleave the apple if he had known that his arrow would defcribe a parabola under the influence of the attraction of the earth. Captain Barclay would not have been more likely to walk a thoufand miles in a thoufand hours, if he had known the place and name of every mufcle in his legs. Monfieur Jourdain probably did not pronounce D and F more correctly after he had been apprifed that D is pronounced by touching the teeth with the end of the tongue, and F by putting the upper teeth on the lower lip. We cannot perceive that the ftudy of gram- mar makes the fmalleft difference in the fpeech of people who have always lived in good fociety. Not one Londoner in ten thoufand can lay down the proper rules for the ufe of *will* and *fhall*. Yet not one Londoner in a million ever mifplaces his *will* and *fhall*. Dr. Robertfon could, undoubtedly, have written a luminous differtation on the ufe of thefe words. Yet, even in his lateft work, he fometimes mifplaced them ludicroufly. No man ufes figures of fpeech with more propriety becaufe he knows that one figure of fpeech is called a metonymy, and another a fynecdoche. A drayman in a paffion calls out ' You are a pretty fellow,' without fufpecting that he is uttering irony, and that irony is one of the four primary tropes. The old fyftems of rhetoric were never regarded by the moft experienced and dif- cerning judges as of any ufe for the purpofe of forming an orator. " Ego hanc vim intelligo" faid Cicero "effe in præceptis omnibus, non ut ea fecuti oratores eloquentiæ laudem fint adepti, fed quæ fuâ fponte homines eloquentes facerent, ea quofdam obfervaffe, atque id egiffe; fic effe non eloquentiam ex artificio, fed artificium ex eloquentia natum." We muft own that we entertain the fame opinion concerning the ftudy of Logic, which Cicero entertained concerning the ftudy of Rhetoric. A man of fenfe fyllogizes in *celarent* and *cefare* all day long without fufpecting it: and though he may not know what an *ignoratio elenchi* is, has no difficulty in expofing it whenever he falls in with it.—('*Lord Bacon*,' in *Critical and Hiftorical Effays contributed to the Edinburgh Review.* By Thomas Babington Macaulay.)

This brilliant paffage has, I have no doubt, appeared to many completely decifive of the queftion which it affirms: and, as fo often happens in like cafes, there is a certain exaggeration againft which it is of truth. It is good againft thofe who confound analyfis and recombination of exifting materials with introduction of them: and who might profefs to fee in agriculture fomething which would have benefited mankind, though plants and animals had not been natural products of the foil. But I now proceed to examine it, againft thofe who affirm that Ariftotle and Bacon are of *no* ufe, and who very frequently fall into the common logical fallacy of fuppofing that their cafe is proved, as foon as it is made out that they are not of *all the* ufe: which Mr. Macaulay himfelf has done, except as againft the exaggerators aforefaid.

We reafon inductively from morning till night, and even in our dreams. True: and how badly we often do it, particularly in fleep. A plain man is then produced, to reafon on Bacon's principles: and Mr. Macaulay has imitated a plain man better than he intended, by making him do it wrongly. Look over the induction, and it will appear that the cafe is not made out; an exclufion is wanting: it may have been the *mixture* of minced pies and brandy which did the mifchief. The plain man fhould have tried minced pies without brandy; but he had drunk the latter daily for years, and it never ftruck him. This is precifely one of the points in which we are moft apt to deceive ourfelves, and for which we moft need to have recourfe to the completenefs of a fyftem of rules; fomething is left taken for granted. The things of courfe, our daily habits, are neglected in the confideration of anything of a lefs ufual character: the plain man left off the minced pies upon trial; but not the brandy: Chriftmas mifchief muft be. referred, he thinks, entirely to Chriftmas fare, if at all.

But even if this omiffion had been fupplied, and the refult found to confirm the conclufion, yet the plain man has ftopped where the plain man frequently does ftop, at what Bacon calls the *Vindemiatio prima*, the rudiments of interpretation. Completenefs is feldom anything but ftudy and fyftem. Philofophy ought to bring him to the refult that daily brandy has made that fpirit ceafe to give the ftimulus which, were its ufe only occafional, would enable his ftomach to bear an unufually rich diet for a fhort time. Our plain friend is precifely in the pofition of a bankrupt who curfes the times, on reafoning ftrictly Baconian as far as it goes, and forgets that a cafual tightnefs in the money market would never have upfet him, if it had not been for the previous years of extravagant living and rafh fpeculation.

But there are many proceffes which are not better performed becaufe men know " how they perform them." Mr. Macaulay here means " becaufe men know the laws of that part of the procefs which nature does for them." That men fhould not know better how to perform for knowing how *they* perform is almoft a contradiction in terms. William Tell knew how to fhoot all the better for knowing which end of the arrow *he* was accuftomed to fit to the ftring: had he wanted this knowledge, his chance of cleaving the apple would have been much diminifhed. But he would not have been improved by knowing that his arrow defcribed a parabola. True, becaufe it did not do fo. The *centre of gravity* of the arrow would defcribe a parabola, if it were not for the refiftance of the air ; or fomething fo near it as to be undiftinguifhable. But, taking the defcription as roughly correct, William Tell did know, inductively, that the arrow defcribes a curve, concave to the earth : and had made thoufands of experiments in connexion of the *two ends* of that curve, which were all that he was concerned with. It is no argument againft the ftudy, as a ftudy, of *induction*, that the amount of ufeful refult which it had recorded in the mind of William Tell in the fhape of habit, would not have been augmented by *deductive* knowledge of an intermediate *ftatus* with which *he* had nothing to do. But let knowledge advance, under both modes of progrefs, and Tell becomes an artillery officer, the rude arrow a truly fhaped and balanced ball, means of meafurement are applied, the true curve is more correctly reprefented than by the parabola, and thirty pounds of iron are thrown to four times the diftance which an arrow ever reached, and with a certainty almoft equal to that of the legend.

But if Captain Barclay had known the places and names of the mufcles, he would not have been more likely to walk a thoufand miles in a thoufand hours. The inftance is far fetched: becaufe the feat confifted in the exhibition of power of endurance acquired by practice. If my denial feem' as far fetched, it is the fault of the propofer. Captain Barclay muft, by habit, by induction, have acquired facility in varying his pace and gefture fo as to eafe the mufcles. Had he been well acquainted with the *difpofition and ufes* of thefe organs to begin with (towards which knowledge of their *places and names* would have contributed) he would have learnt this art more eafily. Though not altogether *ad elenchum*, yet I may fay that in this cafe the effect of fuch knowledge would have been that he would have been *lefs* likely to have performed the feat. Had he directed his attention to fome fcience of obfervation, he would not have needed to have fought fame, or exhauftion of remarkable energy, in fuch a trifling purfuit. And further, in a very common cafe, mechanics has taught what few ever learn by induction, though they have conftant opportunities of doing it: namely, that in walking, the ordinary practice of fwinging the arms is injurious and tiring; that a very trifling amount of it tells ferioufly in a long journey. Here is one ufeful refult, which natural induction does not commonly teach, and there may be many more of the fame kind: the queftion between it and regular ftudy requires the confideration, not only of what is done, and whether it might be done better, but of what is not done.

Next, M. Jourdain did not pronounce D and F more correctly after his attention had been called to the details of the act of pronunciation. None but Moliere ever knew whether he did or not: but all who have watched the progrefs of inftruction know that the bad habits or natural imperfections of children are removed or alleviated by making them practice mechanical pronunciation, with perceptive adoption of rules. In every one of a few detached inftances in which I have feen children at their reading leſſons in France, I have noticed that a return upon the habits of pronunciation is always a part of the exerciſe: and that the letters are pronounced with that diſtinct effort which makes the pupil ſenſible of the action required. I have always attributed to this practice the more uniform ſtandard of pronunciation which prevails among the educated French, as compared with ourſelves.

But the ſtudy of grammar makes no difference in the ſpeech of people who have always lived in good ſociety. If Mr. Macaulay mean merely as to the uſe of *ſhall* and *will*, and the like, it may certainly be ſaid that the perpetual uſe of ſpeech (which is not reaſoning) does enable every one to form the habits of thoſe about him. . But that grammar, as a whole, produces no effect upon the ſpeech of good ſociety, is one ſide of a balanced matter of opinion. Many contend that it has produced, in our generation and the one above it, a very unfortunate effect: they aver that the purity and character of our Engliſh has been deteriorated by Lindley Murray and his ſchool, and that we much want better grammar teaching. On the ſubject of *ſhall* and *will*, it is remarkable that Mr. Macaulay, whom a vigorous faculty of illuſtration, combined with immenſe reading, enables to ſtrew his path with inſtances, has to invent his caſe, and to refer to a treatiſe which Robertſon could have written. But it is ·not enough: if we grant that ſuch a treatiſe would have been *luminous*, we may be ſafe; but would it have been *correct?* And further, knowledge muſt abdicate at once, if we pronounce uſeleſs all that has been clearly explained by thoſe who have not rightly practiſed. Bacon himſelf might have taken *exſors ipſa ſecandi* for his motto.

Next, it is ſaid that no man uſes figures of ſpeech more correctly becauſe he knows that one is *metonymy* and another *ſynecdoche*. True; and in like manner no man conſults his books more eaſily becauſe he has a bookcaſe. But, having the bookcaſe, he arranges his books in it, and then he knows where to find them. Mr. Macaulay dwells throughout upon nomenclature. I might inſiſt upon its ſuperſtructure: but even mere naming is uſeful, when the meaning of the name is clearly underſtood. A mind well ſtocked with underſtood names cannot keep itſelf from being conſtantly in the act of claſſification, which contains induction. The mere involuntary reference of inftance number two to inftance number one, which is made when we remember that the fecond muft have the fame name as the.firft, is comparifon and induction, leads to reflection, cultivates tafte, and gives power. The drayman, who calls out·in a paffion, " You are a pretty fellow !" without knowing that he is uttering irony, is an incomplete picture : there is omitted a wifh relative to the eyes of his opponent, and an adjective which is (in fuch quarrels) fometimes prophetically, but feldom defcriptively, true. The value of the difference between this favage irony and the more elegant form of it which is fo pleafing in the defcription of the plain man's induction quoted above, is not within the comprehenfion of the drayman : the foundation of a better mode of expreffion than undifciplined rhetoric furnifhes, fo far as its adoption is matter of tafte, was laid by thofe who placed irony among the primary tropes. Good tafte is a refult of comparifons, which could not have been made without nomenclature.

Did Cicero declare that fyftems of rhetoric are not of any ufe ? The very quotation appears to mean that thefe fyftems, *præcepta*, have their power; that men get them by obfervation, and put them into practice. The *ea fecuti oratores* refers to what was done in the firft inftance, by the firft eloquent men, *fuâ fponte*. Moft truly does he fay that the art of rhetoric is derived from eloquence, and not *vice verfâ :* moft falfely, as far as can be judged, does he feem to infinuate that it was all done at one ftep ; firft, fome one or more confummate orators, fecondly, a finifhed fyftem, drawn from obfervation of their methods. Perhaps he intended a particular reference to a certain orator then namelefs : the fentence, thus conftrued, contains nothing but matter which Tully is likely enough to have whifpered to Cicero.

A fyftem is a tool, and it muft be employed upon materials which different men furnifh from their different means. But the coat muft be cut according to the cloth, both in fize and quality : no reproach to the fciffors, nor prejudice to their fuperiority over the fharpened wood of the favage, even though practice will enable him to ufe the latter better than any civilized man who is not a tailor can ufe the former. The formation of tools, mental or material, is a cyclical procefs. The firft iron was obtained by help of wood ; one of the firſt uſes of it was to make better tools, to get more iron, with which better tools ſtill were made, and ſo on. And in this way we may trace back any art to natural tools, and to materials which are to be had for the gathering. The aſſertion made by Mr. Macaulay, and many others, that in logic only, of all the abſtract ſciences, our natural means are as good as thoſe which reſult from diligent analyſis, is one which terminates in an iſſue of fact. The inſtances given are contained in the aſſertion that a *man of ſenſe* ſyllogizes in *ceſare* and *celarent* all day long without *ſuſpecting it*, and though he does not know what an *ignoratio elenchi* is, can always detect it when he meets with it.

Mr. Macaulay begins with an indefinite term, a man *of ſenſe :* and the clauſe is deficient in logical perſpicuity. Firſt, what is a man *of ſenſe?* I grant that I ſhould doubt the ſenſe of a man who could not make the inferences deſcribed by *ceſare* and *celarent*. But do men become men of ſenſe by nature, without education ? if yes, I deny the aſſertion that men of ſenſe reaſon (correctly) in *ceſare*, &c. The man of ſenſe who is not educated is as likely to aſſert that *ceſaro* is all that can be obtained, or to invent the form *feſape*, as the plain man to forget to try the mince pies without brandy before he concludes. If no, then the aſſertion is itſelf *ignoratio elenchi :* for the very queſtion is how to make men of ſenſe ; can they not be, *ceteris paribus*, formed better and faſter with ſtudy of logic than without : it being agreed on all hands that *this* man of ſenſe is always a practical logician.

Next, a man of ſenſe reaſons, &c. without ſuſpecting it. Suſpecting what ? that he is reaſoning, or that he is reaſoning in *ceſare ?* I ſuppoſe the latter : that is to ſay, I take it to be meant that a man of ſenſe *may* (not *muſt*, for ſome Ariſtotelians are men of ſenſe) not know that the logicians call the form of reaſoning he uſes *ceſare*. This is eaſily granted : but what is it but the celebrated *ignoratio elenchi* of Locke, who fancied that he raiſed an objection againſt the pretenſions of the logicians, when he declared he never could believe that God had made men only two-legged, and left it to Ariſtotle to make them rational. No one ever denied that men reaſoned before Ariſtotle, and would have reaſoned ſtill if he had never lived.

Mr. Macaulay, probably without ſo much as a new application

to the inkſtand, after falling into the *ignoratio elenchi*, ſingles out this very fallacy as the one which a man of ſenſe is ſure to detect. But if there be a fallacy which is the ſtaple of paralogiſm, it is this one. *Delectat domi*, for ordinary diſcuſſion (eſpecially after dinner) is little elſe; *impedit foris*, for three fourths of public debate, from the Houſes of Parliament downwards, is made up of it. A man who expoſes it in converſation is conſidered a tireſome, and if he do it often, an uncourteous perſon: he " has no converſation," he " harps upon one ſubject," he " won't let you ſpeak."

I have made the above comments upon a very marked paſſage of an eminent writer, in preference to introducing their ſubſtance as a diſſertation of my own, that I might have the advantage of the reader ſeeing that I meet real arguments, inſtead of my own verſion or ſelection. It would probably be difficult to find a better concentration of the ſubſtance of the antagoniſt views, with reſpect to the formal ſtudy of reaſoning, than is contained in my quotation from Mr. Macaulay: and I may ſafely take his adoption of them as proof that theſe views yet require the notice of a writer on logic.

There is one reſult of the theory of probabilities, cloſely connected with induction proper, which it will be adviſable to notice here.

When the ſyllogiſm is declared illegitimate, on account of both premiſes being particular, a probable concluſion of great ſtrength may be admitted in many caſes. This muſt be the more inſiſted on, becauſe it is too common to attend to nothing but the demonſtrative ſyllogiſm, leaving all of which the concluſions are only probable, however probable, entirely out of view.

I take as the inſtance the ſyllogiſm, or imperfect ſyllogiſm, ' Some Xs are Ys, ſome Zs are Ys, therefore there is ſome probability that ſome Xs are Zs.' If the number of Xs and Zs togetner exceed the number of Ys (as in Chapter VIII) there is a certainty that ſome Xs are Zs. Let us then ſuppoſe this is not the caſe.

Let the whole number of Ys in exiſtence be n, and let m and *n* be the numbers of Xs and Zs which are among them. I shall confider two diftinct cafes:—Firft, when the diftribution of the Xs and Zs among the Ys is utterly unknown ; fecondly, when their diftribution is that of *contiguity*, that is, when the Ys being for fome reafon arranged in a particular order, the Xs which are Ys are fucceffive Ys, and the fame of the Zs which are Ys.

For the firft cafe a very rough notion will do, confined to the fuppofition that few Xs and Zs are mentioned, compared with the whole number of Ys. When the Xs and Zs together make a large proportion of the Ys in number, then, if we have no reafon for making them contiguous, or otherwife limiting the equally probable arrangements, it may be faid to be a moral certainty that fome Xs are Zs.

In the firft cafe, if we divide 43 times the product of *m* and *n* by 100 times *n*, it gives us a fufficient notion (not large enough) of the common logarithm of *k*, the odds in favour of fome Xs being Zs being *k* to 1. Say there are 1000 Ys, and that 100 Xs are Ys and 100 Zs are Ys. Then $43 \times 100 \times 100$ divided by 100×1000 is 4·3, which is the logarithm of 20,000. It is then more than 20,000 to 1 that, in this cafe, one or more Xs are Zs. A more exact rule is as follows. To $43mn$ divided by 100*n* add its hundredth part, and to the refult add fuch a fraction of itfelf as $m+n$ is of 2*n*. Thus $43mn \div 100n$ being 4·3, which, with its hundredth part is 4·343, and $m+n$ (200) being the tenth part of 2*n* (or 2000), we add to 4·343 its tenth part, giving 4·777, which is about the logarithm of 60,000, ftill under the mark. It is more than 60,000 to 1 that fome Xs are Zs. When the fractions are very fmall, this rule is accurate enough, if *n* be confiderable. Its refult is, that if *n* be very confiderable, and if a perceptible fraction of the Ys be Xs, and a perceptible fraction Zs, *and if we really have no reafon to make the limitation of contiguity or the like*, then we are juftified in treating it as a moral certainty that fome Xs are Zs. But I fufpect the relation of contiguity, to which I now proceed, better reprefents the actual ftate of the cafe in ordinary argument.

When the Xs which are Ys are contiguous, and alfo the Zs which are Ys, the probability that no Xs are Zs is the fraction having the product of $n-m-n+1$ and $n-m-n+2$ for numerator, and the product of $n-m+1$ and $n-n+1$ for denominator. Thus in the example above proposed, 1000 Ys containing among them 100 Xs and 100 Zs (each set contiguous) we have 801×802 for numerator and 901×901 for denominator. This fraction is about 8-tenths; so that it is now 8 to 2, or 4 to 1, *against* any Xs being Zs.

In order to find the probability against the number of Xs which are Zs exceeding k, add k to both the multipliers in the numerator, which then become $n-m-n+k+1$ and $n-m-n+k+2$. For example, there are 100 Ys, containing 30 Xs and 60 Zs (each set contiguously): what is the chance against the number of Xs which are Zs exceeding 10? The numerator is 21×22: the denominator is 71×41. This fraction is 462 by 2911; whence it is 462 to 2449 *against*, or 2449 to 462 (more than 5 to 1) *for*, the number of Xs which are Zs exceeding 10.

The chances, it is to be remembered, are all *minima :* except when we mean that *m* Xs, *and not more,* are Ys, &c. These questions may serve to give some notion of the manner in which arguments not logically conclusive, may be morally so.

What is called *circumstantial* evidence is a species of induction by probability. The thing required to be found has the marks P,Q,R,S, &c.: this Y has the marks P,Q,R,S, &c.: there is then a certain amount of circumstantial evidence that this Y is the thing we want to find. If it can be shown that there is but one thing which has all these marks, then the circumstantial evidence is demonstrative. But if there were, say 100 Ys, of which 5 have the mark P, 5 the mark Q, &c., then having ascertained one Y which has all the marks, the question is, what chance is there against another Y having them all: the same chance, at least, is there that the Y found is the one sought. Instead however, of attempting the problem in this way, which is never resorted to for want of data (I mean that the resemblance which the rough processes of our minds bear to those of the theory of probabilities does not here exist) I take it as follows. If the possession of the mark P give a certain probability to the Y found being that sought, it is as a witness whose testimony has a certain credibility. Similarly for Q,R,S, &c. Compound these testimonies, when known, by the rule in page 195, and the result is the value of the circumstantial evidence.

CHAPTER XII.

On old Logical Terms.

IN this chapter I propofe to fay fomething on a few terms of the old Logic, which though they keep their places in works on the fubject, and have fome of them paffed into common language, are very little ufed. They relate generally to the fimple notion, and the name by which it is expreffed: and have little of fpecial reference, either to the propofition or fyllogifm. They are moftly derived from Ariftotle, whofe incidental expreffions became or give rife to technical terms, and whofe fingle fentences were amplified into chapters. And here, as in other places, I have nothing to do with the degree of correctnefs with which Ariftotle's meaning was apprehended, nor even with how much was drawn from Ariftotle and how much added to him, but only with the actual phrafes and their ufual meaning.

The words *logic* and *dialectics* * are now ufually taken as meaning the fame thing: the old diftinction is that dialectics is the part of logic in which common and probable, but not neceffary, principles, are ufed. But the diftinction is neither clearly laid down, nor faithfully adhered to, even by Ariftotle himfelf.

The *term* (in this work always called *name*) was divided into *fimple* and *complex*: the fimple term was the mere name, the complex term was what all moderns call the *affirmative propofition*. Thus *man* and *run* were fimple terms : *man runs*, a complex term. Later writers rejected this confufion : and divided the acts of the mind confidered in logic into *apprehenfion*, *judgment*, and *difcourfe*, taking cognizance of notions, propofitions, and arguments. The common meaning of the word *difcourfe*,

---

* Our language is capricious with regard to the ufe of fingular and plural of words in *ic :* thus we have logic and dialectic*s*, arithmetic and mathematic*s*, phyfic and phyfic*s* for medicine and natural philofophy. Some modern writers are beginning to adhere uniformly to the fingular, in which I cannot follow them, for I am afraid an Englifh ear would not bear with *mathematic* as a fubftantive. Would it not better confift with the genius of our language if the plurals were to be always ufed, and the fingulars made adjectives without the termination *al* ?

(which now generally applies to fomething fpoken) is derived from its place in this divifion. The word *argument*, which is now equivalent to *reafoning againft oppofition expreffed or implied*, was originally nothing but the middle term of a fyllogifm.

The fimple term was *univerfal* or *fingular*: univerfal, when of more inftances than one, as man, horfe, ftar; fingular, when of one inftance only, as the fun, the firft man, the pole-ftar, this book. Singular names were called *individuals*, from the etymology of the word, as belonging to objects not divifible into inftances to each of which the name could be applied. I have not dwelt upon the diftinction between fingular and univerfal, becaufe it is ineffective in inference. And moreover, a fingular propofition is only objectively fingular, but ideally plural. 'Julius Cæfar was a Roman': in point of fact, there was but one Cæfar. But take any imaginary repetition of the circumftances of Cæfar's life; fuch, for inftance as occurs to thofe who have thought of the poffibility of the fame courfe of events returning into exiftence after a certain cycle: and then the term Cæfar becomes plural. Or, even without fo forced a fuppofition, we may fay that, if we defcribe Cæfar, we muft defcribe a Roman: that our definition of Cæfar is fo clofe as to fit only one man that ever lived, makes no effential difference in the character of the propofition.

But a further diftinction which was made divided fingular terms into fubjects of univerfal, and fubjects of particular, propofitions. A determinate (or definite) individual, as Cæfar, this man, was the former: a vague (or indefinite) individual, as a certain man, the firft comer, was the latter. The diftinction is that of 'fome man' and 'this one man.'

Certain notions of effence or relation, accompanying the apprehenfion of a name, were called *categories*, or *predicaments*, meaning 'modes of affertion with refpect to' the object named. Ariftotle gave ten categories, and might have given ten hundred. In their ufual Latin form they were *fubftantia, quantitas, qualitas, relatio, actio, paffio, ubi, quando, fitus, habitus*.

The word tranflated by fubftance, ὐσία, means mode of being: and its literal Latin is *effentia*, effence. It is called *fubftance* (that which ftands under) as fupporting *accidents*, prefently explained. It is far too metaphyfical a term to come into common life with-

out fome degradation : and accordingly it there means that of
which a thing is compofed, whether material or not. Accordingly
we have the material fubftance of a coat, the intellectual fubftance of an argument. But, as we ufe the word, its meaning
belongs to the other predicaments. In fact, the fubftance of the
old logicians ftands, as to exiftence, in the fame fituation as *matter* (page 30) with refpect to our fenfible perceptions, or *object*
with refpect to our ideas. The fubftance, it was faid, is *per fe
fubfiftens*, while the accident could not be faid *effe*, but *ineffe*.
The diftinction between the fubftance (mode of being) and the
material fubftance (in the modern fenfe) may be helped by the
diftinction between *fubftantia prima* and *fubftantia fecunda*, the
firft referring to the individual, the fecond to the general term.
Thus the fubftance of John, as John, was *fubftantia prima;* as
man, *fubftantia fecunda.* All thefe very metaphyfical notions
were the ftudent's firft introduction to logic, and were confidered
as of the utmoft importance.

The predicament of *quantity*, derived from the notion of whole
and part, was conceived as either *continuous* or *difcrete*. In continuous quantity, the unit was divifible, in difcrete, indivifible.
Thus ten feet is continuous, ten men difcrete. The diftinction
is precifely that of magnitudinal and numerical.

Quality was fubdivided into 1. *Habit* and *difpofition*, the latter
term being ufed for the imperfect ftate of the former 2. *Power*
and want of it 3. *Patibilis qualitas* and *paffio*, applied to the
ideas of that which is undergone, the firft permanently, the fecond
for a time. 4. *Form* and *figure.*

Relation then, as now, referred to the fuggeftions derived from
comparifon of two things or ideas. It was divided into verbal
and real (*fecundum dici* and *fecundum effe*). Thus the relation of
profit to *profitable* was verbal: that of *father* to *fon*, or of *above*
to *below*, real. The two things related, or *correlatives*, were called
fubject and *term :* fo that of two correlatives, giving two oppofite
relations, the fubject of either was the term of the other. The
fundamentum of the relation was that in which it took its rife,
when it had a beginning.

Action and *paffion*, the production and reception of an effect,
requiring the producing *agent*, and the receiving *patient*, were
divided into *immanent*, or enduring in the agent, and *tranfient*,

or paffing out to another. Actions were *univocal*; or *æquivocal*, according as their effects were of the fame or different fpecies. A few years before the publication of Newton's Principia, it was taught in a work imported into Cambridge that when mice bred mice, the action was univocal, but when *the sun* bred mice (the writer muft have been thinking of Ariftotle and fome of the fchoolmen) æquivocal. There was alfo the *terminus à quo* and the *terminus ad quem* to reprefent the ftate before and the ftate after the action. Thus, when all this nonfenfe was fent to Coventry, the *terminus à quo* was an immenfe quantity of univocally bred learning of the preceding kind; the *terminus ad quem* was the rooting up of the wheat of logic with the tares.

The *where* (as to abfolute pofition), the *when*, and the *fite* (relative pofition) gave no peculiar terms of fubdivifion. The *habitus* (ἔχειν) referring to *poffeffion* generally in the firft inftance, was materialized by fome of the old logicians till it related to *drefs* only, or *habit* in the thence acquired meaning.

The word predicament (and category as well) has been intro-cuced into common language to fignify a fet of circumftances under which any thing takes place. It is then no longer con-fined to the above predicaments, nor is there any occafion that it fhould be.

The *predicables* (κατηγορώμενα) are diftinguifhed from *predica-ments* (κατηγορίαι) in that the former belong to any fimple notion or name, and may be predicated of it: the latter belong to the connexion (when affirmative) between two names. They are faid to be five in number, *genus*, *fpecies*, *differentia*, *proprium*, and *accidens*.

The words *genus* and *fpecies* have preferved their old meaning. If there be a number of names of which each is fubidentical of the one which follows, fay V, W, X, Y, Z : then of any two, fay W and X, X is a *genus* containing the *fpecies* W. Here Z is the *fummum genus*, and V the *infima fpecies*: X is the *genus proximum* of W, Y the *genus remotum*. In what I have called a *univerfe*, which is a *fummum genus*, having for its *infima fpecies* the individual inftance of any name in it, the fuperidentical is the genus, the fubidentical the fpecies. Subcontraries (and contraries) are *oppofite fpecies*; fupercontraries and complex particulars have no ancient name.

The *differentia* is that by which one clafs (be it fpecies or genus, the difference being accordingly termed *fpecific* or *generic*) is diftinguifhed from another. Thus the difference (or one difference) feparating the fpecies *man* from the other fpecies of the genus *animal*, is the epithet *rational*.

The *proprium* (or property) is that which belongs to the fpecies *only*, whether it be to all or only to fome : thus to ftudy, and to fpeak, are equally *propria* of man. But the old commentators give definitions of the property as follows. There are four kinds. 1. That which belongs to the fpecies alone, but not to all. 2. To all the fpecies, but not to that alone. 3. To the fpecies only, and to all of it, but not at all times. 4. To the fpecies alone, to all, and always.

The *accidens* (or accident) is that which may fometimes belong to the individual of a fpecies, but not neceffarily, nor to that fpecies alone. In modern language, the term is limited to what is unufual and unexpe&ted.

The word *caufe* was ufed by the ancients in a wider fenfe than by us : more nearly in the fenfe of the Latin *caufa*, or the Italian *cofa*. Caufes were diftinguifhed into *material*, *formal*, *efficient*, and *final*. The *material* caufe was the very matter of a thing, confidered as a kind of giver of exiftence ; the *formal* caufe was its form, in the fame light ; the *efficient* caufe (our common Englifh word) the agent or precedent ; and the *final* caufe, the ultimate end or obje&t, confidered as a reafon for the exiftence of the thing. Sometimes writers ftill talk of final caufes, and are as unintelligible to moft readers as if they had talked of final beginnings.

The word form was ufed in a wider fenfe than that of figure or fhape, to mean, as it were, law of exiftence, mode, difpofition, arrangement. Mere figure or fhape was only one of the *accidental forms*, as diftinguifhed from *fubftantial forms*, belonging to the fubftance. And *motion* was as widely ufed as *form* : it meant any alteration. Thus, *corruption* was one of the *motions* of matter. Change from place to place, to which the modern word is confined, was *local* motion.

The original ufe of the terms *fubje&t* and *obje&t* is to denote a thing confidered as that which may have fomething inherent in it, or attached to it, or fpoken of it, &c. ; and as that which may

be *objected* to the mind or reason, or made to come in its way.
Thus it was said that matter is the *subject* of those properties
which are the *objects* of the mind in natural philosophy. The
transition to the modern sense of *object*, namely, end proposed, is
natural enough.   In modern times, *subject* and *object* are used*
with respect to *knowledge:* the subject being the mind in which
it is, the object being the external source from which it comes.
For *subjective* and *objective* I have in this work used *ideal* and *objective* (page 29).   *Adjunct* was the technical term for that
which is in the subject.

A *modal* proposition was one in which the affirmation or negation was expressed as more or less probable : including all that is
technically under probability (Chapter IX) from necessity to
impossibility.   The theory of probabilities I take to be the unknown God which the schoolmen ignorantly worshipped when
they so dealt with this species of enunciation, that it was said to be
beyond human determination whether they most tortured the
modals, or the modals them.   Their gradations were *necessary*,
contingent, *possible*, *impossible;* contingent meaning more likely
than not, possible less likely than not.   These they connected
with the four modes of enunciation, A, I, O, E, and when by
some is meant *more than half*, the connexion is good.   The controversy about *modal* forms continues up to this day among
logicians who are not mathematicians : I should suppose that the
latter would never give it a thought, except as a branch of the
theory of probabilities, and except as to the consideration how
the terms by which the non-mathematical logician indicates his
degrees of belief are to be placed upon the numerical scale.   In
like manner he reads the thermometer by graduation, and though
he admits the freezing and boiling point, which have an origin
in nature, he leaves temperate, summer heat, blood heat, &c. to
the fancy of those who choose to employ them.

At the same time it is clear that these modal forms were considered not merely as useful in expression of the nature and amount
of belief, but as suggestive of real branches of inquiry, subservient
to that great *à priori* inquiry into the nature of things to which

* See a full account of these words in Sir William Hamilton's notes to
Reid, p. 806, &c.

mediæval logic was applied. We are not fit to judge of the in-
ftrumental part of this philofophy, unlefs we confider alfo the
materials on which it was founded. In an age in which much
more faith was demanded of the ftudent than now; when he was
much more frequently required to decide in one way or the
other upon a fingle teftimony; when, in addition to the non-
mythic wonders recorded in ancient writers, which there was no
mode of contradicting, all that was known of immenfe regions
and countries refted upon very few accounts, and thofe filled
with ftories quite as ftrange :—the abfence of other means of
diftinguifhing truth from falfehood obliged thofe who thought to
lay much ftrefs upon *à priori* confiderations. It matters little to
us whether we infer the *neceffity* of man being a walking animal
from the non-arrival of exceptions, and thence the *univerfality*
of the rule, or the univerfality from the fuppofed perfect induction
of inftances, and thence the neceffity. But it was of much more
confequence to the old logician : of more *real* confequence. He
did not know but that any day of the week might bring from
Cathay or Tartary an account of men who ran on four wheels
of flefh and blood, or grew planted in the ground like Polydorus
in the Æneid, as well evidenced as a great many nearly as mar-
vellous ftories. As he could not pretend to inductive and demonftrative univerfality, even upon the queftion of the form of his
own race, he was obliged to combine with his argument the an-
tecedent teftimony of his own and other minds, in the manner
which the real doctrine of modals (page 205) fhows to be neceffary in all non-demonftrated conclufions. It is true that he frequently confounded the predifpofition of minds with the confti-
tution of objects ; the teftimony with the thing teftified about.

We fhall never have true knowledge of the fchools of the
middle ages, until thofe who have ftudied both their philofophy,
their phyfics, and their ftate of tradition, will look at their
weapons of controverfy as both offenfive and defenfive, and give
a fair account of the amount of protection afforded by the firft,
in the exifting ftate of the fecond and third. It would alfo be
advifable to confider whether, looking at the power of communication by land and fea, and all the circumftances of literary inter-
courfe, it would have been practicable to place the knowledge of
the earth and its details upon any better footing of evidence.

One leading feature of the fchoolmen, acute as they were, and as to reprefentation of notions, inventive, and which is fhared by many more modern writers who have not difciplined themfelves mathematically, is feen in their employment of quantity: there are inftances of the ftrange ufe, the wrong ufe, and the no-ufe. Moft of them arife from indiftinct apprehenfion of continuity, which obliges them to accept fuch ftages of quantity as are ex-preffed by exifting terms, without any effort to fill up gaps. There is alfo a flovenlinefs of definition in what relates to quantity. Thus dozens of inftances might be given in which the *fome* of the particular propofition is fo defined that we might fuppofe it is ' fome, not all,' inftead of ' fome, it may be all,' and the former is the exprefs definition of fome writers: and it is only when we find in rules that XY does not allow us to infer $X:Y$, nor to contradict $X)Y$, that we afcertain the real intended meaning. "Logicians," fays Sir William Hamilton, "have referred the quantifying predefignations *plurimi,* and the like, to the moft oppofite heads; fome making them univerfal, fome particular, and fome between both." They muft have had curi-ous ideas of quantity who made the propofition ' moft Xs are Ys' either univerfal, or between the univerfal and particular: I fhould fuppofe that thofe who did the latter muft have imagined *fome* to refer to a *minority*.

There is a ftrange notion of quantity revived in modern times, which confifts in making *plurality of attributes* a part of the quan-tity of a notion. It is called its *intenfive quantity,* or its *intenfion,* or *comprehenfion.* It is oppofed to *extenfive quantity,* or *extenfion,* which is the more common notion of quantity, referring to the number of fpecies or of individuals (it may be either, the individual is the real *infima fpecies*) contained under the name. Thus *man* is not fo extenfive as *animal,* but more intenfive; the attribute *ratio-nal* gives greater comprehenfion. But ' man refiding in Europe' is lefs extenfive and more comprehenfive than either. It is faid that the greater the intenfive quantity the lefs the extenfive, but this is not true, unlefs no two of the figns of intenfion be properties of the fame fpecies. Thus, according to fuch ftatements as I have feen, ' man, refiding in Europe, drawing breath north of the equator, feeing the fun rife after thofe in America,' would be a more intenfively quantified notion than ' man refiding in Europe';

but certainly not more *extenſive*, for the third and fourth elements
of the notion muſt belong to thoſe men to whom the firſt and ſecond belong. Thus, in the Port-Royal Logic, one of the earlieſt
modern works (according to Sir W. Hamilton), in which the diftinction is drawn, it is ſaid that the *comprehenſion* of the idea of a
triangle includes ſpace, figure, three ſides, three angles, and the
equality of the angles to two right angles. But the idea of *rectilinear three-ſided figure* has juſt as much extenſion.

The relation between comprehenſion and extenſion exiſts, and
is uſeful: but not, I think, as that of different kinds of *quantity*.
In page 148, where I hold that the propoſition *is contained* in its
neceſſary conſequence, the view is one of extenſion : the ordinary
view is one of comprehenſion. ‘ Every caſe in which P is true,
is a caſe in which Q is true,’ tells us that all the P-caſes are contained, as to extent (number and location of inſtances), among
the Q-caſes. But, as to comprehenſion, every P-caſe contains
all that diſtinguiſhes a Q-caſe from other things. When, in
page 47, it is ſaid that the idea of man is contained in that of
animal, I ſpeak of extenſion : all the inſtances to which the firſt
idea applies are among thoſe to which the ſecond applies. But,
as to comprehenſion, the idea of animal is contained in that of
man: all that defines animal goes to the definition of man, and
other things beſides. In page 50, the “ *is* of poſſeſſion of all
eſſential characteriſtics,” refers to comprehenſion ; the “ *is* of
identity” to extenſion : both poſſeſſing equally the characters
under which the verb may occur in logic. There is no diſtinction
which affects inference : for X)Y has exactly the ſame properties whether we interpret it as expreſſing that Y has all the extenſion of X, and may be more ; or that X has all that Y has in
comprehenſion, and may be more.

In pages 115, &c. we have the mode of repreſenting names
of more or leſs comprehenſion. Thus, P, Q, R, &c. being characteriſtics, the obvious propoſition PQ)P, illuſtrates the theorem
that where the comprehenſion of one name has all that of a ſecond (as PQ has that of P) the extent of the ſecond is at leaſt
as great as that of the firſt. And the ſelf-evident poſtulate in page
115, by which we may diminiſh the extent of a term univerſally
uſed, or increaſe that of one particularly uſed, may be expreſſed in
language of comprehenſion. That is, we may augment the com-

prehenfion of a univerfal, or diminifh that of a particular. Thus, X)Y gives XP)Y, and X.Y gives XP.Y: but X)YP gives X)Y.

It will be eafily feen that comprehenfion has the firft attribute of quantity (page 174): there is *more* and *lefs* about it. But it is not of the *meafurable* kind (page 175). As to extent, 200 inftances bear a definite ratio to 100, which we can ufe, becaufe our inftances are *homogeneous*. But different qualities or defcriptions can never be numerically fummed as attributes, to any purpofe arifing out of their number. Does the idea of *rational animal*, two defcriptive terms, fuggeft any ufeful idea of *duplication*, when compared with that of *animal* alone. When we fay that a chair and a table are more furniture than a chair, which is true, we never can cumulate them to any purpofe, except by abftracting fome homogeneous idea, as of bulk, price, weight, &c. To give equal quantitative weight to attributes, as attributes, feems to me abfurd: to ufe them numerically otherwife, is at prefent impoffible.

The reader will have feen the origin of feveral very common terms, which are ufed in a fenfe coinciding with, or at leaft much refembling, that put upon them by the fchoolmen. But there is one which has diametrically changed its meaning; it is the word *inftance*. The word *inftantia* (and alfo ἔνστασις) implied a cafe *againft*, not *for*; the latter was *exemplum*: fo that *inftance to the contrary* would have been tautology.

I have referred the word *enthymeme* to this chapter, though it is always regularly explained in connexion with the fyllogifm. According to Ariftotle, Ἐνθύμημά ἐστι συλλογισμὸς ἀτελὴς ἐξ εἰκότων καὶ σημείων, an enthymeme is an imperfect fyllogifm from probables and figns: the modern critics reject the word ἀτελὴς, *imperfect*, as interpolated. The word *fign* feems to mean indication, fymptom, or effect, which makes the caufe almoft neceffary or highly probable. But the fchools took the word *enthymeme* to mean a fyllogifm with a fuppreffed and implied premife, fuch as 'He muft be mortal, being a man.' I cannot help fufpecting that Ariftotle *

---

* He fays all that is communicated (λέγεται) of the predicate, will be afferted in words (ῥηθήσεται) of the fubject. Thefe two different tenfes of two different verbs are often both tranflated by *dicitur*. Why did they

made no difference between a fuppreffed premife, clearly intended and diftinctly received, and one formally given. It feems to me that we might as well diftinguifh a written from a fpoken fyllogifm, as to the logical character of the two.

CHAPTER XIII.

On Fallacies.

THERE *is* no fuch thing as a claffification of the ways in which men may arrive at an error: it is much to be doubted whether there ever *can be.* As to mere inference, the main object of this work, it is reducible to rules: thefe rules being all obeyed, an inference, as an inference, is good; confequently a bad inference is a breach of one or more of thefe rules. Except, then, by the production of examples to exercife a beginner in the detection of breaches of rule, there is nothing to do in a chapter on fallacies, fo far as thofe of inference are concerned. Neverthelefs, there are many points connected with the matter of premifes, to which it is very defirable to draw a reader's attention: and above all to queftions in which it is not at firft obvious whether the miftake be in the matter or in the form; or in which it may be the one or the other, according to the fenfe put upon the words.

If there be anything *ridentem dicere verum quod vetat*, writers on logic have in all ages moft grievoufly neglected the prohibition in treating this fubject, and have given the ftudent a prefcriptive right to fome amufement. One reafon of this was, that the

occur? For various reafons, I allow myfelf to fufpect, though not fcholar enough to maintain, that λόγος generally meant communication, paffage from one mind to another by any means, as much at leaft with reference to the receiving, as to the imparting, mind: and that it is here oppofed to ῥῆσις, fpeech, in that fenfe. Throw the verbs back to their primary meanings, and it will be ' That which *is picked up* of the predicate, fhall *flow out* about the fubject.' If my conjecture be correct, the modern enthymeme is here put on the fame footing as the fully expreffed fyllogifm.

Greeks endeavoured to try the new art by inventing inferences the falfehood of which could not be detected by its rules. Thefe, as may be fuppofed, were whimfical efforts of reafoning: neverthelefs, they have been handed down from book to book, unfurpaffed in their way. Another reafon is, that jefts, puns, &c. are for the moft part only fallacies fo obvious that they excite laughter; and the greater number of them can be fhown to break one or another of the rules of logic. Accordingly, they furnifh ftriking examples of thefe rules; the application of which, in ferious terms, has itfelf a tafte of the ludicrous. Boccacio has, by his inimitable mode of narration, made a good ftory the jeft of which could be defcribed as confifting in nothing more than the affumption that what can be predicated of ftorks* in general can be predicated of roafted ftorks: which is what logicians would call the *fallacia accidentis*, or arguing *a dicto fimpliciter, ad dictum fecundum quid.*

The terms *fallacy, fophifm, paradox,* and *paralogifm,* are applied to offences againft logic; but not with equal propriety, *Fallacy* and *fophifm* may technically have been firft applied to arguments in which there is a failure of logic: but it is now very common to apply them alfo to arguments in which there is a falfehood of fact, or error of principle, though logically treated; and if this laft ufe be not correct, writers on logic have fanctioned it in their examples. Many perfons go further, and call the erroneous ftatement itfelf a fallacy: that men are in the habit of walking on their heads, they would fay is a very obvious fallacy. A *paradox* is properly fomething which is contrary to general opinion: but it is frequently ufed to fignify fomething felf-contradictory: thus the newfpaper which recently avowed

---

* A fervant who was roafting a ftork for his mafter was prevailed upon by his fweetheart to cut off a leg for her to eat. When the bird came upon table, the mafter defired to know what was become of the other leg. The man anfwered that ftorks had never more than one leg. The mafter, very angry, but determined to ftrike his fervant dumb before he punifhed him, took him next day into the fields where they faw ftorks, ftanding each on one leg, as ftorks do. The fervant turned triumphantly to his mafter: on which the latter fhouted, and the birds put down their other legs and flew away. "Ah, Sir," faid the fervant, "you did not fhout to the ftork at dinner yefterday: if you had done fo, he would have fhown his other leg too."

its opinion that the repeal of the corn laws would make food both cheap and dear is said to have maintained a paradox. The modern use of the word implies disrespect, but it was not so formerly. Thus in the sixteenth century the opinion of the earth's motion was styled the *paradox of Copernicus* by writers who meant neither praise nor blame, but only reference to the opinion of Copernicus as *an unusual one.* The more precise writers of our day use the word paradox for an opinion so very singular and improbable, that the holder of it is chargeable with an undue bias in favor of singularity or improbability for its own sake. *Paralogism,* by its etymology, is best fitted to signify an offence against the formal rules of inference. It has been frequently abused by mathematical writers, who have signified by it errors of statement, and undue assumptions: but it is not completely spoiled for the purpose, and I shall therefore use it to denote a formal error in inference, as a particular class of fallacy or sophism, words which it would now be difficult to distinguish in meaning. Some have defined *paralogism* to be that by which a man deceives himself, and *sophism* that by which he tries to deceive others: on what grounds I do not know.

The question of a premise being right or wrong in fact or principle, unless indeed it contradict itself, does not belong to logic: nor could it so belong unless logic were made, in the widest sense, that attempt at the attainment of the *cognitio veri* which some have defined it to be. All that relates to the collection of true premises with respect to the vegetable world belongs to botany; with respect to the heavenly bodies, to astronomy; with respect to the relation of man to his Creator, to theology. Even were it within the province of logic, it would be impossible, in less space than an encyclopædia, to enter upon questions connected with the matter of syllogisms. With regard to paralogisms, or *logical* fallacies, (so called, as an error about the measure of space is called a *geometrical* error) the classification under breach of rules would be good in form, but would afford no basis for the treatment of the subject. Those who bring them forward seldom proceed in direct defiance of rule, but in various modes of evasion. These it would be almost impossible to arrange in satisfactory order.

Aristotle made a classification of fallacies, which was of course adhered to by the writers of the middle ages. In this, as in every other place, when I fpeak of Ariftotle and his fyftem, I fpeak of it as underftood by thofe writers. How far they diftinctly comprehended their mafter is a queftion into which I could not enter here, even if I were competent to write on the fubject. It is, however, fufficiently apparent that the logic of Ariftotle is not of the purely formal character which marked the dialectics of the middle ages: there is a much more decided introduction of the attempt to write on the matter of fyllogifm than many perfons think there is. The claffification of fallacies feems to be one proof of this: and the interpretation of that claffification by the middle writers feems to add their teftimony to the affertion: in this part of the fubject they abandon technicalities almoft entirely.

It ought to be efpecially remembered that we are very differently fituated from thofe writers, not as to what is fallacy, but as to what the fpecimens of it produced are likely to be. Out of a world of general principles declared by authority, or declared to be felf-evident by authority, they had to produce logical deductions; and, of courfe, the pure fyllogifm and its rules were to them as familiar as the alphabet. The idea of an abfolute and glaring offence againft the ftructure of the fyllogifm being fupported one moment after it was challenged, would no more fuggeft itfelf to the mind of a writer on logic than it would now occur to a writer on aftronomy that the accidental error (which might happen to any one) of affixing four ciphers inftead of five in multiplying by a hundred thoufand would be maintained after expofure. Accordingly, their formal chapters on fallacies would naturally relate, if not entirely to fallacies of matter, at leaft to thofe in which the fallacy of matter very clofely hinges upon that of form. And fo it is in all the old fyftems which I have examined. The Ariftotelian divifion (or rather felection, for it is far from including everything) lends itfelf eafily to this adaptation.

We, on the contrary, live in an age in which formal logic has long been nearly banifhed from education: entirely, we may fay, from the education of the habits. The ftudents of all our univerfities (Cambridge excepted) may have heard lectures and learnt the forms of fyllogifm to this day: but the practice has been fmall: and out of the univerfities (and too often in them) the very name of logic is a bye-word.

The philofophers who made the difcovery (or what has been allowed to pafs for one) that Bacon invented a new fpecies of logic which was to fuperfede that of Ariftotle, and their followers, have fucceeded by falfe hiftory and falfer theory, in driving out from our fyftem all ftudy of the connexion between thought and language. The growth of inaccurate expreffion, which this has produced, gives us fwarms of legiflators, preachers, and teachers of all kinds, who can only deal with their own meaning as bad fpellers deal with a hard word, put together letters which give a certain refemblance, more or lefs as the cafe may be. Hence, what have been aptly called " the flipfhod judgments and crippled arguments which every-day talkers are content to ufe." Offences againft the laws of fyllogifm (which are all laws of common fenfe) are as common as any fpecies of fallacy : not that they are always offences in the fpeaker's or writer's mind, but that they frequently originate in his attempt to fpeak his mind. And the excufe is, that he meant differently from what he faid : which is received becaufe no one can throw the firft ftone at it, but which in the middle ages would have been regarded as a plea of guilty. The current notions about what logic is, are beautiful and wonderful. I have heard a difputant, an educated man, a graduate, efcape from allowing himfelf to be convinced that he was arguing with a middle term particular in both premifes by declaring that *facts* were better than *fyllogifms* : the form of his argument would have proved that men are plants, becaufe both require air. " I " he faid, " produce you *facts*, like Bacon : you quibble about their combination, like Ariftotle."

The Ariftotelian fyftem of fallacies contains two fubdivifions. In the firft, which are *in dictione*, or *in voce*, the miftake is faid to confift in the ufe of words : in the fecond, which are *extra dictionem*, or *in re*, it is faid to be in the matter.

Of the firft fet fix kinds were diftinguifhed, as follows :—

1. *Æquivocatio* or *Homonymia*, in which a word is ufed in two different fenfes ; giving really no middle term (if the middle term be in queftion) or a term in the conclufion which is not the fame name as that ufed in the premifes. For example, ' All criminal actions ought to be punifhed by law : profecutions for theft are criminal actions ; therefore, profecutions for theft ought to be punifhed by law.' Here the middle term is doubly ambiguous,

both *criminal* and *action* having different senses in the two premises. But here, as in many other cases, the choice lies with the sophist to bring the fallacy under the head to which we refer it or not. It may please him to assert that he means the same thing by *criminal action* in both premises ; in which case, the inference is logical, but one or the other premise must be denied as to the matter. Again, ' Finis rei est illius perfectio ; mors est finis vitæ ; ergo mors est vitæ perfectio.' Here the ambiguity may be thrown either on *finis* or on *perfectio.* The following example can be traced through books for three centuries. 'Every dog runs on four legs ; Sirius (the dog-star) is a dog ; therefore Sirius runs on four legs.' It has been the defect of many old works on logic that *all* their examples have been of that obvious absurdity, which is well enough in one or two instances. Such as ' Nothing is better than wisdom and virtue ; dry bread is better than nothing ; therefore, dry bread is better than wisdom and virtue.' Some of the old examples are ' A mouse eats cheese ; a mouse is one syllable ; therefore one syllable eats cheese.' And again, ' Iste pannus est de Anglia ; Anglia est terra ; ergo, iste pannus est de terra.'

Where the syllogism is formally put, equivocation of the middle term is generally seen with great ease. The most difficult exception is, I think, the old fallacy, in which giving the name of the genus is confounded with giving the name of the species, and thereby, of course, giving the name of the genus. As in 'To call you an animal is to speak truth ; to call you an ass is to call you an animal ; therefore, to call you an ass is to speak truth.' This equivocation will puzzle a beginner as to its form, and the more so from the evident falsehood of the matter. The middle term is " He who says that you are *one* among all animals." He speaks truth ; and the one who calls you an ass or a goose, certainly says that you are *one* among all animals. The equivocation is in the two different uses of the word one ; in the first premise, it is an entirely indefinite *one ;* in the second it is a less indefinite *one.* This *one* is not attached to the quantity of the middle term, which is universal in the first premise, and particular in the second : but is part of the middle term itself.

The manner in which the serious fallacy of equivocation most frequently appears, is in the connection of the old associations of a word which has shifted its meaning with the altered meaning of the same. The word loyal, for instance, originally meaning no more (and no less) than *lawful*, which, as applied to a man, meant one who respected the laws, and had not forfeited any right by misbehaviour, now means attached to the Crown and to the title of the holder of it. In contests for succession, the winner would, of course, assume that *lawful* men were on his side. In more recent times, the term was always self-applied, at elections, by those who supported the party which had the confidence of the Crown for the time being: but on such occasions, abstinence from the fallacy which the French call the *voie du fait* is the utmost which can be expected of human nature.

The word *publication* has gradually changed its meaning, except in the courts of law. It stood for *communication to others*, without reference to the mode of communication, or the number of recipients. Gradually, as printing became the easiest and most usual mode of publication, and consequently the one most frequently resorted to, the word acquired its modern meaning: if we say a man publishes his travels, we mean that he writes and prints a book descriptive of them. I suspect that many persons have come within the danger of the law, by not knowing that to write a letter which contains defamation, and to send it to another person to read, is *publishing a libel*; that is, by imagining that they were safe from the consequences of publishing, as long as they did not print. In the same manner, the well-established rule that the first publisher of a discovery is to be held the discoverer, unless the contrary can be proved, is misunderstood by many, who put the word printer in the place of publisher. I could almost fancy that some persons think rules ought to travel in meaning, with the words in which they are expressed.

A similar change has taken place in the meaning of the word to *utter*, the sense of which is to *give out*, but which now means usually to give out of the mouth in words. As yet, I am not aware that any person charged with the *utterance* of counterfeit coin has pleaded that no one ever uttered coin except the princess in the fairy tale: but there is no saying to what we may come, with good example, and under high authority.

It may almost be a question whether, in the time of Aristotle, successful equivocation, that is, undetected at the moment, would not have been held binding on the difputant who had failed to detect it. The genius of uncultivated nations leads them to place undue force in the verbal meaning of engagements and admiffions, independently of the underftanding with which they are made. Jacob kept the blessing which he obtained by a trick, though it was intended for Efau : Lycurgus feems to have fairly bound the Spartans to follow his laws till he returned, though he only intimated a fhort abfence, and made it eternal : and the Hindoo god who begged for three fteps of land in the fhape of a dwarf, and took earth, fea and fky in that of a giant, feems to have been held as claiming no more than was granted. The great ftrefs laid by Ariftotle on fo many different forms of verbal deception, may have arifen from a remaining tendency among difputants to be very ferious about what we fhould now call play upon words.

Governments permit what would otherwife be equivocation to take a ftrong air of truth, by legiflating in detail againft the principles of their own meafures. The window-tax is a fpecial inftance. A newfpaper calls it a tax upon the light which God's beneficence has given to all. The anfwer would be plain enough, namely, that it is an income tax levied upon a ufe of that light which (how truly matters not here) is afferted to be a fair criterion of income. But this anfwer is deftroyed by the permiffion to block up windows, and thereby evade the tax : which is thus made to fall upon the light ufed, and not upon the means of ufing it which the fize of the houfe affords. According to the principle of this impoft, the blocked window is as fair a criterion of the income of the occupant as the open one, and fhould have been fo confidered.

Among the forms which the fallacy of equivocation frequently affumes, is that of the fophift altering or qualifying the known meaning of a word in his own mind, without giving the other party any notice : fo that there may be, if not two meanings in one mind, yet different meanings in the two minds concerned. A perfon afferts that ' Nobody denies, &c. &c.' Should this go down, the point is gained ; what nobody denies muft be undeniable. But fhould it be contefted (and it will generally be found that the things which nobody denies are matters of fome difference of opinion, while thofe which nobody *can* deny are quite fure to be points of conftant controverfy) the evafion is ready. It is no fenfible perfon, or nobody that underftands the fubject, nobody that is anybody, in fhort: while perhaps it cannot be fettled who does, or who does not, underftand the fubject, until, among other things, the very point in difpute is determined.

There is a wide range of equivocations arifing out of meanings which are fometimes implied and fometimes not. A large clafs of them is made by the ufual, but not univerfal, practice, of giving to the thing the name of that which it is intended to be, whether the attempt be fuccefsful or not. This is now abbreviation or courtefy; but it was the rule. According to old definitions, bad reafoning is reafoning, *fyllogifmus fophifticus* is a fyllogifm, and in an old book now before me, the fruits and effects of demonftration are fcience, opinion, and *ignorance*, the latter containing belief of falfehood derived from *bad* demonftration, which we fhould now call *no* demonftration.

One fallacy of our time, and a very favourite one, is the fettlement of the merit of a perfon, or an opinion, not by arguing the place of that perfon or opinion in its fpecies, but by arbitrary alteration of the boundary of the fpecies, with the intent of excluding the individual in queftion altogether.

It is fomewhat analogous to the proceeding of the landlord who unroofs the houfe to get rid of a tenant. Thus we have had the controverfy whether Pope was a *poet*, not whether he was a good poet or a bad one, but whether he was a poet at all. The difputants, or fome of them, claimed a right to define a poet, and decided that none but verfe-makers of a certain goodnefs (to be fettled by themfelves) were poets. They might juft as well have decided, on their own authority, that none but men of a certain amount of reafoning power were *men*. Had they done this laft, as long as they fixed the amount at a figure which included themfelves under the name, nobody would have thought they materially altered the extent of the term : it is not eafy to fee why they have rights fo arbitrary, over words the objective definitions of which are nearly as well fixed as that of man.

Another form of the fallacy of equivocation is the affuming, without exprefs ftatement, that the meaning of a phrafe can be determined by joining the meanings of its feveral words : which is not always true in any language. When two words come together, it often happens that their dictionary meanings would never enable us to arrive at their known and usual (and therefore proper) compound meaning: though they might help us in explaining how that last meaning arose. A person undertakes to cross a bridge in an incredibly short time : and redeems his pledge by crossing the bridge as one would cross a street, that is, by traversing the breadth. Now, though it be true that, in general, to cross is to go over the breadth, or shorter dimension, yet in the case before us, the phrase is elliptical, and signifies crossing *the river* upon the bridge. Nor can it be said that this common meaning is incorrect : that which is common and well known is, in language, always correct. No reasonable person would say that a French newspaper is wrong in reporting an army to be *à cheval sur la rivière*, because a river is not a horse. This literal (or rather unlettered) mode of interpretation is adopted among gamblers in settling bets : and is of itself enough to raise a strong presumption that their occupation is not that of well-educated men.

It is common enough in controversy, for one side or the other to have fixed meanings of words in his own mind, on which he proceeds without any inquiry as to whether those meanings will be *conveyed* by the words to the other side, or to the reader. It is very difficult to avoid this form of the fallacy, without giving the meanings of the most essential terms, on the first occasions of their occurrence. It is not uncommon to meet with a writer who appears to believe, at least who certainly acts upon, the notion that the right over words resides in him, and that others are wrong so far as they differ from him. I do not only mean that there are many who have an undue belief in their own judgments, both as to words and things : but I speak of those who, though showing a proper modesty in respect to their own conclusions, seem to be unable to do the same with respect to their definitions of words. If all mankind had spoken one language, we cannot doubt that there would have been a powerful, perhaps a universal, school of philosophers who would have believed in the inherent connexion between names and things ; who would have taken the sound *man* to be the mode of agitating the air which is essentially communicative of the ideas of reason, cookery, bipedality, &c. The writers of whom I speak, are more or lefs of this fchool; they treat words as abfolute images of things by right of the letters which fpell them. "The French," faid the failor, "call a cabbage a *fhoe*; the fools! why can't they call it a cabbage, when they muft know it is one?"

Equivocation may be ufed in the form of a propofition; as for inftance, in throwing what ought to be an affirmative into the form of a qualified negative, with the view of making the negative form produce an impreffion. Thus a controverfial writer will affert that his opponent has not attempted to touch a certain point, except by the abfurd affertion, &c. &c. &c. To which the other party might juftly reply, "Your own words fhow that I have made the attempt, though your phrafe has a tendency, perhaps intended, to make your reader think that there is none, or at leaft to blind him to the difference between *none* and *none that you approve of.*"

2. The *fallacia amphiboliæ*, or *amphibologiæ*, differs in nothing from the laft, except in the equivocation being in the conftruction of a phrafe, and not in a fingle term: as in confounding that which is Plato's (property) with that which is Plato's (writing). Or, as in 'Qui funt domini fui funt fui juris; fervi funt domini fui; ergo fervi funt fui juris.' The ambiguities of conftruction in our language, arifing from want of inflexions and genders are tolerably (and intolerably) numerous. The difficulty of determining the emphatic word often gives a doubt as to the meaning. But very often indeed there is a want of the diftinction which the algebraift makes when he writes three-and-four tens as diftinguifhed from three and four-tens: $(3+4).10$ and $3+4.10$. It cannot, for inftance, be faid whether 'I intend to do it and to go there to-morrow' means that it will be done to-morrow or not. It may be either—(I intend to do it and to go there) to-morrow, or—I intend to do it and (to go there to-morrow). The prefumption may be for the firft conftruction: but it is only a prefumption, not a rule of the language. In an inftance cited by Dr. Whateley—"If this day happen to be Sunday, this form of prayer fhall be ufed and the faft kept the next day following," the conftruction is ambiguous, and the intended meaning probably againft the prefumption. There is a book of the laft century, written by a "teacher of mathematics, and writing mafter to Eton College." Were mathematics taught at Eton,

or not? Punctuation may be an assistance; but it so often happens that the author leaves that point to the printer, that it is
hardly safe to rely upon it. Printers punctuate correctly when
the meaning is clear: but when it is ambiguous, they may be
as apt to take the wrong meaning as any other readers.

3, 4. The *fallacia compositionis*, and *fallacia divisionis*, consist
in joining or separating those things which ought not to be joined
or separated. If we may say that A is X and B is Y, so that A
and B is X and Y, we have no right to infer that we may form
the compound and collective names 'A and B,' and 'X and Y,'
and say that 'A and B' is 'X and Y.' Thus two and three are
even and odd: but five is not even and odd. Again, two and
five are four and three; but neither is two four, nor five three.
It must be remembered that the word *all*, in a proposition, is not
necessarily significative of a universal proposition: it may be a
part of the description of the subject. Thus in 'all the peers are
a house of Parliament,' we do not use the words *all the peers* in
the same sense as when we say 'all the peers derive their titles
from the Crown.' In the second case the subject of the proposition is *peer;* and the term *all* is distributive, synonymous with
each and every. In the first case the subject is *all the peers*, and
the term *all* is collective, no more distinguishing one peer from
another than one of John's fingers is distinguished from another
in the phrase, 'John is a man.' The same remarks may be made
on the word *some;* as in 'some peers are dukes,' and 'some peers
are the committee of privileges.' The *all* and *some* of the quantity of the proposition are distributive terms; the all and some of
the subject are collective. Again, all men are a species (of animals) which no number of men are, wanting the rest. *All men*
here make the one individual object of thought of a singular proposition. This amounts to an ambiguity of construction, an
amphibologia, as do most sources of fallacy falling under this head,
which can therefore hardly be considered as anything more than
a case of the last. We want another idiom or the algebraical
distinction, as in 'All (peers) hold of the Crown; (all peers) are
a house of Parliament.'

5. The *fallacia prosodiæ* or *accentus* was an ambiguity arising
from pronunciation, and its introduction seems to lead to very
minute subdivision of the subject, and to ensure the entrance of

none but ludicrous examples. Burgerſdicius does not think it unworthy of himſelf to deſcend to the following, 'Omnis equus eſt beſtia; omnis juſtus eſt æquus, ergo omnis juſtus eſt beſtia. An older writer has 'Tu es qui es ; quies eſt requies ; ergo, tu es requies.' Theſe are mere puns; and the makers of them were fairly beaten by the contriver of 'Two men eat oyſters for a wager, one eat ninety-nine, the other eat two more, for he eat a hundred and won.' But more ſerious fallacies may be referred to this head. A very forced emphaſis upon one word may, according to uſual notions, ſuggeſt falſe meanings. Thus, 'thou ſhalt not bear falſe witneſs againſt thy neighbour,' is frequently read from the pulpit either ſo as to convey the oppoſite of a prohibition, or to ſuggeſt that ſubornation is not forbidden, or that anything falſe except evidence is permitted, or that it may be given *for* him, or that it is only againſt *neighbours* that falſe witneſs may not be borne.

A ſtatement of what was ſaid, with the ſuppreſſion of ſuch tone as was meant to accompany it, is the *fallacia accentus*. Geſture and manner often make the difference between irony or ſarcaſm, and ordinary aſſertion. A perſon who quotes another, omitting anything which ſerves to ſhow the *animus* of the meaning; or one who without notice puts any word of the author he cites in italics, ſo as to alter its emphaſis; or one who attempts to heighten his own aſſertions, ſo as to make them imply more than he would openly ſay, by italics, or notes of exclamation, or otherwiſe, is guilty of the *fallacia accentus*.

To this fallacy I ſhould refer one of very common occurrence, the alteration of an opponent's propoſition ſo as to preſent it in a manner which is logically equivalent, but which alters the emphaſis, either as noticed in page 134, or in any other manner. It is generally not reaſoning, but retort, which is the objeɕt of the alteration: for inference cannot be altered by changing a propoſition into a logical equivalent, but a ſmart repartee may be very effeɕtive againſt 'Some Xs are Ys,' but flat enough againſt 'ſome Ys are Xs.' And even when the proponent miſtakes his own meaning, and miſcalculates his own emphaſis, ſtill, if the miſtake be obvious, there is fallacy in taking advantage of it ; for he who communicates in ſuch incorreɕt terms as ſhow what the correɕt ones are, does, in faɕt, communicate in correɕt terms, to all who see the showing. Of course, respect for logic never stood in the way of a succesful retort from the time of Aristotle till now, nor will on this side of the millenium. A speculator once wrote to a scientific society, to challenge them to an (on his part) anti-Newtonion controversy, relying on it that he could contend in mechanics, though avowedly ignorant of geometry. He was answered by a recommendation to study mathematics and dynamics. His rejoinder was an angry pamphlet, in which, indignant at the unfairness, as he took it to be, of the recommendation, he exclaimed, ' I did not confess my ignorance of dynamics.' Had he been worth the answering, it would have been impossible to resist the reply ' No, but you showed it.' Had he written, as he meant ' It was not dynamics of which I confessed ignorance,' and had an opponent written, as many would have done, ' You say, sir, that you did not confess your ignorance of dynamics : indeed you did not, you contented yourself with an ample display of it,' he would have used the *fallacia accentus*. Nor would he, in my opinion, have been clear of it though he had only taken advantage of a wrong, but evidently wrong, placement of emphasis on the part of the assailant. The use of such a weapon, as to its legitimacy, depends entirely upon the manner in which the question shall be settled how far irony is allowable. Where the answer is in the affirmative, a very obvious fallacy, as a sarcasm, may be permitted. But I may here observe, that irony itself is generally accompanied by the *fallacia accentus* ; perhaps cannot be assumed without it. A writer disclaims attempting a certain task as above his powers, or doubts about deciding a proposition as beyond his knowledge. A self-sufficient opponent is very effective in assuring him that his diffidence is highly commendable, and fully justified by the circumstances.

6. The *fallacia figuræ dictionis*, as explained, means literally a mistake in grammar and nothing else ; as that because *fluvius* is *aqua* it is *humidA*, or that because *aqua* is feminine, so is *poeta*.

All these fallacies *in dictione* come under the head of ambiguous language, and amount to nothing but giving the syllogism four terms, two of them under the same name. The fallacies *extra dictionem* are set down as follows.

1. The *fallacia accidentis* ; and 2. That *à dicto secundum quid ad dictum simpliciter*. The first of these ought to be called that

of *à dicto simpliciter ad dictum secundum quid*, for the two are
correlative in the manner described in the two phrases. The first
consists in inferring of the subject with an accident that which
was premised of the subject only: the second in inferring of the
subject only that which was premised of the subject with an accident. The first example of the second must needs be 'What you
bought yesterday, you eat to-day; you bought raw meat yesterday; therefore, you eat raw meat to-day.' This piece of meat
has remained uncooked, as fresh as ever, a prodigious time. It
was raw when Reisch mentioned it in the *Margarita Philoso-
phica* in 1496: and Dr. Whateley found it in just the same state
in 1826. Of the first, we may give the instance 'Wine is pernicious; therefore, it ought to be forbidden.' The expressed
premise refers to wine used immoderately: the conclusion is
meant to refer to wine however used. This species of fallacy
occurs whenever more or less stress is laid upon an accident,
or upon any view of the subject, in the conclusion, than was
done in the premises. As in the following:—'All that leads to
such philosophy as that of the schoolmen, with their logic, must
be unworthy to be studied, except historically.' The intent of
such a sentence is not formally to propose the false syllogism,
'The schoolmen had that which led them to a false philosophy;
the schoolmen had logic; therefore, logic led them to a false philosophy,' but only to take the chance of the stress thus laid upon
logic producing a disposition to suppose that the logic was in fault.
The premises are really :—

The philosophy of the school-
men (who paid particular atten- } is { a false philosophy.
tion to logic)

Every false philosophy } is { that the guides to which
should be neglected, except
as history.

whence it is rightly inferred that the guides to such a philosophy
as that of the schoolmen (who studied logic) are only of historical
use. And the same thing might equally be inferred of the schoolmen who ate mutton, a practice to which most of them were as
much addicted, no doubt, as to making syllogisms. The art of

the fophift confifts in making the accident which is either un-
fairly introduced, or withdrawn, or fubftituted, have an apparently
relevant relation to the fubject itfelf. Undoubtedly, the fchoolmen's logic has a connexion with their philofophy which the
mutton they ate has not : but as long as it is not *the* connexion
which permits the inference, it is abfolutely irrelevant.

All the fallacies which attempt the fubftitution of a thing in
one form for the *fame thing* (as it is called) in another, belong to
this head : fuch as that of the man who claimed to have had one
knife twenty years, giving it fometimes a new handle, and fometimes a new blade. The anfwer given by the calculating boy
(page 54, note) was, relatively to the queftion, a worthy anfwer,
and took advantage of the common notion that a bean, after
being fkinned, is ftill a bean, as before. More ferious difficulties
have arifen from the attempt to feparate the *effential* from the
accidental, particularly with regard to material objects. The
Cartefians denied weight, hardnefs, &c. to be effential to matter, until at laft they made it nothing but fpace, and contended
that a cubic foot of iron contained no more matter than a cubic
foot of air.

The law, in criminal cafes, demands a degree of accuracy in
the ftatement of the *fecundum quid* which many people think is
abfurd : and it appears to me that the lawyers often help the
popular mifapprehenfion, and give it excufe, by confounding
errors of things with errors of words, after the example of the
world at large. Any error of any kind, provided it be fmall in
amount, paffes for a miftake in words only, by virtue of its fmall-
nefs. By a miftake in words, I mean the addition or omiffion
of words which, whatever they might do under another ftate of
things, do not, as matters ftand, affect the meaning.

Take two inftances, as follows ;—Some years ago, a man was
tried for ftealing a ham, and was acquitted upon the ground that
what was proved againft him was that he had ftolen a portion of
a ham. Very recently, a man was convicted of perjury, 'in the
year 1846,' and an objection (which the judge thought of importance enough to referve) was taken, on the ground that it ought
to have been ' in the year *of our Lord* 1846.' There may, of
courfe, be acknowledged rules, which, as long as they are rules,
muft be obeyed, and which may make the fecond miftake as ne-

ceffarily vitiate an indictment as the firft. But, in difcuffing the policy of the rules, it would feem to me that the two cafes are entirely different. In both, no doubt, the reft of the indictment might, by implication, make good the meaning required: but there feems a great difference between allowing the remainder to correct an error, and allowing it to make good an infufficiency (fuppofing the date, in the fecond cafe, to be really infufficient). In the fecond cafe, the accufed may fee the omiffion as well as another, and may confider of his defence againft every alternative: in the firft, he may be actually led to appear in court with a defence not relevant to what will be brought againft him. The fecond may be a hardfhip, the firft is an injuftice. And this, even on the fuppofition that the reft of the indictment is to be allowed in explanation: for we have no more right to fuppofe that the true parts will correct the erroneous ones, than that the erroneous parts will affect the conftruction of the true ones. But there is good reafon to think that the fufficient defcription of one fentence may fupply what is wanted in the infufficient defcription of another, when infufficiency is all.

But, perhaps, it will be held to be the better rule, that the remainder of the indictment fhould not be allowed in explanation. It will then be admitted by all that a material error, or a material infufficiency, fhould be allowed to nullify the charge. The difference between the law and common opinion entirely relates to what conftitutes a material amount of one or the other. And here it is impoffible to bring the two together: for the law muft judge fpecies, while the common opinion will never rife above the cafe before it. In the two inftances, which by many will be held equally abfurd, a great difference will be feen by any who will imagine the two defcriptions, in each cafe, to be put before two different perfons. One is told that a man has ftolen a ham; another that he has ftolen a part of a ham. The firft will think he has robbed a provifion warehoufe, and is a deliberate thief: the fecond may fuppofe that he has pilfered from a cook-fhop, poffibly from hunger. As things ftand, the two defcriptions may fuggeft different amounts of criminality, and different motives. But put the fecond pair of defcriptions in the fame way. One perfon is told that a man perjured himfelf in the year 1846; and another, that he perjured himfelf in the year of our Lord

1846. As things ſtand, there is no imaginable difference : for there is only one era from which we reckon. The two deſcriptions mean the ſame thing : nor can it even be ſaid that one is complete and the other incomplete ; but only that one is leſs incomplete than the other. The next queſtion might have been, what lord was meant, our Lord Jeſus Chriſt, or our Lord the King? both being phraſes of law. The anſwer will be, that the number 1846 leaves no doubt which was meant. A very good anſwer, certainly ; but equally concluſive as to the ſimple phraſe ' in the year 1846.' The firſt caſe is one in which the two deſcriptions have a real difference of meaning : it is not ſo in the ſecond.

3. The *petitio principii* is one of the logical terms which has almoſt found its way into ordinary life. It is tranſlated by the phraſe *begging the queſtion*, that is, aſſuming the thing which is to be proved. This is alſo called *reaſoning in a circle*, coming round, in the way of concluſion, to what has been already formally aſſumed, in a manner expreſſed or implied. I ſhall reſerve what I have to ſay on the juſtice of this tranſlation, and take it for the preſent as good.

Every collective ſet of premiſes contains all its valid concluſions ; and we may fairly ſay that, ſpeaking objectively of the premiſes, the aſſumption of them is the aſſumption of the concluſion ; though, ideally ſpeaking, the preſence of the premiſes in the mind is not neceſſarily the preſence of the concluſion. But by this fallacy is meant the abſolute aſſumption of the ſingle concluſion, or a mere equivalent to it, as a ſingle premiſe. If the concluſion be ' Every X is Z' and if it be formally known that A and X are identical names, and alſo B and Z, then to aſſume ' Every A is B' as a premiſe in proving ' Every X is Z' would be a manifeſt *petitio principii*, or begging of the queſtion. But even this muſt be ſaid hypothetically ; it is ſuppoſed fully agreed between the diſputants that the two identities are granted. Let it be otherwiſe, and there is no *petitio principii* : it is *then* fair to propound A)B, which, if diſputed, is to be proved, and afterwards to reaſon as in A)B + B)Z = A)Z, X)A + A)Z = X)Z. Strictly ſpeaking, there is no formal *petitio principii* except when the very propoſition to be proved, and not a mere ſynonyme of it, is aſſumed. This of courſe, rarely occurs : ſo that the fallacy to be guarded against is the assumption of that which is too nearly the same as the conclusion required. And then the fallacy is nothing distinct in itself: but merely amounts to putting forward and claiming to have granted that which should not be granted. When this is done, it matters little as to the character of the fallacy, whether the undue claim be made for a proposition which is nearer to, or further from, the conclusion to be proved. When proof is offered, the advancement of the conclusion in other words is of course not *petitio* principii : when proof is not offered, the assumption of that which (with other things proved) would prove the conclusion, is a fallacy of the same character in all cases. There is an opponent fallacy to the *petitio principii* which, I suspect, is of the more frequent occurrence : it is the habit of many to treat an advanced proposition as a begging of the question the moment they see that, if established, it would establish the question. Before the advancer has more than stated his thesis, and before he has time to add that he proposes to prove it, he is treated as a sophist on his opponent's perception of the relevancy (if proved) of his first step. Are there not persons who think that to prove any previous proposition, which necessarily leads to the conclusion adverse to them, is taking an unfair advantage ?

There is another case in which begging the question may be unjustly imputed. It should be remembered that *demonstrative* inference is not the only kind of inference : there is *elucidatory* inference, *recapitulatory* inference, &c. A proposition may have its asserted *explanation* presented as a syllogism, the inference of which, as demonstration, might well be called a result of *petitio principii*. Say ' it never could have been doubted that men would apply science to the production of food.' If there should be any hesitation about this, the explanation of *man* under the phrase which is exclusively characteristic of him, *rational animal*, would remove it : the animal must have food, the rational being will have science. But it would be begging the question to assert that the syllogism of elucidation ' A rational animal is, &c. ; man is, &c ; therefore man is, &c.' is a demonstration. And out of this arises the fallacy of presuming that an author meant *demonstration*, when he can only be fairly construed to have attempted elucidation of what he supposed would, upon that elucidation, be *granted*. The forms of language are much the same in the two cases.

It has been obſerved that Ariſtotle hardly ever uſes the phraſe ἀρχὴν ἀιτεῖσθαι, *principium petere*: it is τὸ ἐξ ἀρχῆς and τὸ ἐν ἀρχῇ, that which is (ought to come) out of, or is in, the principle.   By the word *principium* he diſtinctly means *that which can be known of itſelf*.   He lays down five ways of *aſſuming* that which ought to come out of a ſelf-known principle, of which begging the queſtion is the firſt.   The others are aſſuming the univerſal to prove the particular; aſſuming a particular to help to prove the univerſal; aſſuming all the particulars of which the univerſal may be compoſed; and aſſuming ſomething which obviouſly demonſtrates the concluſion.

Among the earlier modern writers, as far as I have ſeen them, there is ſome diverſity in their deſcription of the *petitio principii*. That the *principium* was meant to be the thing known of itſelf, the ἀρχή of Ariſtotle, as far as the introduction of the word is concerned, ſeems clear enough.   Was it not then by a mere corruption that it was frequently confounded with the concluſion, the ʻquod in *principio* quæſitum fuit?ʼ   Did not the ſame inaccuracy, * which confounds the τὸ ἐν αρχῇ of Ariſtotle with the ἀρχή itſelf, govern the change of the word?   Moſt writers take the fallacy of the *petitio principii* as meaning that in which the concluſion is deduced either from itſelf, or from ſomething which requires proof more, or at leaſt as much, *ignotius aut æque ignotum*.   But ſome, in their definitions, and ſtill more in their examples, ſupport the following meaning, which I ſtrongly ſuſpect to be the true derivation of the phraſe, however the *principium* and *quod in principio* might afterwards have been confounded with one another.   The philoſophy of the time conſiſted in a large variety of general propoſitions (principles) deduced from authority, and ſuppoſed to be ultimately derived from intrinſic evidence, ſelf-known, or elſe by logical derivation from ſuch principles. Theſe were at the command of the diſputant, his opponent could not but admit each and all of them : the laws of diſputation demanded† the aſſent which the geometer requires for his poſtu-

---

* Sir W. Hamilton of Edinburgh (notes on Reid, p. 761,) ſays that *principium* is always uſed for that on which ſomething elſe depends.

† Does a traditional remnant of this convention ſtill linger in the not unfrequent notion that a diſputant is entitled to the conceſſion of his *principia?* We uſed to hear ʻYou muſt grant me my firſt principles, elſe I cannot

lates. Except when, now and then, literary fociety was fhaken to its very foundations by a difpute which affected any of them, as a nominalift controverfy or the like moral earthquake. The moft frequent fyllogifm was one which, having the form *Barbara*, had a *principium* for its major, and an *exemplum* for its minor: as in ' All men are mortal (*principium*); Socrates is a man (*exemplum*); therefore Socrates is mortal. The *petitio principii*, then, occurred, when any one, to prove his cafe, made it an example of a principle which was not among thofe received, without offering to bring the former under the logical empire of the latter. And fome writers define the fallacy as occurring *fi contingat in fyllogifmo principium petere;* where by *principium* they mean the principle which generally occurs in the major premife, and by their inftances they clearly fhow that they mean to include nothing but the fimple fyllogifm of principle and example. They would leave us to infer that if any one fhould happen to conftruct a fyllogifm in which *both* premifes are principles, one or both not received, the inference, though denied by fimple denial of one or both premifes, would not be confidered as technically the *petitio principii,* which with them was, as it were, *petitio principii exemplum continentis.*

It has often been afferted that all fyllogifm is a begging of the queftion, or a *petitio principii* in the modern fenfe, an affumption of the conclufion. That all premifes do, when the argument is objectively confidered, contain their conclufion, is beyond a doubt: and a writer on logic does but little who does not make his reader fully alive to this. But the phrafe, as applied to a good fyllogifm, is a mifapprehenfion of meaning: for its definition refers it to what is affumed *in one premife.* The moft fallacious *pair* of premifes, though exprefsly conftructed to form a certain conclufion, without the leaft reference to their truth, would not be affuming the queftion, or *an* equivalent. But a further charge has been made againft the fyllogifm, namely that very often the conclufion, fo far from being deduced from the principle, is actually required to deduce it: that for inftance, in ' All men are

argue.' Cardinal Richelieu's anfwer to his applicant's *il faut vivre*, namely, *Je n'en vois pas la néceffité,* had fomething of inhumanity in it: but, as applied to the *Mais, Monfieur, il faut fe difputer* of the preceding affumption, it would generally be quite the reverfe.

mortal; Plato is a man; therefore Plato is mortal' we do not know that Plato is mortal becaufe all men are mortal, but that we need to know that Plato is mortal, in order to know that it is really true that all men are mortal.   There is much ingenuity in this argument : but I think a little confideration, not of the fyllogifm, but of how we ftand with refpeĉt to the fyllogifm, will anfwer it.

When we fay that A is B, we do not merely mean that the thing called A is the thing called B : if we fpoke of objeĉts as objeĉts, it would not matter under what name, and ' A is B ' would be no other than ' B is B ' and the very propofition itfelf would be of its own nature a mere identity, an affertion that what is, is. It feems to me that between objeĉts, thus viewed, there can neither be propofitions nor fyllogifms.   A may remind us of a thing as fuggefting one idea to our minds ; B of the fame thing as fuggefting another : and the propofition ' A is B ' then afferts that the two ftates of our mind are from the fame external fource. Our logic, in wholly feparating names from objeĉts, and dealing only with the former, makes a fort of fymbolic reprefentation of the diftinĉtion between ideas and objeĉts.

Now the objeĉtion above ftated to the fyllogifm appears to me to be founded upon thinking of the objeĉt, as if it had no names. Suppofe all things marked, each with every name which can be applied to it.   Undoubtedly then, each one marked *man* will have the mark *mortal* upon him, and fome the mark *Plato*, it may be : and by the time all the marks are put on, and to a perfon who is fuppofed to be immediately cognizant of the fimultaneous exiftence of two or more marks on the fame thing, it would be an abfurdity to attempt any fyllogifm at all.   What coexiftence of marks could there be which he muft not be fuppofed to have noted in making the induĉtion neceffary for a univerfal propofition.   When he colleĉted the elements of ' All men are mortal ' he faw $\frac{\text{Plato}}{\text{man}}$ among the reft and fet it down.   But fuppofe that his knowledge is not acquired, as to different marks, all at once : but that each coincidence of marks is to be a feparate acquifition to his mind.   Then he does not know, by the time he has found out that ' All men are mortal ' whether Plato be mortal or not.   Plato may be a ftatue, a dog, or a book written by a man of that name. *Plato* does not carry *man* with it : his major tells him nothing about Plato, until he has the minor, ' Plato is a man ' and then, no doubt, he has abſolutely acquired the concluſion ' Plato is mortal.' The whole objeᷓion tacitly aſſumes the ſuperfluity of the minor ; that is, tacitly aſſumes we know Plato to be a man, as ſoon as we know him to be Plato. Grant the minor to be ſuperfluous, and no doubt we grant the neceſſity of conneᷓing the major and the concluſion to be ſuperfluous alſo. Grant any degree of neceſſity, or of want of neceſſity, to the minor, and the ſame is granted to the conneᷓion of the major and concluſion.

In the preceding caſe, the ſyllogiſm is looked upon as one of communication, by the authors of the objeᷓion ; while at the ſame time it is tacitly aſſumed that the minor does not communicate : Plato, by virtue of our acquaintance with the name, is taken to be a man.

Moreover, it is to be noted that the propoſition uſed in argument, whether to ourſelves or to others, is very frequently not ſo much the mere attribution of one idea to another, as a declaration that *pro hac vice* the idea contained in the more extenſive term is all that is wanted, and that the differences which conſtitute the ſpecies are not to the purpoſe. Or (page 234) it is the diminution of the comprehenſion which is neceſſary, and the increaſe of extenſion is only contingent. It is ſtripping the complex idea of the unneceſſary parts, to prevent only what is requiſite. Thus any one who will aſſert that, in the Moſaic account, no animal life whatever was deſtroyed by ſlaughter before the deluge, muſt be convinced by being reminded that an antediluvian (Cain) killed Abel who was a man and therefore an animal.

With the *petitio principii* may be claſſed (for it might alſo be referred to other fallacies) caſes of the imperfeᷓ *dilemma*. Suppoſe we ſay ' Either M or N muſt be true : if M be true, Z is impoſſible ; if N be true, Z is impoſſible ; therefore Z is impoſſible.' Now if the disjunᷓive premiſe ought to have been ' either M or N or Z is true,' here would have been almoſt an expreſs *petitio principii*. For example, ſay ' A body muſt either be in the ſtate A or the ſtate B ; it cannot change in the ſtate A ; it cannot change in the ſtate B ; therefore, it cannot change at all.' Now, if the alternative A or B be neceſſary, the correᷓ ſtatement may be ' A body muſt either be in the ſtate A, or in the ſtate B, or in the ſtate of tranſition from one to the other.' Of this kind is the celebrated ſophiſm of Diodorus Cronus, that motion is impoſſible, for all that a body does, it does either in the place in which it is, or in the place in which it is not, and it cannot move in the place in which it is, and certainly not in the place in which it is not. Now, motion is merely the name of the tranſition from the place in which it is (but will not be) to that in which it is not (but will be). It is reported that the inventor of this ſophiſm ſent for a ſurgeon to ſet his diſlocated ſhoulder, and was anſwered that his ſhoulder could not have been put out either in the place in which it was, or in the place in which it was not ; and therefore, that it was not hurt at all.

4. The *ignoratio elenchi*, or *ignorance of the refutation*, is what we ſhould now call anſwering to the wrong point: or proving ſomething which is not contradictory of the thing aſſerted. It may be conſidered either as an error of form or of matter; and it is, of all the fallacies, that which has the wideſt range. Such, for inſtance, as the caſe of a writer I have read, who admits that certain evidence, if given at all, would prove a certain point; and admits that ſuch evidence has been given: but refuſes to admit the point as proved, becauſe the evidence was given in anſwer to objections, and in a ſecond pamphlet. The *pleadings* in our courts of law, previous to trial, are intended to produce, out of the varieties of ſtatement which are made by parties, the real points at iſſue ; ſo that the defence may not be *ignoratio elenchi*, nor the caſe the counter-fallacy, which has no correlative name, but might be called *ignoratio concluſionis*. If a man were to ſue another for debt, for goods ſold and delivered, and if defendant were to reply that he had paid for the goods furniſhed, and plaintiff were to rejoin that he could find no record of that payment in his books; the fallacy would be palpably committed. The rejoinder, ſuppoſed true, ſhows that either defendant has not paid, or plaintiff keeps negligent accounts; and is a dilemma, one horn of which only contradicts the defence. It is plaintiff's buſineſs to prove the ſale, from what *is* in his books, not the abſence of payment from what *is not ;* and it is then defendant's buſineſs to prove the payment by his vouchers.

It is commonly ſaid that no one can be required to prove a

negative, and often that no one *can* prove a negative. There is much confusion about this : for any one who proves a positive, proves an infinite number of negatives. Every thing that can be proved to be in St. Paul's Cathedral at any one moment is fairly proved *not* to be in more places than I can undertake to enumerate. What is meant is, that it is difficult, and may be impossible, to prove a negative without proving a positive. Accordingly, when the two sides of the question consist of a positive and negative, the burden of proof is generally considered to lie upon the person whose interest it is to establish the positive. This being understood, it is *ignoratio elenchi* to attempt to transfer the charge of proving the negative to the other party. But this rule is by no means without exception : there are many departures from it in the law, for example, though not under the most logical phrases. For instance, a homicide, as such, is considered by the law a murderer, unless, failing justification, he can prove that he had no malice. Here, in the language of the law, the homicide, supposed unjustifiable, is in itself a presumption of malice, which the accused is to rebut. It is not true, in point of fact, that such presumption exists on the mere case of homicide, independent of the manner of it : if the law will consult its own records, it will find that, for one homicide with malice of which it has had to take cognizance, there are dozens at least, done in heat of blood, and called manslaughters. But the case stands thus ;—the alternatives are few, so that proving the negative of one, which the accused is called on to do, can be done by proving the affirmative one out of a small number. There are but malice, heat of blood, misadventure, insanity, &c. to which the action can be referred. Of these few things, it is easier for the accused to establish some one out of several, above all when motive is in question (of which only himself can be in possession of the most perfect knowledge) than it is for the prosecutor to establish a particular one. And the principle on which he is called on to establish a negative (or rather another positive) is that the burden of proof fairly lies on the one to whom it will be by much the easiest. The proof of a negative, then, being as easy as, in fact identical with, the proof of one of the positive alternatives, such proof may, from the circumstances, lie upon a disputant, particularly when the number of the alternatives is few. But the *negative proof*, a very different thing, is of its own nature hardly attainable, and therefore hardly to be required. A book has been mislaid; is it in one room or the other? If found in the second room, there is proof of the negative as to the first: and almost any one who can read can be trusted to say, on his own knowledge, that in a certain room there is a certain book. But to give negative proof as to the first room, it must be made certain, first, that every book in the room has been found and examined, secondly, that it has been correctly examined. No one, in fact, can prove more than that he cannot find the book: whether the book be there or not, is another question, to be settled by our opinion of the vigilance and competency of the searcher. Controversialists constantly lay too much stress on their own negative proofs, on their *I cannot find*, even as to cases in which it is palpably not their interest to find.

Somewhat akin to the preceding is the constant fallacy of controversialists, conveyed in their strong assertion of the results of their own arguments. Few can bear to admit that there is a question for others to decide; and after summing up both sides, to separate the points which the reader is to pronounce upon. They must decide for him, and thus act both counsel and judge: probably because their arguments are not so convincing to their own minds as they wish them to be to the reader's. They prove, at the utmost, their own conviction that they have the right side: but the thing to be proved is that such conviction is well founded. They know the maxim *Si vis me flere, dolendum est primum ipsi tibi*, and think it will hold good of the reason, as well as of the feelings: as it will, to some. The consequence is, that the deliberate reader suspects them, and feels inclined rather to differ than agree: he will not dance to a writer who pipes too much. Just as " I'll tell you a capital thing," sets the hearer upon avoiding laughter, and gives him notice to try; so ' I intend to give most unimpeachable proof,' puts the judicious reader upon looking for inadmissible assumptions, and he is seldom allowed by such writers to look in vain. But, if the disputant who begins by declaring his intention to be irresistible, be suspicious, the one who ends by announcing that he is so, is absolutely self-convicted. If it be very clear, why should he say it? Does he tell his reader that he must remember to distinguish the black letters from the white paper, or does he print at the top of the book ' keep this fide uppermoft ?' Thefe things (effential as they are) he really does leave to the reader: but he dares not truft the latter to find out (though he fays it is as clear as black and white) that his arguments are fo ftrong and fo good, that nothing but wilful difhonefty, or hopelefs prejudice, can refift their force.

Another common form of the *ignoratio elenchi*, lies in attributing to the conclufion afferted fome ultimate end or tendency. Thus, an argument in favour of checking the power of the Crown is called Jacobinifm; of an increafe of that power, abfolutifm : though the argument propofed may be found, independently of its propofer's wifhes. This is a cafe in which the refult of the method is juftifiable, though the method is wrong. Many readers will remember the advice given by an old judge to a young one, ' Give your judgments without reafons; moft likely your decifions will be right; and it is juft as likely that your reafons will be wrong.' This advice fhould be followed by many of thofe who judge or decide arguments. The propofer is of a known opinion, which gives him a ftrong bias towards the conclufion of the argument. He is a witnefs (page 205), and the effect upon the mind of the receiver is to be that of the united argument and teftimony. The teftimony is, in the receiver's mind, of a low order; the propofer is a radical, and the receiver is of opinion that a radical would pick a pocket: or elfe, perhaps, the propofer is a tory, and the receiver is of the belief that a tory muft have picked a pocket. Thefe opinions may be right or wrong; but they exift: and there is certainly no formal fallacy in admitting them, as affecting the teftimony, to fubtract from the probability of the truth of the conclufion. But there is a formal fallacy, a decided *ignoratio elenchi*, in throwing all the indifpofition to receive upon the invalidity of the argument.

There is a much more culpable form of the fame fpecies. If fuch a conclufion were admitted, it would lead to fuch and fuch another conclufion, which is not to be admitted. In queftions of abfolute demonftration, this procefs is found: if B be certainly falfe, and if it be the neceffary confequence of A, then A muft alfo be falfe. But it is unfound when it takes the form, ' I believe B to be falfe; I believe it to follow from A; therefore I affume a right to difbelieve A whatever evidence may be offered for it? This fallacy is fufficiently expofed in page 209. There is a tradition of a Cambridge profeffor who was once afked in a mathematical difcuffion ' I fuppofe you will admit that the whole is greater than its part,' and who anfwered, ' Not I, until I fee what ufe you are going to make of it.' This was no doubt the extreme cafe; the more ordinary one arifes in a great meafure from the great fallacy of all, the determination to have a particular conclufion, and to find arguments for it. Obferve a certain perfon who is led on by a wily opponent in converfation : nothing is prefented to him except what his reafon fully concurs in, and no inference except what is indifputable. At a fudden turn of the argument, he fees a favourite conclufion, which he cares more for than for all the reafonings that ever were put together, upfet and broken to pieces. He confiders himfelf an ill-ufed man, entrapped, fwindled out of his lawful goods; and he therefore returns upon his fteps, and finds out that fome of the things which he admitted when he did not fee their con-fequences, are no longer admiffible. Neither he nor the oppo-nent has the leaft idea of the nature of probable arguments, and of their oppofition : both proceed as if the train of reafoning were either demonftration or nothing. The conclufion, formed perhaps upon teftimony, which is more likely to be a guide to truth for the mind in queftion than any appreciation of argument which that mind could make, muft, according to the maxims of the age, be referred to argument, and argument only. The perpetual and wilful fallacy of that mind is the determination that all argument fhall fupport, and no argument fhall fhake, the conclufion. If there were only a diftinct perception of another fource of con-viction, fo ftrong that ordinary argument can neither materially weaken, *nor materially confirm it*, there would be fenfe in the conclufion ; fenfe, becaufe there is truth. Right or wrong, fuch is the fource of moft convictions in, perhaps, moft minds : fuch fource ought therefore to be acknowledged. It would be an ex-cellent thing, if, in any difputed matter, thofe who are better fatisfied by authority of the truth of one fide of the conclufion than of the validity of argument in general, would avow it, keep their own fide, and let others do the fame. But here is the diffi-culty : the perfons who fhould avow fuch a ftate of mind are as much difpofed to make converts as others : they do not like to debar themselves from dissemination of their opinions. Accordingly they propound their best arguments, be they what they may, as what ought to produce all the conviction which themselves feel. On this point see page 194.

The whole class of *argumenta ad hominem*, having some reference to the particular person to whom the argument is addressed, will generally be found to partake of the fallacy in question. Such are *recrimination* and *charge of inconsistency*, as, 'You cannot use this assertion, because in such another case you oppose it.' But if the original argument itself should be a personal attack, then such a retort as the preceding may be a valid defence.

In many such *argumenta ad hominem*, it is not absolutely the same argument which is turned against the proposer, but one which is asserted to be like to it, or parallel to it. But *parallel cases* are dangerous things, liable to be parallel in immaterial points, and divergent in material ones. A celebrated writer on logic asserts, that no one who eats meat ought to object to the occupation of a sportsman on the ground of cruelty. The parallel will not exist until, for the person who eats meat, we substitute one who turns butcher for amusement. There is, or was, a vulgar notion that butchers cannot sit on a jury. Suppose that such a law were proposed, on the ground of the habits arising from continual infliction of death. Would it really be a counter-argument that men who eat meat have the same *animus* and are liable to acquire the same habits. It is contended (justly or not) that a desire to *take life* for *sport* is a cruel desire; to answer that those who eat flesh from which life has been taken by others have therefore also cruel desires, ought to be called arguing *a dicto secundum quid ad dictum secundum alterum quid*. The matter is clear enough. Cruelty of *intention* (the thing in question) must be settled by our judgment of the circumstance in which the sport consists. A person who seeks bodily exercise and the excitement of the chase, and who can acknowledge to himself that his object is gained on the birds which he misses, as well as upon those which he hits, even if thoughtless, cannot be said to act with cruelty of intention. But the sportsman, as he calls himself, who collects his game in one place, merely that he may kill, without exercise, or feeling of skill, is either culpably thoughtless, or else a savage, who delights in the infliction of death. Let any man aſk himſelf, whether in the event of his being called upon to vote for a perfectly abſolute ſovereign, he would feel much concerned to inquire whether the candidate was or was not a ſportſman of the firſt kind : and then let him aſk himſelf the ſame queſtion with reſpect to the ſecond.

The moſt amuſing, and perhaps the moſt common, example of the *ignoratio elenchi*, is the taking exception to ſome part of an illuſtration which has nothing to do with the parallel. The word illuſtration (though it mean *throwing light* upon a thing) is uſually confined to that ſort of light which is derived from ſhowing a proceſs of difficulty employed upon an eaſier caſe. The firſt fallacy may be committed by the illuſtrator. He has before him the ſubject matter of the premiſes, their connexion in the proceſs of inference, and the reſult produced. Either may be illuſtrated ; thus, if it be doubtful whether ſuch premiſes may be employed, the illuſtrator may throw away his mode of connexion, and chooſe another : if the proceſs of inference be doubtful, he may chooſe other premiſes : and ſo on. But he may illuſtrate the wrong point : and this is a fallacy very common to teachers and lecturers. The greateſt difficulty in the way of learners is not knowing exactly in what* their difficulty conſiſts ; and they are apt to think that when *ſomething* is made clear, it muſt be *the* ſomething. I am of opinion that the examples given of ſyllogiſms in works of logic are examples of wrong illuſtration. The point in queſtion is the form, the object is to produce conviction of the form, of its neceſſary validity. If the ſtudent receive help from an example ſtated both in matter and form, the odds are that the help is derived from the plainneſs of the matter, and from his conviction of the matter of the concluſion. If this be the caſe, he has not got over his difficulty. Many learners are puzzled to ſee that ' Every Y is X' is not a neceſſary conſequence of ' Every X is Y.' If the want of con-

---

* Every learner, in every ſubject, ſhould accuſtom himſelf to endeavour to ſtate the point of difficulty *in writing*, whether he want to ſhow the reſult to another or not. I wiſh I had kept a record of the number of times which I have inſiſted on this being done, previouſly to undertaking the explanation, and of the proportion of them in which the writer has acknowledged that he ſaw his way as ſoon as he attempted to aſk the road in preciſe written language. That proportion is much more than one half. Truly ſaid Bacon, that writing makes an exact man.

nexion be eftablifhed by an inftance, as by appealing to their knowledge that every bird is not a goofe, though every goofe be a bird, their knowledge of the propofition is not logical. The right perception may, no doubt, be acquired by reflection on inftances : but the minds which are beft fatisfied by material inftances, are alfo thofe which give themfelves no further trouble.

The illuftration being fuppofed correct, there is more than one fallacious mode of oppofing it. Some perfons will difpute the very method of illuftration of form, in which the fame mode of inferenceis applied toeafier matter; but thefe are mere beginners, hardly even entitled to a name which fuppofes the poffibility of progrefs. Others will deny the analogy of the matter, and thefe there is no means of meeting: for illuftration is *ad hominem,* and the perception of it cannot be made purely and formally inferential : a denier of the force of an illuftration is inexpugnable as long as he only denies. But when he attempts more, when he indicates the point in which the illuftration fails, he very often falls into the error of attacking an immaterial point. If any one were to contend (as fome do) that it is unlawful to take the life of any animal, he might be afked what he would fay if Guy Faux had trained a pigeon to carry the match to the vault, would it have been lawful to fhoot the bird on its way or not? There are not a few who would think it an anfwer to fay that he could not have trained the pigeon, or that pigeons were not then trained to carry.

5. The *fallacia confequentis* (now very often called a *non fequitur*) is the fimple affirmation of a conclufion which does not follow from the premifes. If the fchoolmen had lived in our day, they would have joined with this the affirmation of logical form applied to that which wants it, a very common thing among us. A little time ago, either the editor or a large-type correfpondent (I forget which) of a newfpaper imputed to the clergy the maintenance of the 'logic' of the following as 'confecutive and without flaw.' This was hard on the clergy (particularly the Oxonians) for there was no middle term, neither of the concluding terms was in the premifes, and one negative premife gave a pofitive conclufion. It ran thus,

Epifcopacy is of Scripture origin.

The church of England is the only epifcopal church in England,

Ergo, the church eftablifhed is the church that fhould be fupported.

Many cafes offend fo flightly that the offence is not perceived. For inftance 'knowledge gives power, power is defirable, therefore knowledge is defirable' is not a fyllogifm ; there is no middle term. It is a forites, as follows, 'knowledge is a giver of power, the giver* of power is the giver of a defirable thing, the giver of a defirable thing is defirable, therefore knowledge is defirable.'

It fhould be noted, however, that the copula 'gives' refembles 'is greater than' (page 5) and is an admiffible copula in inferences with no converfion, provided that 'A gives B and B gives C,' implies 'A gives C.' The fame may be faid of the verbs to bring, to make, to lift, &c. And many of thefe verbs are, by the unfeen operation of their having the effect of *is* in inference, often fupplanted by the latter verb in phrafeology. Thus we fay 'murder *is* death to the perpetrator' where the copula is *brings;* 'two and two *are* four' the copula being 'have the value of' &c. But this practice may lead to fallacies, as above fhown : which muft be avoided by attention to the clafs of verbs which communicate their action or ftate, fuch as make, give, bring, lift, draw, rule, hold, &c. &c. All thefe verbs are applied to denote the caufe of the feveral actions : fo, to give that which gives, or to bring that which brings, is to give or to bring. The boy who was faid to rule the Greeks becaufe he ruled his mother, who ruled Alcibiades, who ruled the Athenians, who ruled the Greeks, would have been correctly faid fo to do, if the matters of rule had been the fame throughout.

6. The *non caufa pro caufa.* This is the miftake of imagining neceffary connexion where there is none, in the way of *caufe,* confidered in the wideft fenfe of the word. The idioms of language abound in it, that is, make their mere expreffions of phenomena attribute them to apparent caufes, without intent to affert real connexion. Thus we fay that a tree *throws* a fhadow,

---

* Becaufe power is defirable. See page 115, as to this ftep.

to defcribe that it hinders the light. When the level of a billiard table is not good, the favoured pocket is faid to *draw* the balls. A particular cafe of this fallacy, which is often illuftrated by the words *poft hoc, ergo propter hoc,* is the conclufion that what follows in time follows as a confequence. When things are feen together, there is frequently an affumption of neceffary connexion. There is, of courfe, a prefumption of connexion : if A and B have never been feen apart, there is probability (the amount of which depends upon the number of inftances obferved) that the removal of one would be the removal of the other. It is when there is only one inftance to proceed upon that the affumption falls under this fallacy; were there but two, induCtIve probability might be faid to begin. The fallacy could then confift only in eftimating the probability too high.

As may be fuppofed, the *non caufa pro caufa* arifes more often from mere ignorance than any other fallacy. To take the two inftances that I happened to meet with neareft to the time of writing this page;—Walpole, remarking on the uniform praCtice among the old writing-mafters of putting their portraits at the beginning of their works, remarks that thefe men feem to think their profeffion gives pofterity a particular intereft in their features. Probably they did not think about it : the ufage of the day prevented any man from being chargeable with undue vanity who exhibited his phyfiognomy, and *moft of the writing mafters were themfelves engravers,* and either did their own portraits, or more probably made ufe of their acquaintance with the more celebrated engravers for whom they did the under drudgery, to get themfelves done on eafy terms. Again, Noble (in his continuation of Granger) remarks that Saunderfon had fuch a profound knowledge of mufic, that he could diftinguifh the fifth part of a note. The author did not know, firft, that any perfon who cannot diftinguifh lefs than the fifth part of a note to begin with, fhould be bound over to keep the peace if he exhibit the leaft intention of learning any mufical inftrument in which intonation depends upon the ear; and fecondly, that if Saunderfon were not fo gifted by nature, knowledge of mufic would no more have fupplied the defeCt, than knowledge of optics would give him fight.

The *fallacia plurium interrogationum* confifts in trying to get

one anfwer to feveral queftions in one. It is fometimes ufed by
barrifters in the examination of witneffes, who endeavour to get
yes or *no* to a complex queftion which ought to be partly anfwered
in each way, meaning to ufe the anfwer obtained, as for the whole,
when they have got it for a part.   An advocate is fometimes
guilty of the argument *à dicto fecundum quid ad dictum fimpli-*
citer : it is his bufinefs to do for his client all that his client might
honeftly do for himfelf. Is not the word in Italics frequently
omitted? *Might* any man honeftly try to do for himfelf all that
counfel frequently try to do for him? We are often reminded of
the two men who ftole the leg of mutton ; one could fwear he
had not got it, the other that he had not taken it.   The counfel
is doing his duty by his client ; the client has left the matter to
his counfel.   Between the unexecuted intention of the client,
and the unintended execution of the counfel, there may be a
wrong done, and, if we are to believe the ufual maxims, no
wrong doer.   The anfwer of the owner of the leg of mutton is
fometimes to the point, ' Well, gentlemen, all I can fay is, there
is a rogue between you.'   That a barrifter is able to put off his
forenfic principles with his wig, nay more, that he becomes an
upright and impartial judge in another wig, is curious, but certainly true.
  The above were the forms of fallacy laid down as moft effential to be ftudied by thofe who were in the habit of appealing to
principles fuppofed to be univerfally admitted, and of throwing
all deduction into fyllogiftic form.   Modern difcuffions, more
favourable, in feveral points, to the difcovery of truth, are conducted without any conventional authority which can compel
precifion of ftatement : and the neglect of formal logic occa-
fions the frequent occurrence of thefe offences againft mere rules
which the old enumeration of fallacies feems to have confidered
as fufficiently guarded againft by the rules themfelves, and fufficiently defcribed under one head, the *fallacia confequentis*.   For
example, it would have been a childifh miftake, under the old
fyftem, to have afferted the univerfal propofition, meaning the
particular one, becaufe the thing is true in moft cafes.   The rule
was imperative : *not all* muft be *fome*, and even *all*, when not
known to be *all*, was *fome*.   But in our day nothing is more
common than to hear and read affertions made in all the form,

and intended to have all the power, of univerſals, of which nothing can be ſaid except that moſt of the caſes are true. If a
contradiction be aſſerted and proved by an inſtance, the anſwer
is ‘Oh! that is an *extreme caſe.*’ But the aſſertion had been
made of *all* caſes. It turns out that it was meant only for ordinary caſes; why it was not ſo ſtated muſt be referred to one of
three cauſes;—a mind which wants the habit of preciſion which
formal logic has a tendency to foſter, a deſire to give more
ſtrength to a concluſion than honeſtly belongs to it, or a fallacy
intended to have its chance of reception.

The application of the *extreme caſe* is very often the only teſt
by which an ambiguous aſſumption can be dealt with: no wonder that the aſſumer ſhould dread and proteſt againſt a proceſs
which is as powerful as the ſign of the croſs was once believed
to be againſt evil ſpirits. Where anything is aſſerted which is
true with exceptions, there is often great difficulty in forcing the
aſſertor to attempt to lay down a canon by which to diſtinguiſh
the rule from the exception. Every thing depends upon it: for
the queſtion will always be whether the example belongs to the
rule or the exception. When one caſe is brought forward which
is certainly exception, the aſſertor will, in nine caſes out of ten,
refuſe to ſee why it is brought forward. He will treat it as a
fallacious argument againſt the rule, inſtead of admitting that it
is a good reaſon why he ſhould define the method of diſtinguiſhing the exceptions: he will virtually, and perhaps abſolutely, demand that all which is certainly exception ſhall be kept back,
ſimply that he may be able to aſſume that there is no occaſion to
acknowledge the difficulty of the uncertain caſes.

The uſe of the extreme caſe, its deciſive effect in matters of
demonſtration, may furniſh preſumption as to what it is likely to
be in matters of aſſerted near approach. As in the following inſtance. It ſeems almoſt matter of courſe, when ſtated, to thoſe
who have not ſtudied the ſubject of life contingencies, that the
proper value of a life annuity is that of the annuity made certain
during the average exiſtence of ſuch lives as that of the annuitant.
That if, for example, perſons aged 22 live, one with another,
40 years, an office which receives from every ſuch perſon the
preſent value of forty payments certain, will, without gain or loſs,
in the long run, be able to pay the annuities. If this be (as was

stoutly contended by some writers of the last century) a universal truth, it will hold in this extreme case. Let there be two persons, one of whom is certain to die within a year from the grant (and therefore never claims anything) and the other of whom is certain to live for ever. It is clear that the value of an annuity to both is 0 + the value of a perpetual annuity. But the *average life* of both is eternal: one perpetual duration makes the average of any set in which it is, perpetual. Hence by the false rule the value is *two* perpetual annuities, or just double of the truth.

We might suppose that most persons have no idea of a universal proposition: but use the language, never intending *all* to signify more than *most*. And in the same manner principles are stated broadly and generally, which the assertor is afterwards at liberty to deny under the phrase that he does not *carry them so far* as the instance named. It would not do to avow that the principle is not always true: so it is stated to be *always true*, but not capable of being *carried* more than *a certain length*. Are not many persons under some confusion about the meaning of the word *general?* In science it always has the meaning of *universal*: and the same in old English. Thus the catechism of the church of England asserts that there are two sacraments which are *generally* necessary to salvation: meaning necessary for all of the *genus* in question, be it man, Christian, member of the church, or any other. But in modern and vernacular English, *general* means only *usual*, and generally means usually.

A great deal of what is called evasion belongs to this head, or to that of the *ignoratio elenchi*, as the sophist answers. The advocates, for instance, of the absolute unlawfulness of war never tell, unless pressed, what they think of the case of resistance to invasion. Is the country to be given up to the first foreigner who chooses to come for it? Sometimes the extreme case comes into play: sometimes the assertion that no one will come; which is irrelevant as to the question what would be right if he did come.

Among amusing modern evasions are ‘ There is no *occasion* to consider that’ and ‘ I do’nt consider it in that *point of view.*’ Any one who watches the manner in which men defend their opinions will frequently see “A is B and B is C, therefore A is C’ answered, not by denial of either premise, but by ‘ that is not the proper point of view' or 'I don't fee it in that light.' This fhould be called the confufion between logic and perfpective.

The denial of one univerfal is often made to amount to, or to pafs into, the affertion of the oppofite, or fubcontrary, univerfal. This craving after general truths, the moft manifeft fault of the old logicians in their choice of premifes, did not expire with them. Bacon fays ' the mind delights in fpringing up to the moft general axioms, *that it may find reft.*' Many perfons are defirous of ' fettled opinions,' which is well; unlefs by fettled opinions they mean univerfal, as is often the cafe. That fome are and fome are not is no fettlement: it makes every cafe require examination, to fee under which it falls. And with the above we may couple the tendency to believe that refutation of an argument is proof of the falfehood of its conclufion, and that a falfe confequence muft be a falfe propofition. Hence it arifes that fo many perfons dare not give up any argument in favour of a propofition which they fully believe: they think they abandon the propofition.

It fometimes happens that an affertion is made, which it is difficult to fuppofe can be anything but a cafe of a univerfal propofition: and yet the affertor takes care not to make his propofition univerfal, but perfifts in the particular cafe. A logician in our day has afferted that when Calvin fays that *all* officers of the church fhould be elected by the people, he muft be underftood as fpeaking in reference to *deacons* only, becaufe the affertion is made in the chapter on deacons. If it had been roundly ftated that all univerfal propofitions are to have their univerfes limited by the headings of the works or chapters in which they occur—for inftance, that the affertion that all men are mortal, occurring in a hiftory of England, is to be taken as made of Englifhmen only—there would have been at leaft no ambiguity. But as it is, we are left to furmife whether this be meant, or whether the propofition be to apply to Calvin only, or to Reformers only, or to men whofe names begin with C, &c. The odds are that the *application* of a univerfal propofition will be dictated by the heading of a chapter: but the extent to which a premife is *afferted as true* is not to be judged of by that to which it is *wanted for ufe*: and the lefs, the nearer we go to the day of the old logicians.

T

Wrong views of the quantity of a propofition are as frequent as any fallacies. *Some*, meaning moft, and *fome*, meaning few, are frequently confounded. This is the neceffary confequence of the nature of human knowledge, in which we can but rarely form a definite idea of the proportion which the extent fpoken of bears to the whole. It is part of the value of the mathematical theory of probabilities, that the mind is accuftomed to the view of refults drawn from perfectly definite fuppofed cafes; as ufelefs, it may be, in themfelves, as many of the queftions in a book of arithmetic, but neverthelefs good for exercife. It is not furprifing that fallacies about quantity fhould be capable of moft ftriking expofure in queftions concerning meafurable quantity, that is, in queftions of mathematics : nor that there fhould be claffes of fallacy of which it is difficult to illuftrate the detection by any other inftances. What can be more clear, for example, to ordinary apprehenfions than the broad ftatement that ' of things of the fame kind, that which is fometimes right muft be better than that which is always wrong. But a little confideration will fuggeft that what is always wrong may be as good as that which is fometimes right, if we do not know how to diftinguifh the cafes in which the latter *is* right : and alfo that what is not much wrong, generally, may be more ufeful than that which is moftly very wrong, when it is not abfolutely right. A watch which does not go is right twice a day : but it is not fo ufeful as one which does go, though very badly.

To give an account of all the fallacies which depend upon wrong notions of quantity would require much fpace, and more affumption of mathematical knowledge in my reader than is confiftent with my plan. But I may mention the miftaken ufe of abfolute terms and notions in queftions of degree. There can be, a difputant will fay, but a right and a wrong ; and if this be not right, it is wrong. Many perfons will announce that their watches are *quite right*, abfolutely at the true time, to a fecond : and will end by giving the time which was fhown when they looked, as being accurately that of the inftant at which they announce it. The proverb *Fruftra fit per plura, quod fieri poteft per pauciora* contains an inaccuracy of degree : a bargain which cofts twenty fhillings and is worth fifteen, is not twenty fhillings loft, but only five, though the vexation of the party overreached will feldom fuffer him to fee this.

Proverbs in general are liable to this miſtake. They are often uſed in exactly the ſame manner as the firſt principles of the old logicians. In fact, remembering that theſe firſt principles were bandied from mouth to mouth till they were perfectly *proverbial*, as we now call it, among the learned; and obſerving the application of our modern proverbs, as made by the maſs of thoſe who have not profited by mental diſcipline, we may ſee that the faults of the ſchoolmen are only thoſe of the ordinary human mind. It is hard indeed if there be a purpoſe which a proverb cannot be found to ſerve: it is a univerſal propoſition of no very definite meaning, ſanctioned by uſage, having the appearance of authority, and capable of ſtretching or contracting like Prince Ahmed's pavilion. One only is allowable—*In generalibus latet error:* this deſtroys all the reſt, and then, when cloſely looked at—commits ſuicide.

All miſtakes of probability are eſſentially miſtakes of quantity, the ſubſtitution of one amount of knowledge and belief for another. It is often difficult to convey a proper notion of the degree of force which is meant to be given; and ſtill more ſo to retain it throughout the whole of a diſcuſſion. A perſon begins by ſtating an explanation as poſſible, or probable enough to require conſideration, as the caſe may be. The forms of language by which we endeavour to expreſs different degrees of probability are eaſily interchanged; ſo that, without intentional diſhoneſty (but not always) the propoſition may be made to ſlide out of one degree into another. I am ſatisfied that many writers would ſhrink from ſetting down, in the margin, each time they make a certain aſſertion, the numerical degree of probability with which they think they are juſtified in preſenting it. Very often it happens that a concluſion produced from a balance of arguments, and *firſt preſented* with the appearance of confidence which might be repreſented by a claim of ſuch odds as four to one in its favour, is afterwards *uſed* as if it were a moral certainty. The writer who thus proceeds, would not do ſo if he were required to write $\frac{4}{5}$ in the margin every time he uſes that concluſion. This would prevent his falling into the error in which his partiſan readers are generally ſure to be more than ready to go with him, namely, turning all balances for, into demonſtration, and all balances againſt, into evidences of impoſſibility.

One of the great fallacies of evidence is the diſpoſition to dwell on the actual poffibility of its being falfe : a poffibility which muft exift when it is not demonftrative. Counfel can bewilder juries in this way till they almoft doubt their own fenfes. A man is fhot, and another man, with a recently difcharged piftol in his hand, is found hiding within fifty yards of the fpot, and ten minutes of the time. It does not follow that the man fo found committed the murder : and cafes have happened, in which it has turned out that a perfon convicted upon evidence as ftrong as the above, has been afterwards found to be innocent. An aftute defender makes thefe cafes his prominent ones : he omits to mention that it is not one in a thoufand againft whom fuch evidence exifts, except when guilty.

All the makers of fyftems who arrange the univerfe, fquare the circle, and fo forth, not only comfort themfelves by thinking of the neglect which Copernicus and other real difcoverers met with for a time, but fometimes fucceed in making followers. Thefe laft forget that for every true improvement which has been for fome time unregarded, a thoufand abfurdities have met that fate permanently. It is not wife to tofs up for a chance of being in advance of the age, by taking up at hazard one of the things which the age paffes over. As little will it do to defpife the ufual track for attaining an object, becaufe (as always happens) there are fome who are gifted with energies to make a road for themfelves. Dr. Johnfon tells a ftory of a lady who ferioufly meditated leaving out the claffics in her fon's education, becaufe fhe had heard Shakfpeare knew little of them. Telford is a ftanding proof (it is fuppofed by fome) that fpecial training is not effential for an engineer.

The difpofition to judge the prudence of an action by its refult, contains a fallacy when it is applied to fingle inftances only, or to few in number. That which, under the circumftances, is the prudent rule of conduct, may, neverthelefs end in fomething as bad as could have refulted from want of circumfpection. But upon dozens of inftances, fuch a balance would appear in favour of prudence as would leave no doubt in favour of the rule of conduct, even in the inftances in which it failed. The fallacy confifts in judging from the refult about the conduct of one who had only the previous circumftances to guide him. ' You acted unwifely, as is proved by the refult,' is a paralogifm, except when it implies ' You did, as it happens in this inftance, take a courfe which did not lead to the defired refult.' Take a ftrong cafe, and the abfurdity will be feen. A chemift makes up a prefcription wrongly, and his cuftomer leaves him for another : this other, fo it may happen, makes it up ftill more wrongly, and poifons the patient. Who would venture to fay that he acted unwifely, as is proved by the refult, in leaving the tradefman whom he knew to be carelefs, for another of whom he knew no harm. The only way in which blame can be imputed, is when it can be faid ' You acted unwifely, in not finding out, as you might have done, that the refult which has happened is the one which was likely to happen.' One refult proves very little as to the fuperior wifdom of the courfe which produced it ; feveral may give a prefumption of it, and the greater the number, the greater the prefumption.

So little is this thought of, that the common phrafe, ' I acted for the beft,' meaning originally ' I acted in the manner which under the circumftances, appeared likely to lead to the beft refults,' very often lofes its proper meaning, and is ufed as fynonymous with ' I acted with good intentions.'

Thefe, and many other points, I can only flightly touch on : I will proceed to notice a few other caufes of error.

And firft, of equivocations of ftyle. I have before referred to fuch a phenomenon as the alteration of a good fyllogifm into a bad one, to make the fentence *read better.* But nothing ever reads well (for a continuance) except the natural current of a writer's thought. I fhould like it to be the law of letters, that every book fhould have inferted in it the printer's affidavit, fetting forth the number of verbal erafures in the manufcript, fair copies being illegal. It would be worth at leaft one review.

There is a wilful and deliberate equivocation, which it is fuppofed the age demands. It is the ufe of fynonymes, or fuppofed fynonymes, to prevent the fame word from occurring twice in the fame paffage. So far is the neceffity of this practice recognized, that there are few printing-offices in London, the *readers* of which do not *query* the fecond introduction of any word which prominently appears twice. And then the author obeys the hint, ftrikes out one of the offenders, fticks in a dictionary equivalent, and would have been content if the printer's reader had done it for him. And fo he writes a *good ftyle.* To be fure, he does not say what he meant, exactly ; for synonymes are seldom or never logical equivalents : but what is that to elegance of expression?

The demand for non-recurrence of words arises from the public (I beg its pardon) not knowing how to read. If, when a word occurs twice, the proper emphases were looked for, and observed, there would be nothing offensive about the repetition. It is the reader who makes one and one into two, by giving both units equal value. Take this sentence from Johnson, (the first I happened to light on, in the preface to Shakspeare), and read it first as follows :—" He therefore indulged his natural *disposition:* and his *disposition*, as Rymer has remarked, led him to comedy :" and then as follows—" He therefore indulged his *natural disposition ;* and *his* disposition, as Rymer has remarked, led him to comedy." This reading is what the context requires, and the ill effect of the repetition is next to nothing. Take the next sentence :—" In tragedy he often *writes*, with great appearance of toil and study, what is written *at last* with little felicity : but in his comic scenes he seems to produce, without *labour*, what *no* labour can improve." These were the first instances I found, from a chance opening of the *Elegant Extracts*, purposely chosen as a miscellany. The laws of thought generally dictate this rule, that the first occurrence of a word is the more emphatic of the two : the lesson of experience is, that a writer who prevents recurrence by the use of the dictionary of synonymes, is a good style-maker for none but a bad reader, and may very possibly be a good arguer for none but a bad logician. Of course, I should not deny that recurrence of both word and emphasis is a defect, if it be frequent.

The confusion between the means and the end, and putting one in the place of the other, is well enough known in morals : but there is a corresponding tendency to forget the distinction between the principle which is to be acted on, and the rule of action by which adherence to that principle is secured. A reference to the derived rule is in all respects as good as one to the first principle, between parties who understand both, and the connexion between them. But those who understand the rule only, are apt to forget that a rule may or may not be the true expression of a principle, according to the circumstances in which it is proposed to apply it. If, indeed, it were of universal appli-

cation, thofe who do and thofe who do not underſtand the principle might be on the fame footing as to fecurity: but there are
few fuch rules.

The preceding caution may be applied in all departments of
thought, in law and in logic, in morals and in arithmetic. It is
impoffible, for inftance, to ftate the rule of three in fuch a manner as eafily to include the cafes in which it ſhall apply, and exclude thofe to which it does not. To fay that it muſt be uſed
where the fourth quantity, the one fought, is to be a fourth proportional to the three which are given, though correct, ſtill leaves
it open to inquiry what are the cafes in which this condition is to
be fatisfied: and many cafes might be, and are propofed, in which
the inquiry is not eafy to a beginner. In law, there are not only
rules, but rules for their application. To an unlearned ſpectator,
particularly in the courts of equity, in which the advocate addreſſes
a judge, and not a jury, the argument takes that technical form
which makes many perfons think that the whole law is, at beſt,
only arbitrary rule. It may be that fome of thofe who there
addreſs the court can make nothing better of it: and juſt as there
are arithmeticians, and good ones too, who are but the flaves, and
never the maſters, of their proceſſes, fo there may be advocates,
and even judges, who have not one element of the legiſlator in
them. But there are enough of a higher ſpecies.

The great art of ufing rules is to apply them in aid, and not
in contravention, of the principles which they are intended to
embody. A rule may have exceptions, it is faid; but this is
hardly a correct ſtatement. A rule with exceptions is no rule,
unleſs the exceptions be definite and determinable: in which cafe
the exceptions are excluſions *by another rule.* The parallel is
perfect between rules and propoſitions (page 143). Thus, ' All
Europe, except Spain and Portugal' is a univerfal propoſition;
but 'All the ſtates of Europe except two' is a particular one. A
rule which applies to all ſtates except Spain and Portugal is a
rule: but a rule which applies to all except two (unknown) is no
rule. When it is ſtated, in ordinary language, that every rule is
fubject to exception, it is meant, for the moſt part, that the circumſtances under which adherence to the rule gains the object,
are thofe which moſt frequently occur, and that the circumſtances
under which adherence to the rule would defeat the object are

rare. If this were remembered, much confufion would often be faved. We want a word which fhall fo far exprefs *rule*, that it fhall imply that which will generally fucceed, without the notion of *obligation* which accompanies that of *rule*, and which perpetually mifleads. We want, in fact, the *rule nifi* of the courts, which is to be a rule unlefs caufe be fhown againft it : and which will, in moft cafes, be ultimately made abfolute, but is not abfolute from the beginning.

The common miftake is, that the rule *nifi* is an abfolute rule, and that therefore it may be fubftituted for its leading object or firft principle, and that even the very words which exprefs that object gained, may be taken as equally expreffive of fatisfaction of the rule, and *vice verfâ*. For inftance, it is commonly ftated that the rule by which a difcoverer is determined, is publication ; that he who firft publifhes the difcovery, is *to be held* the difcoverer ; one lapfe more, and it is faid that he *is* the difcoverer ; yet one more, and it will be faid that the publication is the difcovery. The very remarkable circumftances attending the recent difcovery of the planet *Neptune*, involving points of peculiar intereft and delicacy, have caufed this rule to be much difcuffed, and have brought out every variety of ftatement of it. The thing to be determined is the *actual truth of the queftion*, the real hiftory of the human mind with regard to it. No one has a right under any rule, no matter what its authority, nor by whom impofed, to fubftitute the thing which is not, for the thing which is, or the lefs probable for the more probable. If philofophers were to attempt, by a law of their own framing, to fubftitute the conventional refult for the real one, the common fenfe of mankind would difpute their authority, and reverfe their decifion. The firft rule (*nifi*) is undoubtedly that the firft printer is the firft publifher, the fecond, that the firft publifher is the difcoverer. Thefe will, unlefs caufe be fhown againft them, be made abfolute in every cafe. A notion which is very prevalent, namely, that the firft publifher has *therefore* the rights of the difcoverer, is as incorrect as that the firft *printer* is therefore the firft *publifher*. To take the current language, one would fuppofe that printing one hundred copies would be held better than circulating one thoufand in manufcript, and that even though the firft publifher could be proved to have plagiarifed, he has ftill the rights of difcovery.

Juſt as (page 244) early notions make laws of literal interpretation ſuperſede thoſe of intended meaning, ſo, in the earlier
ſtages of law, rules are often made to over-ride the principles on
which they profeſs to be founded, and to defeat truth and common
ſenſe. There is more excuſe here than there would be in a
queſtion of ſcience, for peace and convenience are main objects
of law, and it may be that rigid adherence to a rule, as a rule, at
a certain avowed ſacrifice of truth and juſtice, may be the only
practicable means of preventing a larger ſacrifice of both. In old
times, the rule of affiliation, *Pater eſt quem nuptiæ demonſtrant*,
was held ſo abſolutely, that the huſband of the mother would be
the legal father, though the two had been confined in two different jails a hundred miles apart for twelve months preceding the
birth of the child. The modern law has made this rule to be no
more than it ought to be, namely, one which muſt hold unleſs
the contrary be proved.

It is not uncommon, in diſputation, to fall into the fallacy of
making out concluſions for others by ſupplying premiſes. One
ſays that A is B; another will take for granted that he muſt believe B is C, and will therefore conſider him as maintaining that
A is C. But it may be that the other party, maintaining that
A is B, may, by denying that A is C, really intend to deny that
B is C. In religious controverſy, nothing is more common than
to repreſent ſects and individuals as *avowing* all that is eſteemed
by thoſe who make the repreſentation to be what, upon their premiſes, they ought to avow. All parties ſeem more or leſs afraid
of allowing their opponents to ſpeak for themſelves. Again, as
to ſubjects in which men go in parties, it is not very uncommon
to take one premiſe from ſome individuals of a party, another
from others, and to fix the logical concluſion of the two upon
the whole party: when perhaps the concluſion is denied by all,
ſome of whom deny the firſt premiſe by affirming the ſecond,
while the reſt deny the ſecond by affirming the firſt. Any ſect
of Chriſtians might be made atheiſts by logical conſequence, if it
were permitted to join together the premiſes of different ſections
among them into one argument. This is a fallacy which, however common, could eaſily be avoided, and would be, if thoſe
who uſe it cared for anything but victory. But there is another
form of the ſame, which every one is ſubject to, and which it is

not so easy to perceive. It is that of drawing upon our former selves for the premises which are to guide us for the time being. Conclusions remain in our minds long after the grounds on which they were formed are abandoned : and it may happen that one premise of an argument will still have force, when the very reasons on which the second premise is now admitted are contradictory of those which once induced us to admit the first. Thus many who have learnt to advocate the legal toleration of opinions which they still believe, by force of education, to be absolute crimes against society, are logically the advocates of toleration of crime ; whereas, the arguments which they have learned to think valid for the first premise, ought, if worth anything, to teach them to deny the second. I have myself heard from one mouth in one conversation (of course not in one part of it) that all sins against the Creator are sins against society, that all sins against society ought to be punished by society, that certain opinions then named are sins against the Creator, and that it is the height of injustice to punish any one for his opinions.

In printed controversy, the statement of the opposite opinion or assertion may be made by description without *citation* (by chapter or page), by description with citation, or by *quotation* with or without description. The first is not allowable. The presumption is strong that a person who opposes an opinion, imputes an error, or makes a charge, upon the writings of another, is bound at least to cite, in a manner which cannot be mistaken, the part of those writings to which he refers. There are writers who refer descriptively and even commentatively, putting the reference of citation, and thus (as Bayle says Moréri constantly does) lead the reader to suppose that the words of their paraphrase and comment are those of the passage itself. I do not see that quotation is obligatory, though highly desirable : but the reader must remember, when there is only citation, that it is not the author cited who speaks, but the person who brings him forward. It is a man's own account of his own witness : with the advantage of an apparent offer of enabling the reader to go and verify the statement for himself. If the citer be honest, the passage in question exists : if judicious, it is to the effect stated. Consequently, whenever the citer's honesty or judgment is expressly in question, no mere citation is admissible.

When citations are few they ought perhaps to be quotations : when they are many, it may be impracticable to make them fo. But extenfive citation ought to be encouraged. Lazy readers do not like it : they are not pleafed to have a power of verification offered of which they do not mean to avail themfelves ; and they would rather, in cafe of being mifled, have to throw the blame upon the author than upon their own non-acceptance of the offered means of verification. Accordingly, they exprefs their difguft at " pages loaded with references." But the more diligent readers confider every citation as a boon. At the fame time it is to be remembered that there are writers who, relying on the common difinclination to verify, add a large number of citations, and give the appearance of a ftrong body of authorities, which are often nothing to the purpofe, and fometimes not taken from actual examination, but copied from other writers.

Perhaps the greateft and moft dangerous vice of the day, in the matter of reference, is the practice of citing citations, and quoting quotations, as if they came from the original fources, inftead of being only copies. It is in truth the reader's own fault if he be taken in by this, or by the falfe appearance of authority juft alluded to ; for it is in his own power to certify himfelf of the truth : though there may be difficulty when the citations are many, or when fome of them are from very rare books. Honefty and policy both demand the exprefs ftatement of every citation and quotation which is made through another fource. If a perfon quote what he finds of Cicero in Bacon, it fhould be ' Cicero (cited by Bacon) fays, &c.' It has happened often enough that a quoter has been convicted of altering his author, and has had no anfwer to make except that he took the paffage from fome previous quoter.

Quotations are frequently made with intentional omiffion and alteration. But no rule ought to be more inflexible than that all which is within the marks of quotation ought to be a literal tranfcript of the book quoted. Sometimes the omiffion is made becaufe part of the fentence is unneceffary, *as the quoter thinks.* But this is juft the point which he has no bufinefs to decide without letting his reader know that he *has* decided it, which is eafily done by the recognized mark of omiffion (. . . . .) If a perfon would quote the Æneid for the antiquity of Carthage, he has no bufinefs to write down, as from Virgil, 'Urbs antiqua fuit Carthago :' it fhould be ' Urbs antiqua fuit ..... Carthago,' if he decide upon omitting 'Tyrii tenuere coloni.' In this cafe, not only may the omiffion make the propofition appear more categorical than it is in the original, turning it from 'There was an old city, Carthage,' rather towards 'Carthage was an old city;' but a reader may choofe to think that the omitted words qualify the epithet, or even offer proof deftructive of it. What if he fhould deny the antiquity of Tyre? The *omiffion* may (or may not) be right, but the omiffion without notice, or *fuppreffion*, is certainly wrong.

Moreover, it is dangerous to truth to fhorten without notice, inafmuch as thofe who quote the quotation will be apt to do the fame thing; that is, thinking they have the whole paffage, to fhorten it further. What this may end in, no one can predict: but miftakes have been brought about in this way quite as abfurd as any that ever were made. It may reafonably be fuppofed that many very ludicrous errors arife thus. A good many years ago, I fucceeded, by means of a fhortened quotation, put away until it was wanted, in arriving at, and publifhing, the conclufion that Archimedes was once fuppofed to have been an anceftor of Henry IV. of France. The real purport of the fentence was that he was fuppofed to have been an anceftor of the Sicilian martyr St. Lucia, on whofe day Henry IV. was born. It has happened that A has been faid to have afferted in a fecond book, that B related the death of C, when the truth is that A faid in the firft book that B died many years before C (See the *Companion to the Almanack* for 1846, page 27). I do not fpeak of omiffions made becaufe the part omitted would prove more than the quoter likes : this of courfe is fraud.

Unjuftifiable as unnoted omiffions may be, ftill more fo are additions and alterations. Writers have fometimes inferted gloffes of their own, into the text which they quote, either as addition or alteration. Explanatory *additions* may eafily be made within brackets [ ], which are underftood marks of fuch a thing : but alterations are intolerable. But why, the reader may afk, are fuch things infifted on? Is not the fimple rule, *Be honeft*, enough to include thefe and hundreds of things like them, without detail? To this I reply that within a twelvemonth before the time I write this, a clergyman, a man of high education and character both, publifhed a fermon in which he gave a verfe from the Bible within marks of quotation, in which he wilfully ftruck out one word, and inferted another, without notice : and his fermon went through feveral editions, either without detection, or without that detection leading to fuccefsful remonftrance. I do not fuppofe there was difhonefty here; but rather the following reafoning ;—' I am fure it was meant; therefore I may ftate that it was faid.' Such reafoning is one of the curfes of our literature.

There is one alteration within the marks of quotation which may at firft feem reafonable : it is alteration of grammar to bring the quoted phrafes into connected Englifh with the quoter's context. As when a man fays " I know" and another perfon, quoting him, fays " He knows." But it is furely juft as eafy to put down He fays " I know." There is often an alteration of emphafis in this adaptation of grammar, and generally an introduction of irony : and it is the *premier pas* to fomething worfe. As far as I have feen, thofe who do it as a matter of courfe, are apt fometimes to put their own paraphrafes under marks of quotation. A writer fhould fuit his own grammar to that of his quotation, and not the converfe.

Omiffion of context, preceding or following the quotation, may alter its character entirely : and this is one of the moft frequent of the fallacies of reference, both intentional and unintentional. The only way to infure full confidence is to give the egg in its fhell : that is, to begin at a point which clearly precedes the immediate fubject of quotation, and to continue until the matter is as clearly paft : to give a fentence preceding and a fentence following the matter quoted for its own fake, diftinguifhing the latter. This is not always conclufive : becaufe the fubject may be refumed in a fentence or two, or in another part of the book. But it will inform the reader, in moft cafes, whether he is or is not likely to differ from the quoter as to the meaning of the part quoted. And this refers particularly to quotations of opinion : thofe of fact may often be more briefly treated with fafety.

In quoting ancient authors, in cafes where the text is not notorious, the various readings fhould be given, efpecially when it is an author whofe text has an indifferent reputation for accuracy. Or if this cannot be done, the edition fhould be cited. Shameful things have occurred in controverfy, by omiffion of a part of the ordinary text, which the quoter *chofe to confider* as an interpolation, without choofing to confider that the reader ought to have liberty to judge for himfelf on that point.

Among the cafes of indirect citation, fhould be included that in which a book is mentioned as exifting, not on the authority of the writer's own eyes, but on that of a catalogue. The number of nonexifting books which are entered in catalogues and copied, as to their titles, into other works, is greater than any one who has not examined for himfelf would fuppofe poffible. In thofe who know this, confidence is deftroyed; and this fometimes affects queftions of opinion. I am told that Dugald Stewart, who had a ftrong notion of the practical impoffibility of prefenting Euclid in a fyllogiftic form, never would believe that it had been done by Herlinus and Dafypodius. Such a work is entered in catalogues : but I muft fay that the ftate of catalogues is fuch that Stewart or any one elfe had full right to doubt of any work, upon no other than catalogue evidence. The work does exift, and I have a copy of it. But, feeing how matters ftand, no one has a right to declare that an old book ever was written, without informing his reader on what fort of evidence he relies.

CHAPTER XIV.

On the Verbal Defcription of the Syllogifm.

IN page 75, I have made a firft attempt to exprefs the relations of propofitions in language which will make fyllogifms capable of verbal defcription, and the inference of their conclufions matter of felf-evidence. It is defirable that this fhould be more fully done, and I accordingly renew the attempt, with the beft words of defcription which I can find or make. Any one who can fuggeft words which better convey the meaning to himfelf, will find it eafy to fubftitute them for thofe which I have ufed.

The conditions to be satisfied are, that the words should have as much imported meaning as possible, that every word and its contrary should have the connexion of contrariety well marked, and that the verbal descriptions should be capable of being easily formed from the symbolic notation. As may be supposed, these conditions are to some extent contradictory of each other: the sacrifice of either to the others is then to be made to the most advantageous effect.

There are two ways in which it may be necessary to describe the syllogism. First, the one hitherto used throughout this work, in which one concluding term is referred to the other by the intervention of the middle term: what X is of Y, and what Y is of Z, determine what X is of Z. Secondly, that in which the two terms are referred to one another by comparison of both with the middle term: what X and Z severally are of Y determine what X is of Z.

In the first mode, the middle term is mentioned, and its description is middle in the sentence; while the reference term is understood in the predicate of each description. Thus when we say ‘a subcontrary of a supercontrary is a subidentical,’ it is that a subcontrary of a *supercontrary (of* Z) is a subidentical (of Z); and the *supercontrary of* Z is the middle term.

In the second mode, the middle term is understood in the subject, and the concluding terms in the predicate, of the description of the syllogism. Thus when we say ‘genus and species are genus and species,’ it means that two terms which are severally genus and species of the middle term (one entirely containing, the other entirely contained in, the middle term) are genus and species to one another (the first genus, the second species).

Now it will be very easily seen, that the way to change the first description into the second is as follows. Say the description runs thus, ‘ P of Q is R.’ If Q be its own correlative, as happens when Y and Z are convertibly connected, then ‘ P of Q ’ merely becomes ‘ P and Q :’ but if Q have another, Q^0, for its correlative, then ‘ P of Q ’ becomes ‘ P and Q^0.’ Again, if R be its own correlative, its plural takes its place : but if R have R^0 for its correlative, it becomes ‘ R and R^0.’ Thus ‘ subcontrary of supercontrary is subidentical ’ of the first mode, becomes ‘ subcontrary and supercontrary are subidentical and superidentical ' meaning that C_i and C' of the middle term are D_i and D' of each other. But ' fubcontrary of fuperidentical is fubcontrary' becomes ' fubcontrary and fubidentical are fubcontraries.'

I need hardly fay that ' P of Q is R ' with refpect to X in terms of Z, muft be read ' Q^0 of P^0 is R^0 ' with refpect to Z in terms of X. This rule we have already ufed.

It is thus fhown that it is only neceffary to dwell on the firft mode; and now arifes the queftion what words are to be employed in defcribing the eight ftandard propofitions. After a good deal of confideration, I prefer to denote the univerfal relations by pofitive terms, and their contrary particulars by the correfponding negative ones: not without full perception of the facrifice which enfues of the firft condition above mentioned to the third.

The words *genus* and *fpecies* immediately fuggeft themfelves to denote the relation of Y to X and X to Y in X)Y. Thefe are to be underftood as employed up to their limit; or the genus and fpecies may be coextenfive. For two names which have nothing in common, as in X.Y, I propofe to fay that they are *externals* of each other. And for two names which have nothing out of one or the other, as in x.y, that they are *complements* of each other. Remember that complemental does not mean *only juft complemental* (which is contrary), but may be contrary or fupercontrary.

In X:Y, I call X a *non-fpecies* of Y, and Y a *non-genus* of X. Thefe words have not as much as I could wifh of imported meaning, nor are there any pofitive terms which I can propofe to fupply their places. They appear as fynonymous with *not entirely contained in* and *not containing the whole*. In XY, let X and Y be *non-externals ;* and in xy, let X and Y be *non-complements*. Accordingly, in defcribing what X is with refpect to Y, we have as follows, fhowing the fubftitutions which occur in reading the fyllogiftic fymbols into this language.

A_i, fpecies	O_i, non-fpecies.
A', genus	O', non-genus.
E_i, external	I_i, non-external.
E', complement	I', non-complement.

If we confider *genus* and *complement* as *larger terms*, and *fpecies* and *external* as *fmaller ones*, and if we put down each univerfal followed by its two weakened particulars, writing firft that which is of the fame accent, we have

Univerfal.	Firft weakened form.	Second weakened form.
A' Genus	I' non-complement	I_1 non-external.
A_1 Species	I_1 non-external	I' non-complement.
E' Complement	O' non-genus	O_1 non-fpecies.
E_1 External	O_1 non-fpecies	O' non-genus.

Thus it appears that the primary weakened form of a larger name contains a larger name, and of a fmaller a fmaller: and the contrary for the fecondary forms. The words primary and fecondary do not refer to importance, but only to order of derivation: thus A_1 was in our table X)Y, weakened into XY, before it became y)x, weakened into yx or xy.

The rules for forming particular fyllogifms by weakening univerfal premifes may now be repeated. In a univerfal fyllogifm, fubftitute for the *firft* premife and for the conclufion their *primary* weakened forms, or for the *fecond* premife and for the conclufion their *fecondary* weakened forms. In a ftrengthened fyllogifm, fubftitute for the *firft* premife its *fecondary* form, or for the *fecond* premife its *primary* form.

I now write down the whole body of fyllogifms, that the reader may exercife himfelf in the independent comprehenfion of their meaning, and in affent to their inferences; deducing the particular fyllogifms from the univerfals only.

Univerfal and particular Syllogifms.

Symbol.	Defcription of X with refpeft to Z.
$A_1A_1A_1$	Species of fpecies is fpecies.
$I_1A_1I_1$	Non-external of fpecies is non-external.
$A_1I'I'$	Species of non-complement is non-complement.
A'A'A'	Genus of genus is genus.
I'A'I'	Non-complement of genus is non-complement.
$A'I_1I_1$	Genus of non-external is non-external.
$A_1E_1E_1$	Species of external is external.
$I_1E_1O_1$	Non-external of external is non-fpecies.
$A_1O'O'$	Species of non-genus is non-genus.

$\begin{cases} A'E'E' \\ I'E'O' \\ A'O_1O_1 \end{cases}$ Genus of complement is complement.
Non-complement of complement is non-genus.
Genus of non-species is non-species.

$\begin{cases} E_1A'E_1 \\ O_1A'O_1 \\ E_1I_1O' \end{cases}$ External of genus is external.
Non-species of genus is non-species.
External of non-external is non-genus.

$\begin{cases} E'A_1E' \\ O'A_1O' \\ E'I'O_1 \end{cases}$ Complement of species is complement.
Non-genus of species is non-genus.
Complement of non-complement is non-species.

$\begin{cases} E_1E'A_1 \\ O_1E'I_1 \\ E_1O_1I' \end{cases}$ External of complement is species.
Non-species of complement is non-external.
External of non-species is non-complement.

$\begin{cases} E'E_1A' \\ O'E_1I' \\ E'O'I_1 \end{cases}$ Complement of external is genus.
Non-genus of external is non-complement.
Complement of non-genus is non-external.

Strengthened Syllogisms.

$A_1A'I'$ Species of genus is non-complement.
$A'A_1I_1$ Genus of species is non-external.
$A_1E'O'$ Species of complement is non-genus.
$A'E_1O_1$ Genus of external is non-species.
E_1A_1O' External of species is non-genus.
$E'A'O_1$ Complement of genus is non-species.
E_1E_1I' External of external is non-complement.
$E'E'I_1$ Complement of complement is non-external.

No person could propose to himself a better exercise in the acquisition of command over language, than practising the demonstrations of these relations, or more properly their reduction into specific showing, as to the matter of the inference, in what its extent consists. For instance, 'the complement of a non-complement is a non-species': How, and by how much? The non-complement leaves something which is neither in the term understood, nor in that non-complement. This, the complement of that non-complement must fill up: and by this then, at least, the complement of the non-complement is not in the term understood, of which it is therefore so far non-species.

In the preceding view, I have particularly confidered the connexion between contrary forms, and the adaptation of language to that connexion. But in the firft derivation of the fimple fyllogifms (page 88) the univerfals were related, not to their contraries, but to their particular concomitants. I now proceed to the confideration of this view, and to the juftification, on felf-evident principles, of the affertion that there is a real and ftriking affinity between the univerfal fyllogifm and its concomitants, as $A_iA_iA_i$ and $O'A_iO'$, $E'E_iA'$ and $E'I'O_i$, &c.

The complex propofitions D_i, D', and C_i contain each a univerfal which, in common language, is generally confounded with it, and a particular, the exiftence of which is therefore for the moft part fuppofed in thought to accompany the univerfal. The remaining univerfal, E', is differently circumftanced : if we fay that X and Y complete the univerfe, we fhould generally mean that they only juft complete it, and fhould not think of the fuper-contrary relation, or of their overcompleting it. To be contained but not to fill ; to contain with room to fpare, or to *overfill*; to exclude and be excluded without completion ; and to exclude and be excluded with completion (or to complete and be completed without inclufion) ;—are our moft ufual ideas of the relations of the extent of names.

The reduction of the complex propofition to the fimple univerfal, when done by removal of the concomitant *particular*, is in all cafes a lowering of the quantity, by the removal of an excefs, as follows :—

D_i means that X is contained in Y, and more is contained.
D' means that X contains Y, and contains more.
C_i means that X excludes Y, and excludes more.
C' means that X completes Y, and *more than completes.

Drop the fecond claufes, and D_i, &c. are reduced to A_i, &c. Drop the firft claufes, and it would feem as if we had ftill the

---

* The alteration of grammar here feen is in deference to the word *complete*, the beft I can get. In this propofition, the verb refers to *the univerfe*, and it is X(joins in completing the univerfe)Y and joins in completing more (than the univerfe).

complex propofitions ; for *more* will contain its tacit reference to
that which it *is more than*. Let this tacit reference be dropped,
and then we have, inftead of the whole complex propofition, only
its particular. And this abandonment is actually made in common language, by what would be called perhaps a lax, but is a
very logical, ufe of the word *more*. ' There are more than fifh on
the dry land,' would be perfectly intelligible, and not as implying
that there were any fifh : ' he was actuated by more than the
motive, &c.' very often means ' other than the motive' &c.

Now, in the complex fyllogifm, as we have feen (page 81), the
exceffive part of the conclufion (whence comes its fecond claufe,
its additive *more*) is the fum of the exceffive parts of the premifes.
If one of the complex premifes be deprived of its affertion of
excefs, or lowered into a fimple univerfal, the conclufion ftill
remains, though not *à fortiori*, neceffarily. This being done, *the
valid excefs of the conclufion depends upon the excefs of the remaining premife;* and the concomitant particular fyllogifm, confidered
as part of the mixed complex fyllogifm, is the expreffion of this,
without the reft. Finally, the excefs may be ufed in the lax, or
non-correlative, fenfe, and then the concomitant fyllogifm ftands
by itfelf.

For example, $O_1A'O_1$ may be read thus :—Confider O_1 as
concomitant of A' in D'. ' X contains more than [fomething
that is not in] Y ; Z contains X ; therefore, Z contains more
than [fomething that is not in] Y.' If ' more than Y ' mean ' Y
and more,' this would be $D'A'D'$. Again, $O'E_1I'$ is ' more than
X [fomething not X] is contained in Y ; Y excludes Z ; therefore, X excludes more than Z [fomething not in Z].' If ' more
than X ' were ' X and more,' &c. : this would be $D_1E_1C_1$. And
fo on for other cafes.

I now proceed to what I may call the *quantitative* defcription
of the fyllogifm : by which I mean the expreffion of its cafes in
terms of the quantities *only* of its names and propofitions, leaving
the alternative of affirmation and negation to be fettled by the
law of thefe quantities. My reafon for the prefentation of the
fyftem in fo many different points of view will be obvious enough :
that which claims to be complete, muft fhow itfelf to contain juft
the fame, and no more, as to refults, whatever may be the principle which is chofen as the bafis of conftruction.

Every propofition, in fpeaking of two names, fpeaks of their contraries, and (page 63) of the four terms, two direct and two contrary, two are univerfal and two are particular. Since univerfal and particular are themfelves properly contraries, (for 'Every X' is 'Xs, *known* to be all' and 'Some Xs' are 'Xs, *not known* to be all') let us fignify the univerfal and particular forms of the *propofition* by V and v. Again, fpeaking of a *name*, let its mode of entry, univerfal and particular, be denoted by T and t. Writing down V(or v) applied to T(or t), T(or t) we can make eight varieties, which give us the eight ftandard forms applied to one order, fay XY; as follows :—

$$A_1 = V(Tt) \mid A' = V(tT) \mid E_1 = V(TT) \mid E' \doteq V(tt)$$
$$O' = v\,(Tt) \mid O_1 = v\,(tT) \mid I' = v\,(TT) \mid I_1 = v\,(tt)$$

Thus I' or xy, may be defcribed as the particular in which both terms are univerfal: for X and Y are both univerfal in xy, or x:Y, or y:X. And $v(TT)$ defcribes it thus.

If, underftanding the order to be XY, YZ, XZ, we write down any three propofitions, we make an attempt at a fyllogifm, valid or not, as the cafe may be: as in

$$V(Tt).\dot{v}(tt).V(tT) \text{ or } VvV(Tt,tt,tT)$$

which muft be A_1I_1A'. It will affift the memory to obferve that fub-fymbols have VT or vt at the beginning, fuper-fymbols vT or Vt. Alfo, that affirmatives have an even number of capitals (*none** or *two*) and negatives an odd number (*one* or *three*). A univerfal and its particular concomitant have the fame entries of T and t, and contranominals have inverted modes of entry of thefe letters. The convertibles have T in both places, or t: the inconvertibles have T and t.

Firft, it is unneceffary to write down the term-letters of the conclufion, for they muft be taken from the premifes, in every cafe in which the conclufion is the ftrongeft that can be drawn from the premifes; and our fyftem has no others (nor, indeed,

---

* The reader muft here follow the mathematician in confidering o as an even number.

has the Aristotelian any other except *Bramantip*). Thus, TT,tt being the term letters of the premises, strike out the second T and the first t, which refer to the middle term, and Tt must belong to the conclusion. To prove this, observe that we know that t in the premise cannot give T in the conclusion : therefore T cannot give t; for if, the term being Z, T gave t, then, putting z properly in its place, t would give T, which it cannot. Again, we know that the valid forms, as to propositions, are VVV, VVv, vVv, Vvv; so that v occurring once only, must come third, and V must come in the first pair. Further, in the four term letters of the premises, VVV, vVv, Vvv, require Tt, or tT, to come in the middle, while VVv alone requires TT, or tt. Observe these laws, and every formation which can take place under them leads to a valid syllogism. Putting dots to represent a blank place, we form the eight universal syllogisms by filling up the blanks in VVV(.. t,T ..) and VVV(.. T,t..); the eight strengthened syllogisms from VVv(.. T,T ..) and VVv(.. t,t ..); the eight particulars which begin with a universal from Vvv(.. t,T ..) and Vvv(.. T,t..); and the eight particulars which begin with a particular from vVv(.. t,T ..) and vVv(.. T,t..). And, under the rules just given, we have no other cases.

Taking the preceding as a basis, we might make the rules of accentuation follow from it. For, since the first blank in our symbol, and the first concluding term, must agree, and since accents depend only on the first two letters in the symbol of a proposition, we may proceed as follows. Let K and L, each of them, mean T or t, as the case may be, but with the proviso that what it means in either place it shall mean in the other. Then, in VVV(KT,tL,KL) and in vVv(KT,tL,KL), in which symbols of conclusion are introduced, we see that the first and third accents must agree, which is part of the direct rule. As to the first and second accents, they agree in the first instance above, if K be t, which puts an even number of capitals in the first symbol VKT, or an affirmative proposition at the commencement : they differ if K be T, which puts a negative proposition first. In the second instance, they agree if K be T, which puts an affirmative first, &c. I leave it to the reader to deduce the other cases of this rule, the inverse rule, and also that premises give an affirmative, or a negative, conclusion, according as they have like or unlike signs. And thus it will appear, that the symbolic rules given in chapter V, are really expressions of the general rules of quantity.

It will be observed that the concomitant syllogisms of a universal have the same term letters as that universal, and only change VVV into Vvv, or vVv. Also, that the inverted syllogisms of page 96 only invert the order of all the term-letters, and the letters of the premises, when different.

Thus, $E_1A'E_1$ being VVV(TT,tT), its concomitants I'A'I' and E_1O_1I', are vVv(TT,tT) and Vvv(TT,tT). But the inverted form $A_1E_1E_1$ is VVV(Tt,TT). Contranominals have different quantities in all the term-letters. The weakened forms of a universal change the first premise letter and the first term letter, or the second of both. Thus, $E_1E'A_1$ being VVV(TT,tt), its weakened forms, $O_1E'I_1$ and E_1O_1I', are vVv(tT,tt) and Vvv(TT,tT).

The forms of the numerical syllogism (page 161) may be recovered by few and easy rules, in which the premises as they stand determine the conclusion, as follows :—Let ξ be designated as the number of X, and ξ' as that of x; and so on. Let a term of the conclusion be called *direct* when it is in the premise, and *inverse* when its contrary is in the premise. Then,

1. In every case, the conclusion has the sum of the quantities mentioned in the premises, as part of the expression of its quantity.

2. For every inverse term in the conclusion, the number of its direct term appears in the quantity of the conclusion, subtracted. Thus, x in a premise, with X in the conclusion, must have $-\xi'$ in the concluding quantity. But the direct terms of the conclusion never introduce anything into the concluding number.

3. When the entrances of the middle term are similar (YY, or yy), the terms of the two forms of conclusion are both direct and both inverse, with subtraction of the number of the middle term in the former, addition of the number of its contrary in the latter. Thus, yy gives $-\eta'$ in the direct, $+\eta$ in the inverse form.

4. When the entrances of the middle term are dissimilar (Yy,

or yY), each form of conclufion has one direct and one inverfe term ; and no number from the middle term enters the concluding quantity.

Thus, the conclufions from $mx\mathrm{Y}+n\mathrm{YZ}$ are immediately written down as

$$(m+n-n)\mathrm{x}\mathrm{Z} \text{ and } (m+n+n'-\xi'-\zeta)\mathrm{X}\mathrm{z}:$$

while thofe from $mx\mathrm{Y}+n'\mathrm{yz}$ are at once

$$(m+n'-\xi')\mathrm{X}\mathrm{z} \text{ and } (m+n'-\zeta')\mathrm{x}\mathrm{Z}$$

There are relations exifting between the forms of the fyllogifm which I have not confidered. For inftance, the defcription of X with refpect to Z being that it is a fpecies, (A₁), the defcription of its contrary, x, is that it is a fupercontrary, (E'). If then we give the name of *contradefcriptives* to A₁ and E' we find that A' and E₁, I₁ and O', I' and O₁, are alfo contradefcriptives. The arrangement of fyllogifms by contradefcriptives, and the laws of connexion thence refulting, will be an eafy exercife for the ftudent.

APPENDIX.

I.

Account of a Controversy between the Author of this Work and Sir William Hamilton of Edinburgh ; and final reply to the latter.

THIS appendix contains an account of a controversy in which some of the matters treated in the preceding work involved me with Sir William Hamilton, Professor of Logic and Metaphysics in the University of Edinburgh. It has produced four publications (to which I shall refer as I, II, III, IV) namely:

I. 'Statement in answer to an assertion made by Sir William Hamilton, Bart. . . . . by Augustus De Morgan, ... (London, octavo, R. and E. Taylor, pp. 16, published April 30, 1847.)

II. 'A letter to Augustus De Morgan, Esq. . . . . on his claim to an independent rediscovery of a new principle in the theory of syllogism. From Sir William Hamilton, Bart. Subjoined, the whole previous correspondence, and a postscript in answer to Professor De Morgan's "Statement,"' (London and Edinburgh, octavo, Longman and Co., Maclachlan and Co. pp. 44, exclusive of 'Prospectus' hereinafter mentioned: received by me May 22, 1847.)

III. Letter from me to Sir W. Hamilton, dated May 24, published in the *Athenæum* Journal of May 29.

IV. Letter from Sir W. Hamilton to me, dated June 2, published in the same Journal of June 5.

There are two questions involved, one concerning my character, the other purely literary. The former stands thus. March 13, Sir W. Hamilton informed me by letter that (the Italics are his own words) *to* him *it is manifest that* for a certain *principle* I was *wholly indebted to* his *information,* and *that if I should give* it *forth as a speculation of* my *own* (which I had done to himself, and meant to do, as he knew, and have since done, in print) I should, even *though recognizing always* his *priority, be guilty both of an injurious breach of confidence towards* him *and of false dealing towards the public.* This *hypothetical charge,* and *derogatory supposition of which he may formerly have surmised the possibility* (such are his subsequent qualifications of it) is unreservedly retracted at the *beginning* and *end* of II: but it is frequently insinuated in the *middle,* by proposing things as difficult to be explained otherwise, by hint that *others* may believe it, by *hopes* that they will not, by charges of falsehood, &c. &c. For the *formal* charge is substituted imputation of lapfe of memory, intellectual confufion, &c. The following is the programme of the firft intended argument, (II. p. 4.)

'I confefs, that, for a time, I regarded your pretenfion, as an attempt
'at plagiarifm, cool as it was contemptible.

'From this view, feeling, information, reflection turned me; and I
'now, Sir, tender you my fincere apology, for admitting, though founded
'on your own ftatements, an opinion fo derogatory of one, otherwife fo
'well entitled to refpect.

'In itfelf, this view was, to me, painful and revolting.—The cha-
'racter, too, which you bear among your friends, I found to be wholly
'incompatible with a fuppofition fo odious. You are reprefented as an
'active and able man, profound in Mathematics, curious in Logic, wholly
'incapable of intentional deceit, but not incapable of chronological mif-
'takes. Your habitual confufion of times is, indeed, remarkable, even
'from our correfpondence. Your dates are there, not unfrequently of
'the wrong month, and not always, even of the right year. With much
'acutenefs, your works fhow you deficient in architectonic power, the
'concomitant of lucid thinking; and, that you are not guiltlefs of intel-
'lectual rafhnefs is fufficiently manifeft, from your pretention to advance :
'Logic, without having even maftered its principles.'

With regard to the fubfequent infinuation of a retracted charge, my explanation (believing as I do, that Sir W. Hamilton always fpeaks fubjective truth) is that his mind infenfibly fell back to its old bias as he felt that the fubftitute for his charge wanted ftrength : my conclufion is, that it is unneceffary henceforward to notice any thing he may fay or write on my character : and my determination is to act accordingly.

Sir W. Hamilton's pamphlet contains about a fcore and a half of quotations, on which hang fundry jokes and fneers, fome of them at mathematicians in general, and myfelf as one of the body. On thefe I fhall only fay that my notions of the common fenfe of controverfy, and my determination to perfift, generally, in the tone of refpect to my opponent's learning and character which I have hitherto preferved, would, were there nothing elfe, prevent my adopting the habit of which they are fpecimens. But as no man willingly ftands an unreturned fire of facetiæ without defiring to prove that his forbearance does not arife from want of ammunition, I will permit myfelf (difclaiming the *animus* under which fuch things are ufually written) juft to fhow that quotation, application, allufion, fneer, joke, and fling at an opponent's ftudies, are all among the weapons which I could have employed, if I had thought them worthy of my antagonift, or of thofe whom I want to convince.

I might, for inftance, have written fomething like the following ;—

Among the affets of the old logicians, difcovered when the fchools were fwept out, there was found, as is well known, the queftion *Utrum chimæra bombinans in vacuo poffet comedere fecundas intentiones :* a very good title, as Curll would have faid, wanting nothing but a treatife written to it. Now whether it be *comedere,* or whether the fchoolmen invented *comedere,* Sir W. Hamilton, on whom their mantle has fallen, has written the treatife, and fuccefsfully maintained the affirmative. His notion that his communication could give any hint, is clearly and aptly defcribed by *chimæra*, his ftyle by *bombinans*, his proof by *vacuum :* and the *fecond intentions*, above noticed, chewed up and given forth with his firft ones, are a practical example of the poffibility of the Q.E.I. He, or rather the bombinating chimæra which has perfonified itfelf in his form, as the ἅλος ὄνειρος did in that of Neftor, is thus both retractor and detractor. But though the tranfition from flops to folids generally indicates convalefcence, yet, as here made manifeft, the paffage from liquid to dental may be only the growing weaknefs, the perifcence, of the cafe.

I affert the following documents to be all that are relevant with refpect to the literary part of the controverfy. They are given at the end of this appendix.

A is an extract from a communication of mine to the Cambridge Philofophical Society, made before I received any communication what-foever from Sir W. Hamilton. I affert it to contain a diftinct an-nouncement and ufe of the principle of *quantification of the middle term*, be that middle term fubject *or predicate.* On this point the reader is to judge.

B is a communication from Sir W. Hamilton to me. The reader is to judge firft, whether it contain anything which is intelligible with refpect to any fyftem of fyllogifm ; fecondly, whether, if it fhould fo con-tain anything, that fomething would have been information to me who had written A, on fome matter afterwards found in C.

C is the relevant part of an addition made by me to A, when the latter came before me in proof. The reader is to judge firft, whether C contain anything more than an application of A ; fecondly, if fo, whe-ther that fomething more is derived from anything intelligibly hinted at in B.

The only bare fact on which Sir W. Hamilton and myfelf are at iffue is this. I affert and maintain that the matter of C was written in my poffeffion before I received B : Sir W. Hamilton holds me mif-taken, and thinks he can prove from the correfpondence that in this point my memory has failed. This I continue to treat as irrelevant : for we are both agreed that the *corpus delicti*, if *delictum* there be, lies in C containing fomething not fubftantially contained in A, but fuffi-ciently hinted at in B. Any reader who thinks that C does contain fomething fuggefted by B which is not in A, may declare againft the correctnefs of my memory ; any one who thinks the contrary, will hold it of no confequence whether my memory on the difputed fact be good or bad. With the firft reader I have no cafe : with the fecond I have all I think worth caring about.

Sir. W. Hamilton maintains my letters to be effential parts of the cafe. They *may* become fo, as foon as it is *pointed out* what C contains which is hinted at in B, and not contained in fubftance or principle, in A. When Sir W. Hamilton *points out*, by citation from C, what he alleges to have been taken, and by citation from B, what he thinks it has been taken from, and when I thereupon fail to produce equivalent knowledge from A or elſe to expoſe the irrelevance of his citation from B—then thoſe letters may become of importance. This he has not done, though ſpecially challenged to do ſo: and when I come to diſcuſs III and IV, it ſhall appear that he admits he has not done it.

I now give the beſt account I can of the origin of the diſpute, pre-miſing, that up to this 3d of September, 1847, I do not abſolutely know what the ſyſtem is which I am charged with appropriating. There is *a* ſyſtem which I think is moſt probably the thing in queſtion: but a ſyſ-tem containing a defeſt of ſo glaring a charaſter, that I will not attribute it to Sir W. Hamilton, who deſcribes his own as "adequately teſted and matured" until he expreſſly claims it, or until I have the moſt indu-bitable proof.

In the common, or Ariſtotelian propoſition, the quantities of the ſub-jeſt and predicate are determined, the firſt *by expreſſion* or *implication*, the ſecond *by the nature of the copula* (ſee page 57 of this work). And the only quantities conſidered are *all* and *ſome ;* the latter meaning any-thing that *not none* may mean, ſome, it may be all but not known to be all, perhaps not more than one. The matter contained in A ſuggeſted itſelf to me in the ſummer of 1846, and was forwarded to Cambridge with the reſt of the memoir on the 4th of Oſtober.

I will now introduce Sir W. Hamilton's deſcription of the various kinds of quantity (II p. 31, 32).

' Your "Statement" is chiefly plauſible from a wretched confuſion ' of diſtinſt things. This confuſion, with which you delude yourſelf, ' and many of your readers, is of two independent ſchemes of logical ' quantification ; the one, aſſerting *an increaſe* in the expreſſly *quantified* ' *terms*, the other, *a minuter diviſion of the forms of quantification itſelf.* ' To diſintricate this entanglement, we have ſimply to conſider, in' their ' contraſts, the *three* following ſchemes of quantification :—

' The *firſt* ſcheme is that which logically—confines all expreſſed ' quantity to the *Subjeſt*, preſuming the *Predicate* to be taken—in ne-' *gative* propoſitions, always determinately in its *greateſt* and *leaſt* ex-' tenſion (univerſally and ſingularly), in *affirmative* propoſitions, always ' indeterminately in *ſome part* of its extenſion (particularly).

' The *ſecond* ſcheme is that which logically —extends the expreſſion ' of quantity to *both* the propoſitional terms; and allows the *Predicate* to ' be of *any quantity*, in propoſitions of *either quality*. This not only ' ſupplies a capital defeſt, but affords a principle on which Logic ob-' tains a new and general development.

' The *third* ſcheme is that which logically—admits *more expreſſed* ' *quantities* than a determinately leaſt or greateſt extenſion (quantity ſin-' gular and univerſal), and an indeterminately partial extenſion (quantity ' particular.) This, though it correſts, *perhaps*, an omiſſion, yields no ' principle for a general logical development.

' The firſt doſtrine is the common or Ariſtotelic ; the ſecond is mine; ' and in the third—in ſo far as you have gone, and apart from the con-' ſideration of right or wrong—I do not queſtion your originality.

' Now, the ſecond and third ſchemes are both oppoſed to the firſt, ' but in different reſpeſts ; conſequently the ſecond and third may, each

' of them, combine with itfelf, *either* the whole other, *or* that part of
' the firft to which it is not itfelf oppofed. More is impoffible.

' Let the following be noted:*—*Your* OLD *view (that in the body of
' the Cambridge Memoir) is a combination of the* THIRD *fcheme of quan-*
' *tification with the* FIRST; *your* NEW *view (that in its Addition) is a*
' *combination of the* THIRD *fcheme of quantification with the* SECOND: *and*
' *the confufion, of which you are* NOW *guilty, is the recent and uniform,*
' *and perverfe identification, in your* PRESENT " Statement," *of the* SECOND
' *fcheme with the* THIRD.

' Before, however, proceeding to comment on your confufion of the
' fecond and third fchemes, I may alfo relieve a confufion in the term
' *definite* and its reverfe, *indefinite,* as applied to logical quantification.

' In the *firft,* common, or Ariftotelic meaning, *definite,* or more pre-
' cifely *predefinite* (διορισ‌τός, προσ‌διορισ‌τός,) is equivalent to *expreffed,*
' *overt,* or, more proximately, to *defignate* and *pre-defignate;* in this
' fenfe, *definite quantity* denotes *expreffed,* in oppofition to merely *under-*
' *ftood, quantity.*

' In the *fecond* meaning, that which I have always ufed, (and certain
' ancients, I find, were before me,) *definite* is equivalent to determinately
' *marked out;* a fenfe in which *definite quantity* is *extenfion undivided*
' *or indivifible, univerfal* or *fingular* (this including any collected plu-
' rality of individuals) as oppofed to *particular quantity.*

' In the *third* meaning, which you have ufurped, *definite* is equivalent
' to *numerically fpecified;* and in this fenfe, a *definite* is an *arithmetically*
' *articulate* quantity, as oppofed to one arithmetically inarticulate.—
' This your meaning of the word I did not, before the appearance of
' your " Statement," apprehend; for of courfe I prefumed you to ufe it
' in its firft or common meaning, from which you never hint that you
' confcioufly intend to deviate.'

Three fchemes of quantity are here mentioned.

Firft, the ordinary one.

Secondly, that in which the *ordinary quantities, all* and *fome,* are applied in every way to both fubject and predicate.

Thirdly, that in which numerically definite quantity is applied to
fubject or predicate or both: the effential diftinction of this cafe is *nume-*
rical definitenefs: it really contains the fecond fyftem, when numerical
quantity is algebraically expreffed. Of thefe, it appears, Sir W. Hamilton claims the fecond, or rather, the application of fuch a fcheme to
the fyllogifm. What then is it? I fuppofe it to be the following. My
order of reference is XY.

---

* Let the following alfo be noted:—My *old* view (that in the body of the Cambridge paper) is entirely on the *firft* fcheme, except in one *digreffive* fection and one
fubfequent paragraph (from both of which A is quoted) in which the *fecond* and *third*
are combined: my next view (that in the addition) is alfo a combination of the *fecond*
and *third* fchemes: and my " Statement" contained alfo a uniform, but not recent,
identification of the fame *fecond* and *third* fchemes, which I never feparated in thought
until I faw this paragraph. Any one who can form an opinion of the way in which
the fubject would prefent itfelf to the mind of a mathematician, will fee that the fecond
fcheme would prefent itfelf concomitantly with, and as an effential part of, the *alge-*
braical form of the third. A. De M.

All X is all Y means that X and Y are identical: it is my D. *All X is some* Y *is* A$_1$. *Some X is all Y* is A'. *Some* X is *some* Y is I$_1$. As to negative propositions, *All X is not all Y* is E$_1$. *Some X is not all Y* is O$_1$. *All X is not some Y* is O'. *Some X is not some Y* is true of all pairs of terms one of which is plural. In its indefinite form, it is what I have in Chapter VIII. called *spurious*.

The propositions of this system are then the complex D, or A$_1$+A', the six Ariftotelian forms A$_1$, A', E$_1$, O$_1$, O', I$_1$, and the spurious form, which may be called U. In looking over (Sept. 5) Sir W. Hamilton's pamphlet, I happened to light on the affertion (incidentally made) that his system gives *thirty-fix* valid moods in each figure. On examining the preceding system, I find this to be the case. I should not have published the refults, had not Sir W. Hamilton made it necessary for me to comment on them. I shall denote the proposition U, or 'Some Xs are not some Ys' by X :: Y; and I shall, supposing each case to be formed in the first figure, then transpose it into my own notation.

1. There are *fifteen* forms in which D enters. Whenever D is either of the premises, the other premise and conclusion agree. Thus we have A$_1$DA$_1$, DUU, &c. &c.

2. *Fifteen* Ariftotelian forms A$_1$A$_1$A$_1$, A'A'A'; A$_1$E$_1$E$_1$, E$_1$A'E$_1$; A$_1$O'O', O$_1$A'O$_1$; A'O$_1$O$_1$, O'A$_1$O'; A'I$_1$I$_1$, I$_1$A$_1$I$_1$; E$_1$I$_1$O', I$_1$E$_1$O$_1$; A'A$_1$I$_1$; A'E$_1$O$_1$, E$_1$A$_1$O'.

3. *Six* more U syllogisms A'O'U, O$_1$A$_1$U; A'UU, UA$_1$U; I$_1$O'U, O$_1$I$_1$U.

The two things to be confidered are;—the introduction of the identical proposition; and that of the spurious one, as I call it.

It is, I suppose, a fundamental rule of all formal logic, that every proposition must have its denial, its contradiction. Now D has no simple contradiction in this system: that O' and O$_1$ both contradict it (and also E$_1$) is true: but the mere contradiction is the disjunction 'O'. or O$_1$'. A person who can show that one or the other of these is true, has demonstratively contradicted D, even though it could be proved impossible to determine which of the two it is.

The proposition U is ufually spurious. But if we introduce it, we must introduce its contradictory alfo. Now if either X or Y be plural names, it must be true: consequently, the contradiction of U is 'X and Y are singular names, and X is Y.' When a syllogifm having the premife U is introduced, either that premife may be contradicted, or it may not. If it may, there is no form to do it in: if it may not, then it is a spurious proposition, and cannot, by combination with others, prove anything but a like spurious conclusion.

Let X :: Y denote 'Some Xs are not some Ys,' and X$_1$Y$_1$ denote 'there is but one X and one Y, and X is Y.' Then either X :: Y or X$_1$Y$_1$ must be true, and one only. A logical system which admits one and not the other, which contains an affertion incapable of contradiction without going out of the system, can hardly be said to be "adequately tested and matured," and is not felf-complete. The proposition X$_1$Y$_1$ includes in itfelf the conditions of D, and is a kind of *singular form* of D.

I prefume, from the number of Sir W. Hamilton's moods, *thirty-fix*, as above obtained, that the contradiction neither of D nor of U finds a place. Admit them, and the contradiction of U alone (call it V) demands fixteen new moods in each figure. I will now proceed.

In my publication, fpeaking now of (A) what was fent to Cambridge before I communicated with Sir W. Hamilton, I had no quantification intermediate between the ordinary one, and the numerical one applied to either fubject or predicate, as wanted in the *canon of the middle term* there given. Look at the laft of the feven fyllogifms in the *fecond extract*, where *both the predicates*, being of the middle term are quantified, and the condition of validity is quantitatively ftated. But for ' $Y_1 + Y_2$ lefs than 1' fhould be read ' $y_1 + y_2$ greater than 1.' The equivalence of this to ' $Y_1 + Y_2$ lefs than 1' is a miftake. In the *firft extract*, the general canon is given which is afterwards ufed in C.

Up to the time when Sir W. Hamilton publifhed his letter in reply to my ftatement, (II), I never had feparated the idea of his fecond fcheme of quantification from that of the third.

Thus then we ftood on October 3, when I fent my paper to Cambridge. Sir W. Hamilton had been teaching the application of the ordinary quantities to both fubject and predicate: I had arrived at the algebraical reprefentation of the numerical quantification of terms, whether fubject or predicate matters not, as long as they were middle terms.

1846, *October* 6. My communication (containing A) was in the hands of Dr. Whewell (as he informs me) for tranfmiffion to the Cambridge Society: I never faw it again till the next February. *October* 7, Sir W. Hamilton wrote to me, in anfwer to an application of mine on the *hiftory* of the fyllogifm, further informing me that he taught an extenfion and fimplification of its *theory*, which he offered to communicate. *November* 2, (the offer having been accepted) Sir W. Hamilton forwarded the communication B, which I give entire; confifting of a letter, and the *Requifites* which he had furnifhed to his ftudents, for a prize Effay. *December* 28, he wrote again, forwarding a printed *Profpectus* of his intended work on logic. This is not material; for, on receiving it, I thought certain, what from the previous communication I had thought poffible, that Sir W. Hamilton was in poffeffion of the theory of numerically definite fyllogifms (but this was a miftake of mine, as will prefently appear). I accordingly, to preferve my own rights, immediately forwarded (as will prefently be ftated more in detail) an identifying defcription of the fheets of paper on which my numerical theory was written, and an account of both my fyftems (in letters dated *December* 31, 1846, and *January* 1, 1847). Of this, Sir W. Hamilton (who has publifhed both letters) is my witnefs. 1847, *February* 27, I dated the addition to the proof fheet of my Cambridge paper, which was defpatched to Cambridge the next day. This addition contains C, which itfelf contains (in fubftance) all that part of my letter of *January* 1 which refers to the difputed point. *March* 13, Sir W. Hamilton wrote the letter containing the charge of plagiarifm; having been for two months prevented by illnefs from refuming the fubject.

All fubfequent correfpondence referred to proceedings, and not to the fubject matter of the charge.

Many days before the middle of October, I had applied the fyftem of quantification in the manner fhewn in C. Sir W. Hamilton thinks my memory has failed here: I know better. My memory does not depend upon a date, but upon the opening of the Univerfity College Seffion, which takes place in the middle of October. But it matters nothing, for the notion of the complete quantification of a predicate, *when wanted becaufe it is the middle term,* will prove the poffeffion of that procefs as well as quantification *in all cafes whether wanted or not.* On receiving B, I looked with curiofity at 2°, on which, in fact, Sir W. Hamilton grounds his declaration of having made a communication. He demands of his pupils,

'The reafons why common language makes an *ellipfis* of the *expreffed quantity,* frequently of the *fubject,* and more frequently of the *predicate,* though both have always their quantities in thought.'

On looking at this, and feeing mention of the quantities which the terms have in thought, *in common language,* I took it for granted that the common quantities were fpoken of: namely, that of the fubject from the tenor of the propofition, that of the predicate from the nature of the copula. I never fhould have imagined that in the common language of common people, there were any other quantities, even if, in their minds, the *predicate* have thefe. Had this been all, I fhould have paffed it over, as referring to common quantities, and making common people a little more of logicians, as to the predicate, than I have found them to be That this *common language* meant the language of any fcientific fyftem, I had not the leaft idea: ftill lefs that it referred to the language of the writer's own unprinted fyftem, current only between himfelf and his hearers. And, though I gained a fufpicion that Sir W. Hamilton might have (which he had not) adopted numerical quantification, it was not from this paffage, which by itfelf was nothing, but from what is now coming, which made this paffage ambiguous.

On looking further into B, (which fee) I found that Sir William's fyftem, whatever it might be, noted defects in the *converfion of propofitions,* and a *general canon of fyllogifm.* Now I had two fyftems, each of which had its own way of adding to the converfions, and each its own canon of fyllogifm. In my firft fyftem (which has now grown into Chapter V) the permanent introduction of the *contranominals* is a completion of converfion: and the reduction, by the remarks in pages 96, &c. of all fyllogifms to univerfal affirmative premifes, was the canon of fyllogifm. In the fecond, feen in A and C, which has grown into Chapter VIII, there is the univerfality of fimple converfion, and the canon of the middle term. Sir W. Hamilton may deny (I believe he does) that thefe are canons: let it be fo; but I took them for canons, and thought of them when I faw the word *canon* in his fummary. And then the queftion was, had Sir W. Hamilton one of thefe fyftems, or a third one? I had been throughout our correfpondence well pleafed with the idea that I had hit upon fomething in common with Sir W. Hamilton; and in my anfwer to communication B I faid,

'I am not at all clear that I shall *not* have to claim only secondary
'originality on several points. When I see "defects of the common
'doctrine of conversion" and a "supreme canon" of categorical syllo-
'gism, I must wait for further information. .... I think I may yet be
'able to flatter myself that I have followed you in some points unknow-
'ingly.'

The reader will observe that this instructive communication is supposed to tell me, that in my thoughts the predicate has all kinds of
quantity: though in truth *both have their quantities* is not English for
either may have any one of two species of quantity. Sir W. Hamilton
has expressed (perhaps) the dictum which is to have taught me new
quantification, in terms of that new quantification unknown. By *both
have quantities* he seems to assert that he meant *both have all quantities*.
That both have their quantities, is true in the common system: these
words, which express a truth of the common system, Sir W. Hamilton
declares to be a sure mode of communicating the difference between his
system and the common one. This may do in his own lecture room,
in which he has the *arbitrium et jus et norma loquendi* in his own
hands. A distinctively unmeaning phrase may, in virtue of his explanations, pass current between him and his pupils: and a private bank,
of course, must receive its own notes. But they are not lawful tender
anywhere: nor good tender out of the neighbourhood.

I shall now proceed to the letters in the *Athenæum* (III and IV).
These contain the issues raised by the pamphlets: my short letter contains the strength of my case: I am to presume that my opponent's
letter contains the strength of his answer, and I think it does so. At
least I can see nothing stronger in his pamphlet.

MR. DE MORGAN.

I take this mode of acknowledg-
ing the receipt of your printed
letters to me. I promised you an
answer, if you would bring forward the grounds of your assertion
that I had acted with breach of
confidence and false dealing. But
you now admit that your grounds
are no grounds ; you declare your
conviction that (though chargeable
with confusion, want of memory,
&c. &c.) I have acted with good
faith ; and you offer a proper retraction and apology. You state in
various places and manners, that
though you are satisfied of my integrity, all may not be so ; and,
thereupon, you call for an answer.
But I think that others will be

SIR W. HAMILTON.

In reply to your letter in the last
number of the *Athenæum* :—you
were not wrong to abandon your
promise "of trying the strength of
my position ;" for never was there
a weaker pretension than that, by
you, so suicidally maintained. You
would, indeed, have been quite
right had you never hazarded a
second word ; for every additional
sentence you have written is another mis-statement, calling, sometimes, for another correction.

x

quite fatisfied with your own anfwer to your own charge.

There is nothing left which I
care to difcufs with you. Our
views of logic, their coincidences,
their differences, their firft dates,
my memory, &c. I am content to
leave to thofe who will read my
ftatement and your letters, with
two remarks.

There is no ftrength in an abandoned pofition. My pamphlet was
publifhed in defence of my own *charaƈter :* when Sir Wm. Hamilton
retraƈted his charge of breach of confidence and falfe dealing, there was
nothing to which I ftood engaged, nothing I cared to write feparate
pamphlets on, efpecially when the approach of this prefent publication
was confidered. Any one who reads page 9 of my pamphlet, in which
the promife was made, will fee that it has reference to what I there call
" the infamy which would attach to any one who had deferved the
terms he ufed for the conduƈt he defcribed." I certainly forgot to fay
" unlefs you retraƈt :" but as he had already refufed to retraƈt (though
he had propofed to *fufpend* the charge, provided I would then undergo
an examination) it did not enter into my head to provide for fuch a contingency. The affertions about weaknefs, misftatement, &c. are for the
reader's judgment. I did not, in this letter, allude exprefsly to Sir W.
Hamilton's various infinuations that the old charge might be true : both
becaufe, at the firft hurried reading, I did not become aware of their
extent ; and alfo, becaufe I wifhed to take time before I made up my
mind as to the way of treating what I faw of them.

MR. DE MORGAN.	SIR W. HAMILTON.
1. As foon as the queftion of character was difpofed of, it was your bufinefs to fhow that my *Addition,** written after I communicated with you, contained fome principle not contained in my *Memoir,†* written before I communicated with‡ you. This you do not do. You affert, and you defcribe, and you fum up ; but you do not quote,—except a few words, which are not in that part of my *Memoir* which I declared to contain the principles ufed in. my *Addition.*	You do not deny, that your correfpondence afferts a claim to the principle communicated to you by me ; but you complain that I have not fhown that your *Addition* involves a new doƈtrine, uncontained in *that part !* [from the overt contradiƈtions of its *other* parts I had] of your Memoir which you declared to contain the principles ufed in your *Addition.* And this you can fay, when I explicitly ftated that " *throughout the whole paper* (the Memoir) not only is there much in contradiƈtion—there

* Here given in C. † Here given in A, fo far as relevant.
‡ Sir W. Hamilton's part of this is B.

is abfolutely *nothing* in (more then fortuitous) conformity with the theory of a quantified predicate" (L. p. 34). This, too, you can fay whilft before your eyes, *unan-fwered*, there was lying " my *for-mal requeft*, that you would point out *any paffage* of your *previous writings* in which this doctrine (that afferted in your ' *Statement*,' of a quantification of the middle term, be it fubject or predicate) is contained" (Ibid)—for *I could find none; and none has by you been indicated.*

I do deny, in one fenfe, that my correfpondence afferts a claim to the principle communicated by Sir W. Hamilton: for I deny that he *communicated any principle.* I prefume of courfe that the Profpectus and letter fent on the 28th of December are out of the queftion: fince I gave the fyftem on which the charge was made by return of poft. Sir W. Hamilton has very properly confined himfelf, in his pamphlet, to his communication (B) of *November 2*, as containing the communication which he afferts me to have ufed. Let the reader look through it and afk himfelf what *new principle* is communicated, and where.

Sir W. Hamilton afferts that he has fhown my *Addition* to contain a new doctrine, not contained in *one definite part* of my *memoir*, by the contradictions of its *other parts.* Let P, Q, R, be parts of a memoir; and S an addition. By fhowing that P and Q contradict one another, Sir W. Hamilton thinks he fhows that S contains a doctrine not involved in R. The fact is, that all my memoir except 'Section iii. *On the quantity of propofitions*' and one other paragraph (from both which A is taken) belongs to the fyftem of Chapter V. in this work: while Section iii., the other paragraph, and the *addition*, belong to Chapter VIII. Let the reader take notice that Sir W. Hamilton (who, by the way, feems to confider ' I explicitly ftated' as a fufficient anfwer to 'you have not fhown') *does* find *fomething* in my memoir *in conformity with the theory of a quantified predicate.* He fays it is *fortuitous*: but it did not feem to him requifite to bring it forward, and point out its *fortuitoufnefs.* This point is for the reader to judge of. "How dare you," he fays, "rob me of my quantified predicate." "Good Sir," I anfwer, "I had it before I knew you." "What if you had," he replies, "it is enough if I inform you that it was only by accident."

Sir W. Hamilton cannot find either in the *memoir* or the *addition* (he fays here only in the *previous writings*, but in his pamphlet (p. 34) he ftates it of both *memoir* and *addition*), any thing about the doctrine of quantification of the middle term, whether it be fubject or predicate, which doctrine he fays *is repugnant to all that is there taught.* It is

true that in the next sentence he refers to *previous writings*, as cited. I
will therefore conclude that Sir William included the *addition* by mistake, and meant the *memoir* only. Whether my Section iii. (A) is or
is not full of quantification of the middle term, without reference to
whether that middle term be subject or predicate, I am quite content to
leave to the reader. Sir W. Hamilton says he cannot find it. This I
believe, and wonder at : but it does not follow that it is not there. Let
the reader look.

Again, when Sir W. Hamilton asserted that C contains something
which I got from him, and which is therefore not in A, I repeat that
he ought to have pointed out *what it is*. His assertion that he *cannot
find* it in A neither proves that it *is not* in A, nor that it *is* in C.

This is the pinch which obliged him to write forty-four pages of *ac-
cusation* in answer to sixteen of *defence :* and this is the point on which
the question will finally turn. I am tediously often obliged to bring
the whole matter to its A B C; but what else can I do with an opponent who writes an *ignoratio elenchi* of forty-four pages long.

Sir W. Hamilton is not good at finding. Immediately after what
he has quoted from himself as above, comes the following passage ;—

' In regard to your third assertion, that "*perfectly definite quantifica-
' tion destroys the necessity of distinguishing subject and predicate ;*" this
' is altogether a mistake. It is not " definite quantification," (in what-
' ever sense the word *definite* be employed), but the quantification of *both
' the terms* which " destroys the necessity of distinguishing subject and
' predicate ;" and this by showing, that propositions are merely *equations,*
' and enabling us to convert them *all—simply*.'

I now quote from myself. Of the two sentences now coming, Sir
W. Hamilton quotes the first, *omits the second,* which shows that my
phrase '*perfectly* definite' means *definite in both terms,* and then makes
the preceding remark.

' In fact, perfectly definite quantification destroys the necessity of dis-
' tinguishing subject and predicate. To say that some 20 Xs out of 50,
' are all to be found among 70 Ys, or that 20 out of 50 Xs are 20 out of
' 70 Ys, is precisely the same thing as saying that 20 out of 70 Ys are
' 20 out of 50 Xs.'

In a writer of whom dishonest intention might be concluded, we
should know how to explain the omission of the second sentence. But
there is no dishonesty in Sir W. Hamilton : the omission must be
referred to the same disposition which prevents him from seeing quantification of the middle term in A. What I take that disposition to be,
matters nothing to my reader. Perhaps this sentence alone will enable
some to detect that I had not any idea of the second system of quantifi-
tion independently of the third.

MR. DE MORGAN.

2. All the alleged inconsistencies which you find in my letters,
&c. will not help you till you have

SIR W. HAMILTON.

You say, that my exposure of
your inconsistencies is unavailing,
except " I show that my commu-

done this:—and even then, you will have to fhow that your communication was intelligible.

In glancing over my letters and the mafs of notes which you have written on them, I fee that I have feveral times ufed inaccurate language, as people do in hurried letters. Still more often you have mifunderftood me. If my occafional inaccuracy and your occafional mifunderftanding fhould be held to furnifh fome excufe for you when you precipitately charged me with difhonourable conduct, I fhall be better pleafed than not:

nication was intelligible." You forget that it is for *you* to explain how, having "*fubfcribed to,*" as having "*rightly underftood,*" twenty-two fentences of my profpectus (L. pp. 19, 16), you could fubfequently declare that communication to be *unintelligible!!* (L. p. 59). I have now no doubt, however, that you then "fubfcribed to" more fentences than, by you, were "rightly underftood." Indeed, had you only betimes avowed that all you had "fubfcribed to, as rightly underftood," was to you really unintelligible, and that the repetition of my doctrine was in your mouth mere empty found, two pamphlets might have eafily been fpared.

Firft, the *profpectus* is not the "communication." The communication is that of November 2 (B). Let the reader look at it, and fee whether it be intelligible communication of new principle.

In my pamphlet I have feveral times fpoken of *the* communication, though there were two. This was natural enough, inafmuch as there was *one* communication (that of Nov. 2), on which the charge was made againft which that pamphlet was a defence. Sir Wm. Hamilton has never ventured to maintain that I derived anything from the communication of Dec. 28, containing the profpectus, to which I replied on the evening I received it, as prefently mentioned. But he makes, in various places of which the above is one, a mixture of the two communications.

Secondly, I have looked carefully at pages 19 and 16 of Sir Wm. Hamilton's letter, and at all the reft of our correfpondence, without finding that I have ever admitted that I fubfcribed to any part of the profpectus as by me "rightly underftood." Page 59 is no doubt a mifprint for 39. I have neither found, nor have I the flighteft remembrance of, any fubfcription of mine to any thing Sir Wm. Hamilton ever wrote as "rightly underftood."

I repeat the account given in my pamphlet of the manner in which I fubfcribed to this profpectus;—

'The next communication is dated *Dec.* 28, and confifted of—1. A 'letter. 2. A printed profpectus of Sir William Hamilton's intended 'work on logic. Nothing turns on this, for the fimple reafon that my 'anfwer contained the moft exprefs and formal proof that, come by it 'how I might, I was *then* in the moft complete *written* poffeffion of all 'I have fince publifhed.... The profpectus which accompanied this letter

' is very full on the *refults* which Sir William Hamilton can produce
' from his principles; but gives nothing, I think, certainly nothing intel-
' ligible to me, on thofe principles themfelves.

' As foon as I faw thefe *refults*, I inftantly faw that many of them
' agreed with my own. I had then no doubt that we poffeffed fomething
' in common; and I faid fo very diftinctly in my reply. As the reader
' will prefently fee, this firft impreffion has not been confirmed. Feeling
' it now time to fecure whatever of independent difcovery might belong
' to me, I anfwered Sir William Hamilton in two letters, dated Decem-
' ber 31 and January 1. In thefe letters—

' 1. I returned the printed profpectus with the refults underlined
' which *my* fyftem would produce.

' 2. I ftated that I had a fyftem written on certain fheets of paper,
' which I defcribed as to number, fize, &c., adding the head words of
' each page. I felt inclined to get the fignature of fome good witnefs put
' upon thefe papers; but at the fame time I felt reluctant that Sir Wil-
' liam Hamilton fhould fee, if it ever became neceffary to produce thefe
' papers, that I had been taking *precautions* againft him. I therefore de-
' termined to make himfelf my witnefs.

' 3. I ftated diftinctly the firft principles of *both* my fyftems, and the
' fyllogiftic formulæ to which they lead.'

Thirdly, I fubftantiate the above, fo far as the fubfcription is concerned, by quoting two paffages from Sir W. Hamilton's publication
of my letter of December 31.

' I received your obliging communication *this morning* and am now
' fully fatisfied that I have, in one of my views of fyllogifm, arrived at
' your views in fubftance, or fomething fo like them, that I could fub-
' fcribe in my own fenfe to a great part of your paper ...... This
' chapter [meaning the one on the fheets of paper above referred to]
' I might exprefs in your words wherever they are underlined in the
' profpectus which I return, hoping you will fend another.'

Where are thofe words "rightly underftood" which Sir W. Hamilton attributes to me three times in one paragraph?

He muft have been quoting from memory. Seeing *his refults*, I
found they were alfo *my refults;* fo I told him that I could "fubfcribe"
(and I cannot find I have ufed this word more than once, and it is in
page 19 referred to by Sir William) " in my own fenfe to a great part
of " his " paper." If words can fpeak meaning, I here tell him that I
fubfcribe in *my own* fenfe, leaving it to the future to fhow whether I
fubfcribe in *his*, that is, whether I *underftand him rightly*.

[I was reading this for the prefs, when I found out the words which,
applied in one fenfe hypothetically to *one* of his refults, Sir W. Hamilton
has transferred in a different fenfe to all. One of his refults, fpeaking
of the moods, is the eftablifhment of ' Their numerical *equality* under
all the figures,' the Italics being his. I could not make out the Englifh
of this. The others I *underftood* in the grammatical fenfe. For example, ' The abrogation of the fpecial laws of fyllogifm' is intelligible:
I did not know whether my fenfe of thefe words, that is to fay, my

abrogation of thofe laws, was the fame as Sir W. Hamilton's; ftill that he did abrogate certain laws was clear. But numerical equality of moods I could only underftand as referring to the numerical quantities which I fuppofed (the reader will remember that I fent back the profpectus by the next poft, and had little time to look at it) Sir W. Hamilton's fyftem to contain. It means, I find, that there are the fame number of moods in all figures: but to attribute *numerical equality* to different things is a mode of faying that there is *the fame number of them in different fets* to which I was unaccuftomed. Having however, as I thought, divined what the Englifh of this might mean, I underlined it, adding (as Sir W. Hamilton ftates in one of the foot-notes, which I never remarked till now) thefe words, " If I underftand this rightly I may underline it I think." I meant, " If I can make out *the words.*" This *underftand rightly*, Sir W. Hamilton actually takes from this fentence, joins it to my "fubfcription" mentioned in *another document,* and reprefents me as declaring that I have "*fubfcribed to as rightly underftood*" twenty-two *fentences,* &c., and himfelf as quoting from one paffage.]

But, had I *betimes* avowed my non-underftanding, two pamphlets might have been fpared. Where are we now? I did avow my not underftanding the firft communication, and my fubfcribing to the fecond *in my own fenfe.* To which Sir W. Hamilton fubfequently anfwered to the effect that I fpoke falfe, that I did underftand the *firft, for* that I had fent him, in letters written immediately after the fecond was received, his "fundamental doctrine" and "many of its moft important confequences." What have I been contending for all along, except that the doctrine of his *firft* communication *was* to me mere empty found, and that all I was able to produce when I received the *fecond,* was my own? But Sir W. Hamilton actually gives me a right to fay, with reference to the *fecond,* the more developed and more intelligible communication, that I did not underftand it, infifts upon my faying it, and reproaches me for not faying it. Well then, to ufe a Scottifh phrafe, the lefs I lie when I fay I did not underftand the *firft,* which is the point at iffue. So that, as to the matter of our controverfy, Sir W. Hamilton admits that there was (fortuitous he calls it) entrance of the theory of the quantified predicate in my writings prior to his communications; and as to the conduct of it, he admits that I *did not* underftand his communication; and in the face of fact, reproaches me with maintaining that I *did* till after the pamphlets were written: when it was of the effence of my ftatement, firft, that I did not underftand, fecondly that neither I nor any one elfe could have underftood, fave only the pupils to whom the requifites were addreffed.

MR. DE MORGAN.

Your copious and flafhing criticifms on my intellect (by which you avenge yourfelf for the retraction of your afperfion on my integrity), I will profit by fo far as I

SIR W. HAMILTON.

I difregard your mifreprefentation that " I avenge myfelf for the retraction of my afperfion on your integrity by my copious and flafhing criticifms on your intellect."

discover them to be true: the rest shall amuse me;—and the whole will be good for the printer. Take one retort from me on the same terms. You have much skill in forming new words; and, as is fair, you put your own image and superscription on your own coinage. I think you have got into the habit of assuming the same authority over that already existing portion of our language which is commonly said to belong to the Queen—and that you need an interpreter. If I can arrive at your meaning by the time I write the preface to my work on logic, I will state your claim, accompanied by your own words; if not, I can still state your own words. Till then, I have nothing more to say.

When your (excusable) irritation has subsided, you will see that I *could only* secure you from a verdict of plagiarism by bringing you in as suffering under an illusion. What, however, is all in all;—my criticisms will not, I think, be found untrue.

If guilty of *lese majesty* by reference to the Queen's English, have I not my accuser as abettor? For you not only passed my mintages (*quantify* and *quantification*) as current coin; but, in borrowing, actually " thanked me for the words" (L. p. 22). However, my verbal innovations are, at least, not elementary blunders. I do not, for example, confound a *term* with a *proposition,* the *middle* with the *conclusion* of a syllogism.

Sir W. Hamilton unconsciously adapts his language to a very true supposition, namely, that he has, in his pamphlet, made himself the jury in this case. He is unfortunate about the mintage. I say to him ' You make new words well, but I am afraid you alter the old ones.' To which he replies ' Why, you thanked me for my new words.' So I did, and so I do again: but what has that to do with the *lese majesty* part of my insinuation.

Sir W. Hamilton says that I have somewhere (where he does *not* say) used *term* for *proposition, middle* for *conclusion, collectively* for *distributively.* This may be; such slips of the pen are common enough. He sets them down as blunders of ignorance. I am not afraid the reader will follow him. He ought to have said where they occur, that is, when he first mentioned them, in his pamphlet. Till I put these letters together, I was satisfied, on Sir Wm. Hamilton's statement, that I had done all these enormities: but now, after the case of " *rightly understood*" which I have just had to discuss, I do not feel so well satisfied.

Sir W. Hamilton.

Finally, I beg leave to remind you.—There is now evidence in your possession that for *seven years,* at least, the doctrine of a quantified predicate has been puclickly taught by me; whilst, on your part, there is a counter assertion or innuendo, which, as you cannot prove, it concerns *your character* formally to annul.

I never denied that Sir W. Hamilton had taught *a* doctrine of the quantified predicate. By the time I wrote my pamphlet, I was pretty fure that it was not the fame as mine. Sir W. Hamilton's anfwer confirmed me in this, as appears in page 300.

I now come to mention a part of the difcuffion which I fhould perhaps have omitted, if I had not pledged myfelf in my pamphlet to give an account of a certain offer which I there made to Sir William Hamilton, in the event of that offer not being accepted. It is a curious inftance of that difpofition to hold a correfpondent or an opponent capable of folving enigmas, and bound to do it, which appears in his prefuming that (fee B, paragraph 2°) an obfcure reference to what is done in *common language* would enable me to guefs at the *uncommon language* of his fyftem and his lectures. I infert it, alfo, as a fpecimen of the various mifunderftandings and mifapprehenfions which Sir W. Hamilton imputes to me, referring to a matter which readers will feparately comprehend. Had I fpace or inclination to deal with them all, I believe I could ferve them all in the fame way.

Oct. 7, 1846, I learnt from Sir Wm. Hamilton that his doctrine had obtained confiderable publicity through the *notes and effays* of his ftudents. In my reply, referring to this fyftem, and to his offer of communicating it, I afked if he had a pupil whom he could truft with the communication; the anfwer was B, prefently given. But, *Dec.* 28, in fending the profpectus, Sir W. Hamilton informed me that, before forwarding it (the firft communication in which that he had other than Ariftotelian quantification was intelligibly announced) he had waited for a reply from Mr. ———. 'That gentleman' continues Sir W. Hamilton, in words fome of which I place in Italics ' was a pupil of ' mine fix years ago, and obtained one of the higheft honours of the clafs ; ' he was therefore *fully competent to afford you information,* which I ' begged him to do, in regard to my logical doctrines as they were taught ' fo far back. I knew-him to be a graduate of your College, and he tells ' me that he was for three years a pupil of your own. If you are ftill ' interefted in the matter, you can therefore obtain from him as an ' acquaintance, what information you wifh, more agreeably than from a ' ftranger. When he *attended me,* befides the twofold wholes in which ' the fyllogifm proceeds, *the quantification of the predicate,* and the effect ' of that on the doctrine of converfion, on the doctrine of fyllogiftic ' moods, on the fpecial fyllogiftic rules, &c., were *topics difcuffed,* and ' partly *given out for exercifes. They were, in fact, then mere common-* ' *place.*'

Jan. 13, 1847, Mr. ——— called on me at Univerfity College, after an evening lecture of mine, put his *notes* into my hands, and has fince ftated (in which I have no doubt he is correct, though I do not remember it) that he informed me he *was doubtful whether they contained exactly what I wanted,* and that he would gladly furnifh any additional information. Now I conceived, as I thought it was intended by Sir W. Hamilton I fhould do, that the notes of one of the beft ftudents, even if not exactly what I wanted, were fure to contain *fomething* of the *mere commonplace* (by which I took to be meant the ordinary matter of the lectures) which was *difcuffed,* and *given out as exercifes* to thofe who attended. But in these notes I found nothing on quantification (I had now this key word, which did not appear in the main communication B) differing from what is usual; and after expressing this in my pamphlet, I proceeded as follows:

'But if there really be anything in which Sir William Hamilton has
'preceded me, I shall be, of all men except himself, most interested in
'his having his full rights. And I make him this offer, and will take his
'acceptance of it as reparation in full for his suspicions and assertions.
'With the consent of the gentleman to whom these notes belong, which
'I am sure will not be refused to our joint application, I will forward to
'him a copy of their table of contents, having more than a hundred and
'fifty headings. From these Sir William Hamilton shall select those
'which are, in his opinion, sure to contain proof of his priority on any
'point which I have investigated. Of these I will have copies made and
'sent to him: and will print in the work on Logic which I am preparing
'(and in some one part of it) the parts which he shall select as fit to
'prove (or to show that he could prove, let him call it as he likes) his
'case, or the germs of his case (as he pleases, again). Provided always,
'that the matter shall not run beyond some eight or a dozen octavo pages
'of small print. And I on my part propose that I shall be allowed to
'print, to one-half the amount selected by Sir William Hamilton, of ad-
'ditional extract: but if this be refused I will not insist on it. With this
'I will put a heading fully descriptive of the reason and meaning of the
'insertion, and such distinct reference and account at the beginning of
'the preface as shall be sure to call the reader's attention to it. So that
'my book shall establish the claim, if it can be established from the notes
'of one of the best students. If this offer be not accepted, an account of
'it will take the place of any other result. If Sir William Hamilton, or
'any one else, can propose anything to make this offer fairer, I shall pro-
'bably not be found indisposed to accept the addition. And though, I
'will frankly say, my present conviction is that the acceptance of the
'offer would alone cause my work to knock Sir William Hamilton's
'assertions to atoms, yet I will pledge myself, in any case, to abide by it.

Had our places in this discussion been changed, I should have taken care that no reader of my answer should have been left in ignorance of so fair an offer on the part of my opponent: more especially if that opponent had been accused by me of fraud and falsehood, in a manner which I felt obliged formally to retract. But Sir Wm. Hamilton does not notice *the offer*, even by an allusion: and refers to the notes in the following way:

'In regard to Mr ———— and his Notes, I beg leave to say, that in
'my relative letters, neither to that gentleman nor to you, did I ever
'refer to his *Notes* of my lectures, but exclusively to his *personal infor-
'mation* in regard to them. And for a sufficient reason. The Paragraphs
'on Logic dictated to, and taken down by, my students, on which I after-
'wards prelect, were written so far back as the year 1837, and prior to
'many of my new views, and to the *whole* doctrine of a quantified predi-
'cate. These views, as developed, were, and are, introduced in a great

'meafure as corrections of the common doctrine; in the older Notes
'efpecially, they may, therefore, not appear in the dictated and numbered
'Paragraphs at all; whilft, frequently, (particularly at firft,) they were
'given out as data, on which, previous to farther comment, the ftudents
'were called on or excited to write expofitory Effays. I diftinctly recol-
'lect, that in the Seffion during which Mr. ——— attended my courfe of
'Logic (1840-1) it was required, on the hypothefis of a quantified pre-
'dicate,—to ftate in detail, the valid moods of each fyllogiftic figure; and
'I, further, diftinctly recollect, that Mr.——— was one of thofe who
'effayed this problem. If wrong on this point, I fhall admit that my
'memory is as treacherous as yours. It was, indeed, quite natural, that
'Mr. ——— fhould give, and that you fhould receive, his Notes; but,
'of courfe, you could have fought or obtained no perfonal information
'from him, in reference to the point in queftion, without mentioning the
'fact. .... Were it, however, requifite to give proof from *Notes* of fo
'manifeft a fact, I doubt not that fcores of ftudents would be willing to
'place theirs at my difpofal.

On the appearance of Sir W. Hamilton's pamphlet, Mr. ———
wrote him a very ftraightforward letter, of which he fent me a copy,
with permiffion to both of us to ufe it. The general tenor is that Sir
W. Hamilton is correct in his ftatements of what he had taught (which
ftatements I never impugned as to fact; I did not know what they
meant). On the point in queftion Mr. ——— fays (the Italics are
mine);—

'During the Seffion in which I attended your lectures (1840 and
'1841) your new fyftem, bafed on the thorough going quantification of
'the predicate (the fecond of the three fyftems mentioned in page 31 of
'your publifhed letter) and its confequences in making all propofitions
'fimply convertible &c. *was not developed by you in your ordinary feries*
'*of Lectures. I believe it was not touched upon in them, but it was partly*
'*explained to the clafs verbally,** and then given out as a fubject for Ef-*
'*fays.* When the Effays were given in they were read aloud in the clafs,
'and commented upon by you, and in fo doing you fully explained the
'fyftem as "a full extenfion and thereby a complete fimplification of the
'fyllogiftic theory."

'Thefe facts which were ftrongly fixed in my memory, becaufe I
'believe on that occafion I happened to be the only Effayift who had
'rightly apprehended and worked out the thefis, will account for the
'circumftance that my notes, which were originally taken in fhorthand,
'although containing a full Report of all your ordinary Lectures, are
'completely filent on the fubject."

The reader may find out, if he can, where Sir W. Hamilton referred to *perfonal information* as diftinguifhed from *notes*, or to his
teaching of his new fyftem, as a matter diftinct from that of his ordinary
lectures: and muft judge what his fuccefs is in faying what he means

---

* I think this fhould be *extempore:* meaning that Sir W. Hamilton ufually reads his
lectures.

to fay. And he may find out further, how I was to guefs that the *mere commonplace* of the *topics difcuffed* in Sir William's teaching was to come, after an interval of fix years, from his old pupil's *perfonal information,* and not from the full and (as I found them) excellent *notes* which he made at the time.

I fhould add that Mr. ———, fubfequently to the printed controverfy, anfwered every query which I put to him on Sir W Hamilton's fyftem, but did not feel juftified (as in a like cafe I fhould not either) in anfwering pofitively as to the minute details of it, after laying it by for years.

I have mentioned one or two inftances in which, as feems to me, Sir W Hamilton has a ftrange idea of the fenfe of his own words: I will now take one of the cafes in which he has dealt as ftrangely with mine. The way in which we ufe language, is one of the means which the reader has, for forming his judgment on the whole of this difpute: and he muft decide which of us is incapable of giving to the phrafes of the other their proper fignification.

When I returned to Sir W. Hamilton his profpectus, with thofe parts underlined which I could interpret in my own fenfe, the more important parts relating to logical mood and figure were not thus underlined. In the accompanying letter, I ufed thefe words, ‘To mood and figure, I have attended but little; what I get on thefe points will be from your hint, or from your book.’ The whole letter was on what I had done in the way of inveftigation, not of elementary reading: and I may fafely fay that it is clear I meant that I had not made mood and figure, as conftituent parts of a theory of fyllogifm, fubjects of *inveftigation,* with a view to new properties. But Sir W. Hamilton, in two places, makes me avow ignorance of the ordinary fyftem of mood and figure. In a foot-note to the above, he fays, “And yet, though con-‘ feffedly to feek in the very alphabet of the fcience, Mr. De Morgan ‘ would be a logical inventor! What is here acknowledged in terms, is ‘ fufficiently manifefted from miftakes.”* And in his pamphlet (II. p. 9), he reprefents me as ‘ no proficient—no thorough ftudent,—in the fcience;’ and refers to this paragraph of mine as the ground of the affertion. It would have been ftrange, if, avowing ignorance of the ordinary doctrine of mood and figure, I had faid that what I fhould get on thefe points muft be from Sir W Hamilton's hint or unpublifhed book, when any ordinary treatife would have given it: fo ftrange, that this claufe ought, I think, to have fuggefted the obvious meaning. Is Sir W. Hamilton's interpretation a *fair* one? I do not doubt that he meant it to be fair. What I afk is, has he the power to read fairly as well as the will?

The two preceding cafes (that of the *notes* and that of the *avowed ignorance*) are fpecimens of Sir W. Hamilton's *give and take,* of the

---

* Sir W. Hamilton fhould have cited a few: but when he declares I have made elementary blunders, he does not give fo much as a reference. The plan is a fafe one.

manner in which he expects to be underftood, and of that in which he claims a right to underftand. They are alfo, of courfe, fpecimens of my own.

In (A), the fymbols A, E, I, O, are the A_1, E_1, I_1, O_1, of this work: and a, e, i, o, are the A', E', I', O'.

(A) *From the paper as fent to Cambridge before I had any communication whatfoever from Sir William Hamilton (without any corrections).*

Section III. *On the quantity of propofitions.*

" The logical ufe of the word *fome*, as merely 'more than none,' needs no further explanation. Exact knowledge of the extent of a propofition would confift in knowing, for inftance in ' fome Xs are not Ys' both what proportion of the Xs are fpoken of, and what proportion exifts between the whole number of Xs and of Ys. The want of this information compels us to divide the exponents of our proportion into o, more than o not neceffarily 1, and 1. An algebraift learns to confider the diftinction between o and quantity as identical, for many purpofes, with that between one quantity and another: the logician muft (all writers imply) keep the diftinction between o and *a*, however fmall *a* may be, as facred as that between o and 1—*a*; there being but the fame form for the two cafes. We fhall now fee that this matter has not been fully examined.

" Inference muft confift in bringing each two things which are to be compared into comparifon with a third. Many comparifons may be made at once, but there muft be this procefs in every one. When the comparifon is that of identity, of *is* or *is not*, it can only be in its ultimate or individual cafe, one of the two following:—' This X is a Y, this Z is the very fame Y, therefore this X is this Z; or elfe ' This X is a Y, this Z is not the very fame Y, therefore this X is not this Z.' And collectively, it muft be either ' Each of thefe Xs is a Y; each of thefe Ys is a Z; therefore each of thefe Xs is a Z;' or elfe ' Each of thefe Xs is a Y, no one of thefe Ys is a Z, therefore no one of thefe Xs is a Z.'

" All that is effential then to a fyllogifm is that its premifes fhall mention a number of Ys, of each of which they fhall affirm either that it is both X and Z, or that it is one and is not the other. The premifes may mention more: but it is enough that this much can be picked out; and it is in this laft procefs that inference confifts.

" Ariftotle noticed but one way of being fure that the fame Ys are fpoken of in both premifes; namely, by fpeaking of all of them in one at leaft. But this is only a cafe of the rule: for all that is neceffary is *that more Ys in number than there exift feparate Ys fhall be fpoken of in both premifes together.* Having to make *m*+*n* greater than unity, when neither *m* nor *n* is fo, he admitted only that cafe in which one of the two *m* or *n*, is unity and the other is anything except o. Here then are two fyllogifms which ought to have appeared, but do not,

Moſt of the Ys are Xs	Moſt of the Ys are Xs
Moſt of the Ys are Zs	Moſt of the Ys are not Zs
∴ Some Xs are Zs	∴ Some of the Xs are not Zs

And inſtead of moſt, or $\frac{1}{2}+a$, of the Ys, may be ſubſtituted any two fractions which have a ſum greater than unity. If theſe fractions be m and n, then the middle term is *at leaſt* the fraction $m+n-1$ of the Ys. It is not really even neceſſary that all the Ys ſhould enter in one premiſs or the other: for more than the fraction $m+n-1$ of the whole may be repeated twice.

"And in truth it is this mode of ſyllogiſing that we are frequently obliged to have recourſe to; perhaps more often than not in our univerſal ſyllogiſms. '*All* men are capable of ſome inſtruction; all who are capable of any inſtruction can learn to diſtinguiſh their right and left hands by name; therefore all men can learn to do ſo.' Let the word *all* in theſe two caſes mean only *all but one*, and the books on logic tell us with one voice that the ſyllogiſm has particular premiſes, and *no concluſion can be drawn*. But in fact idiots are capable of no inſtruction, many are deaf and dumb, ſome are without hands: and yet *a* concluſion is admiſſible. Here m and n are each very near to unity, and $m+n-1$ is therefore near to unity. Some will ſay that this is a probable concluſion: that in the caſe of any one perſon it means there is the chance m that he can receive inſtruction, and n that one ſo gifted can be made to name his right and left hand: therefore $m \times n$ (very near unity) is the chance that this man can learn ſo much.

"But I cannot ſee how in this inſtance the probability is anything but another ſort of inference from the demonſtrable concluſion of the ſyllogiſm, which muſt exiſt under the premiſes given. Beſides which, even if we admit the ſyllogiſm as only probable with regard to any one man, it is abſolute and demonſtrative in regard to the propoſition with which it concludes.

"But this is not the only caſe in which the middle term need not enter univerſally: this however is matter for the next Section. I now go on to another point."

Extract II.

"I now take the two caſes in which particular premiſes may give a concluſion: namely

$$I_{\prime\prime} \quad XY + XY = XZ \qquad XY + Y:Z = X:Z \qquad O_{\prime o}$$

on the ſuppoſition that the Ys mentioned in both premiſes are in number more than all the Ys. If Y_1 and Y_2 ſtand for the fractions of the whole number of Ys mentioned or implied in the two premiſes, and y_1 and y_2 for the fractions of the ys implied or mentioned, we ſhall by a repetition of the procefs on YX+YZ=XZ (the other being obtained in the courfe of the procefs) arrive at the following refults or their counterparts : remembering that $Y_1 + Y_2$ is greater or lefs than 1, according as $y_1 + y_2$ is lefs or greater.

Defignation.	Syllogifm.	Condition of its exiftence.
I_{II}	$YX + YZ = XZ$	$Y_1 + Y_2$ greater than 1
O_{Io}	$YX + Y{:}Z = XZ$	....
i_{oo}	$Y{:}X + Y{:}Z = xz$	....
O_{oi}	$X{:}Y + yz = X{:}Z$	$Y_1 + Y_2$ lefs than 1
i_{ii}	$yx + yz = xz$	....
O_{Oi}	$X{:}Y + yz = X{:}Z$	....
I_{oo}	$X{:}Y + Z{:}Y = XZ$	....

(B) *Communication received on the 4th or 5th of November from Sir William Hamilton, being the pretext for his charge that I have, with injurious breach of confidence towards himfelf, and falfe dealing towards the public, appropriated his " Fundamental Doctrine of Syllogifm" privately communicated to me : and, after the retraction of that charge, noticed in pages 297, 8, for the affertion that I have done the fame thing unconfcioufly.*

" 16, Great King Street,
November 2nd, 1846.

"DEAR SIR,—I have been longer than I anticipated in anfwering your laft letter. I now fend you a copy of the requifites for the prize Effay, which I gave out to my ftudents at the clofe of laft feffion. It will fhow you the nature of my doctrine of fyllogifm, in one of its halves. The other, which is not there touched on, regards the two wholes, or quantities in which a fyllogifm is caft. I had intended fending you a copy of a more articulate ftatement which I meant, at any rate, to have drawn up; but I have not as yet been able to write this. I will fend it when it is done. From what you ftate of your fyftem having ' little in common with the old one,' and from the contents of your Firft Notions, we fhall not, I find, at all interfere, for my doctrine is fimply *that of Ariftotle, fully developed.*

It will give me great pleafure if I can be of any ufe, in your inveftigations concerning the hiftory of Logical doctrines. I have paid great attention to this fubject, on which I found, that I could obtain little or no information from the profeffed hiftorians of Logic ; and my collection of Logical books is probably the moft complete in this country. But, as I mentioned to you in my former letter, it is only in fubordinate matters that in *abftract* Logic there has been any progrefs.

" I remain, dear Sir, very truly yours,

" W. HAMILTON."

Eſſay on the new Analytic of Logical Forms.

Without wiſhing to preſcribe any definite order, it is required that there ſhould be ſtated in the Eſſay,—

1°. What Logic *poſtulates* as a condition of its applicability.

2°. The reaſons why common language makes an *ellipſis* of the ex-*preſſed quantity*—frequently of the *ſubjeƈt*, and more frequently of the *predicate*, though both have always their quantities in thought. [*This paragraph is the one on which Sir W. Hamilton principally relies*].

3°. *Converſion of propoſitions*—on the *common* doƈtrine.

4°. Defeƈts of this.

5°. *Figure* and *Mood* of Categorical ſyllogiſm, and Reduƈtion,—on *common* doƈtrine (General ſtatement).

6°. Defeƈts of this (General ſtatement).

7°. The one *ſupreme Canon* of Categorical Syllogiſms.

8°. The evolution, from this canon, of all the *ſpecies* of Syllogiſm.

9°. The evolution, from this canon, of all the *general laws* of categorical Syllogiſms.

10°. The error of the *ſpecial laws* for the ſeveral Figures of Categorical Syllogiſm.

11°. *How many Figures* are there.

12°. What are the *Canons* of the *ſeveral Figures*.

13°. *How many moods* are there in all the Figures: ſhowing in concrete examples, through all the Moods, the *uneſſential* variation which Figure makes in a ſyllogiſm.

(Thoſe which follow 13° were wrong numbered.)

15°. What relation do the Figures hold to *extenſion* and *comprehen-ſion*.

16° Why have the *ſecond* and *third* Figures *no determinate major* and *minor* premiſes and *two* indifferent concluſions: while the *firſt* Figure has a *determinate major and minor premiſe*, and a ſingle *proximate concluſion*.

17°. What relation do the Figures hold to *Deduƈtion* and *Induƈtion*.

N.B. This Eſſay open for competition to all ſtudents of the claſs of Logic and Metaphyſics during the laſt or during the enſuing ſeſſion.

April 15th, 1846.

(C) *Extraƈt from the Addition to my Paper, taken, as can be ſhown, from the papers which I gave the means of identifying in January laſt, and which papers (though I hold it immaterial) I aſſert to have been written before I received any logical communication from Sir William Hamilton. (To be compared with the extraƈts given in A).*

"Since this paper was written, I found that the whole theory of the ſyllogiſm might be deduced from the conſideration of propoſitions in a form in which *definite quantity* of aſſertion is given both to the ſubjeƈt and the predicate of a propoſition. I had committed this view to paper, when I learned from Sir William Hamilton of Edinburgh, that he had for fome time paft publicly taught a theory of the fyllogifm differing in detail and extent from that of Ariftotle. From the profpeétus of an intended work on logic, which Sir William Hamilton has recently iffued, at the end of his edition of Reid, as well as from information conveyed to me by himfelf in general terms, I fhould fuppofe it will be found that I have been more or lefs anticipated in the view juft alluded to. To what extent this has been the cafe, I cannot now afcertain; but the book of which the profpeétus juft named is an announcement, will fettle that queftion. From the extraordinary extent of its author's learning in the hiftory of philofophy, and the acutenefs of his written articles on the fubjeét, all who are interefted in logic will look for its appearance with more than common intereft.

"The footing upon which we fhould be glad to put propofitions, if our knowledge were minute enough, is the following. We fhould ftate how many individuals there are under the names which are the fubjeét and predicate, and of how many of each we mean to fpeak. Thus, inftead of ' Some Xs are Ys,' it would be, ' Every one of *a* fpecified Xs is one or other of *b* fpecified Ys.' And the negative form would be as in ' No one of *a* fpecified Xs is any one of *b* fpecified Ys.' If propofitions be ftated in this way, the conditions of inference are as follows. Let the *effeétive number* of a propofition be the number of mentioned cafes of the *fubjeét*, if it be an affirmative propofition, or of the *middle term*, if it be a negative propofition. Thus, in ' Each one of 50 Xs is one or other of 70 Ys,' is a propofition, the effeétive number of which is always 50. But ' No one of 50 Xs is any one of 70 Ys ' is a propofition, the effeétive number of which is 50 or 70, according as X or Y is the middle term of the fyllogifm in which it is to be ufed. Then two propofitions, each of two terms, and having one term in common, admit an inference when 1. They are not both negative. 2. The fum of the effeétive numbers of the two premifes is greater than the whole number of exifting cafes of the middle term. And the excefs of that fum above the number of cafes of the middle term is the number of the cafes in the affirmative premifs which are the fubjeéts of inference. Thus, if there be 100 Ys, and we can fay that each of 50 Xs is one or other of 80 Ys, and that no one of 20 Zs is any one of 60 Ys;—the effeétive numbers are 50 and 60. And 50+60 exceeding 100 by 10, there are 10 Xs, of which we may affirm that no one of them is any one of 20 Zs mentioned.

"The following brief fummary will enable the reader to obferve the complete deduétion of all the Ariftotelian forms, and the various modes of inference from *fpecific particulars*, of which a fhort account has already been given.

"Let *a* be the whole number of Xs; and *t* the number fpecified in the premifs. Let *c* be the whole number of Zs; and *w* the number fpecified in the premifs. Let *b* be the whole number of Ys; and *u* and *v* the numbers fpecified in the premifes of *x* and *z*. Let $X_t Y_u$ denote that each of *t* Xs is affirmed to be one out of *u* Ys: and $X_t : Y_u$ that each of *t* Xs is denied to be any one out of *u* Ys. Let $X_{m,n}$ fignify *m*

Xs taken out of a larger specified number n; and so on. Then the five possible syllogisms, on the condition that no contraries are to enter either premises or conclusion, are as follows :—

$$1.\ X_tY_u + Z_wY_v = X_{t+w-b,\,t}\ Z_w = Z_{t+w-b,\,w}X_t.$$
$$2.\ X_tY_u + Y_vZ_w = X_{t+v-b,\,t}\ Z_w = Z_{t+v-b,\,w}X_t.$$
$$3.\ Y_uX_t + Y_vZ_w = X_{u+v-b,\,t}\ Z_w = Z_{u+v-b,\,w}X_t.$$
$$4.\ X_tY_u + Z_w\!:\!Y_v = X_{t+v-b,\,t}\!:\!Z_w$$
$$5.\ Y_uX_t + Z_w\!:\!Y_v = X_{u+v-b,\,t}\!:\!Z_w.$$

"The condition of inference expresses itself; in the $X_{m,\,t}$ of the conclusion, m must neither be o nor negative. The first case gives no Aristotelian syllogism; the middle term never entering universally (of necessity) into any of its forms, under any degree of specification which the usual modes of speaking allow. The other cases divide the old syllogisms among themselves in the following manner : they are written so as to show that there is sometimes a little difference of amount of specification between the results of different figures, which changes in the reduction from one figure to another. The Roman numerals mark the figures.

2.	$t = a,\ v = b$	$Y)Z_w + X)Y_u = X)Z_{u,\,w}$	*Barbara* I.
	$t = a,\ v = b$	$X)Y_u + Y)Z_w = Z_{a,\,w}X$	*Bramantip* IV.
	$t < a,\ v = b$	$Y)Z_w + X_tY_u = X)Z_{u,\,w}$	*Darii* I.
	$t < a,\ v = b$	$X_tY_u + Y)Z_w = Z_{t,\,w}X_t.$	*Dimaris* IV.
3.	$u = b,\ v = b$	$Y)X_t + Y)Z_w = Z_{b,\,w}X_{b,\,t}$	*Darapti* III.
	$u < b,\ v\quad b$	$Y_uX_t + Y)Z_w = Z_{u,\,w}X_{u,\,t}$	*Disamis* III.
	$u = b,\ v < b$	$Y)X_t + Y_v)Z_w = Z_{v,\,w}X_{v,\,t}$	*Datisi* III.
4.	$t = a, v = b, w = c$	$Y\,.\,Z + X)Y_u = X\,.\,Z$	*Celarent* I.
	$t = a, v = b, w = c$	$Z\,.\,Y + X)Y_u = X\,.\,Z$	*Cesare* II.
	$t = a, v = b, w = c$	$X)Y_u + Z\,.\,Y = Z\,.\,X$	*Camestres* II.
	$t = a, v = b, w = c$	$X)Y_u + Y\,.\,Z = Z\,.\,X$	*Camenes* IV.
	$v = b, w = c$	$Y\,.\,Z = X_tY_u = X_t\!:\!Z$	*Ferio* I.
	$v = b, w = c$	$Z\,.\,Y + X_tY_u = X_t\!:\!Z$	*Festino* II.
	$t = a, v = b,$	$X)Y_u + Z_w\!:\!Y = Z_w\!:\!X$	*Baroko* II.
5.	$u = b, v = b, w = c$	$Y\,.\,Z + Y)X' = X_{b,\,t}\!:\!Z$	*Felapton* III.
	$u = b, v = b, w = c$	$Z\,.\,Y + Y)X_t = X_{b,\,t}\!:\!Z$	*Fesapo* IV.
	$v = b, w = c$	$Y\,.\,Z + Y_uX_t = X_{u,\,t}\!:\!Z$	*Feriso* III.
	$v = b, w = c$	$Z\,.\,Y + Y_uX' = X_{u,\,t}\!:\!Z$	*Fresison* IV.
	$u = b,\quad\ w = c$	$Y_v\!:\!Z + Y)X' = X_{v,\,t}\!:\!Z$	*Bokardo* III.

I conclude by submitting to the reader what I began with, namely, that until Sir William Hamilton produces something from C, intelligibly hinted at in B, and neither substantially contained in the matter, nor immediately deducible from the principles, of A, he has no right whatever to assert that I have borrowed from him consciously or unconsciously. I have not found any person who thinks that such a thing can be produced : and I leave every reader to form his own opinion whether it can be done or not.

APPENDIX II.

On some forms of inference differing from those of the Aristotelians.

I THINK it desirable to state all I know of any attempt to deal with the forms of inference otherwise than in the Aristotelian method. Since the time of Wallis, three well known mathematicians have written on the subject, Euler, Lambert, and Gergonne : there may have been others, but I have not met with them.

Euler's ' Lettres à une Princesse d'Allemagne sur quelques sujets de Physique et de Philosophie' (3 vols. 8vo. Petersburg 1768-1772, according to Fuss) contain the representation of the syllogism by *sensible terms*, namely, areas. There was a Paris edition by Condorcet and Lacroix, in 1787, as is stated by Dr Henry Hunter, who published an English translation from it and from the original edition, London, 1795, 2 vols. 8vo. Euler makes use of circles to represent the terms. In a tract published (or completed) in 1831, in the Library of Useful Knowledge, under the name of ' the Study and Difficulties of Mathematics' I fell upon this method before I knew what Euler had done, using, for distinction, squares, circles, and triangles, as in Chapter I. of this work. The author of the "Outlines" presently mentioned, has what I consider a very happy improvement on Euler. The proposition ' some X is Y,' is represented by the latter as the circle of X, partly inside and partly outside the Y. The author of the "Outlines" puts a broken segment of the circle of X inside the circle of Y, leaving it unsettled whether the rest of the circle is united to the broken piece, or transferred elsewhere.*

But Euler had been preceded in the publication of this idea by Lambert, in his ' Neues Organon, &c.' Leipzig, 1764, 2 vols. 8vo. In this work, the terms are represented by lines, and identical extents by parts of the lines vertically under one another, as in page 79. The whole notion is represented by continuous line, the part left indefinite in particular propositions by dotted line. Some of the contranominal forms are more distinctly mentioned than is usual, but there is no introduction that I can find of any form of inference which is not Aristotelian.

* I should say that Euler does not use the numerical, but the magnitudinal notion, (see page 48 of this work).

In the seventh volume of the *Annales de Mathématiques* (Nismes, 1816 and 1817, 4to.). there is a paper by the editor, M. Gergonne, entitled *Essai de dialectique rationelle.* I did not see this paper, nor Lambert's work, until after my memoir in the Transactions of the Cambridge Society had been published. The second would have given me no hint: the first might have done so. There is the idea, and some formal use, of a complex proposition: but the division is erroneous. The subidentical, identical, and superidentical forms are there; these are not easily missed: the others which Gergonne uses are, the *complete exclusion* (the *contrary or subcontrary* of my system, which, disjunctively, are only the common universal negative) and *partial inclusion with partial exclusion* (the *complex particular, or supercontrary,* of mine). The use of contraries is expressly* forbidden, the old conversion by contraposition formally declared *false,* and the particular proposition asserted to be incapable of being made universal. But M. Gergonne's complex propositions, such as they are, are used in a manner resembling that in chapter V, of this work, though requiring a separate *tâtonnement* for many things the analogues of which appear as connected results of my system. Accordingly, I am bound to attribute to M. Gergonne the first publication of the idea of *a* complex syllogism, and of the comparison of the simple one with it. But numerical statement is not hinted at.

Sir William Hamilton's system dates, as to its publication in lectures, from 1841, as far as has yet appeared. What I have to say of it will be found in another appendix.

In 1842, there was published anonymously 'Outline of the laws of thought'; London and Oxford (Pickering, and Graham) *octavo in twos* (small). The author is the Rev. Wm. Thomson, tutor of Queen's College, Oxford. It is a very acute work, and learned. The system of propositions is extended by the introduction of both the common quantifications of the predicate into the *affirmatives only,* which introduces the propositions U and Y, as the author calls them, or "All Xs are all Ys," and "Some Xs are all Ys."

The memoir in the Cambridge Transactions in which I gave the first account of what has since grown into Chapters IV, V, VIII, and X, of this work, is described as to date in the preceding appendix. With reference to the subject of chapter V, I may note the following defects of that memoir: 1. That only one arrangement of X and Z as premises being taken, only half the system is given, and many correlative arrangements are not obtained (see page 140). 2. That owing to my not seeing distinctly that each universal proposition has *two* weakened forms, the syllogisms A_1A^1I' and $E'E'I_1$ are considered as a class apart. 3. That much of the power of forming easy rules is not gained, by the order of reference being made XY, ZY, XZ, instead of XY, YZ, XZ. The former appears at first the more natural order, and is certainly

---

* I am told that some works on logic used in the Irish colleges formally announce that the truth of the [ordinary] laws of syllogism depends upon the exclusion of contraries: but I have not met with any of them.

more eafily defcribed; namely, to refer each of the concluding terms to the middle term, with which both are compared. I obferve, fince, that M. Gergonne adopts this laft order of reference : but the other is by an immenfe deal more convenient in its refults, as I think I have fhown.

With refpect to the numerical quantification, what I did in the *Memoir* and *Addition* is given in full in the preceding appendix. Sir William Hamilton, who diftinctly renounces all claim to the " arithmetically articulate" fyftem, and doubts whether it afford any bafis for a logical developement, ftates that he had formerly obtained the " ultra-total quantification" (page 317) and thrown it away as a cumbrous and ufelefs fubtlety, without publifhing it, as I underftand, in any way. To his reply, he appends a note which I think it defirable to republifh at length, as a document in the hiftory of this fpeculation, and that I may make that hiftory complete (II. p. 41).

' I have avoided, in the previous letter and poftfcript, all details in
' regard to the *third* fcheme of quantification (p. 32); becaufe that fcheme
' except in fo far as it is confounded with the *fecond*, has no bearing in
' the controverfy; and I admit that whatever Mr. De Morgan has
' therein accomplifhed, he has accomplifhed independently of me. Fur-
' ther, I fhall not deny him any claim of priority to whatever he may
' have ftated in our correfpondence, in reference to this third fcheme.
' Finally, I fhall acknowledge, for I think it not improbable, that his
' fyllogifm (p. 19) fuggefted a reconfideration, on my fickbed, of a cer-
' tain former fpeculation, in regard to the ultratotal quantification of the
' middle term in both premifes together;—a fpeculation determined by
' the vacillation of the logicians, touching the predefignations *more, moft*,
' &c. but which I had laid, afide, as a ufelefs and cumbrous fubtlety.

' Ariftotle, followed by the logicians, did not introduce into his doc-
' trine of fyllogifm, any quantification between the abfolutely univerfal
' and the merely particular predefignations, for valid reafons.—1°, Such
' quantifications were of no value or application in the one whole (the
' univerfal, potential, logical), or, as I would amplify it, in the two cor-
' relative and counter wholes (the logical,— and the formal, actual,
' metaphyfical,) with which Logic is converfant. For all that is out of
' claffification, all that has no reference to genus and fpecies, is out of
' Logic, indeed out of Philofophy; for Philofophy tends always to the
' univerfal and neceffary. Thus the higheft canons of deductive reafon-
' ing, the *dicta de Omni et de Nullo*, were founded on, and for, the
' procedure from the univerfal whole to the fubject parts; whilft, con-
' verfely, the principle of inductive reafoning was eftablifhed on, and for,
' the (real or prefumed) collection of all the fubject parts as conftituting
' the univerfal whole.—2°, The integrate or mathematical whole, on
' the contrary, (whether continuous or difcrete) the philofophers con-
' temned. For whilft, as Ariftotle obferves, in mathematics genus and
' fpecies are of no account; it is, almoft exclufively, in the mathemati-
' cal whole, that quantities are compared together, through a middle
' term, in neither premife, equal to the whole. But this reafoning, in

' which the middle term is never univerfal, and the conclufion always
' particular, is,—as vague, partial, and contingent,—of little or no value
' in philofophy. It was accordingly ignored in Logic ; and the prede-
' fignations *more moft*, &c., as I have faid, referred, to univerfal, or,
' (as was moft common) to particular, or to neither, quantity. This
' difcrepancy among Logicians long ago attracted my attention ; and I
' faw, at once, that the poffibility of inference confidered abfolutely,
' depended, exclufively on the quantifications of the middle term, in both
' premifes, being, together, more than its poffible totality—its diftribution,
' in any one. At the fame time I was impreffed—1°, with the almoft
' utter inutility of fuch reafoning, in a philofophical relation : and 2°,
' alarmed with the load of valid moods which its recognition in Logic
' would introduce. The mere quantification of the predicate, under the
' two pure quantities of *definite* and *indefinite*, and the two qualities of
' *affirmative* and *negative*, gives (abftractly) in each figure, *thirty fix*
' valid moods ; which, (if my prefent calculation be correct,) would be
' multiplied, by the introduction of the two hybrid or ambiguous quan-
' tifications of *a majority* and *a half*, to the fearful amount of *four hun-*
' *dred and eighty* valid moods for each figure. Though not, at the
' time, fully aware of the ftrength of thefe objections, they however
' prevented me from breaking down the old limitation ; but as my fu-
' preme canon of Syllogifm proceeds on the mere formal poffibility of
' reafoning, it of courfe comprehends all the legitimate forms of quanti-
' fication. It is ;—*What worft relation of fubject and predicate, fubfifts*
' *between either of two terms and a common third term, with which one,*
' *at leaft, is pofitively related ;—that relation fubfifts between the two*
' *terms themfelves :* in other words ;—*In as far as two notions both*
' *agree, or one agreeing, the other difagrees, with a common third notion :*
' *—in fo far, thofe notions agree or difagree with each other.* This canon
' applies, and proximately, to all categorical fyllogifms,—in extenfion
' and comprehenfion,—affirmative and negative,—and of any figure. It
' determines all the varieties of fuch fyllogifms ; is developed into all
' their general, and fuperfedes all their fpecial, laws. In fhort, without
' violating this canon, no categorical reafoning can, formally, be wrong.
' Now, this canon fuppofes that the two extremes are compared together,
' through the *fame common middle ;* and this cannot but be, if the
' middle, whether, fubject or predicate, in both its quantifications to-
' gether, exceed its totality, though not taken in that totality in either
' premife.

   ' But, as I have ftated, I was moved to the reconfideration of this
' whole matter ; and it may have been Mr. De Morgan's fyllogifm in
' our correfpondence (p. 19), which gave the fuggeftion. The refult
' was the opinion, that thefe two quantifications fhould be taken into
' account by Logic, as authentic forms, but then relegated, as of little
' ufe in practice, and cumbering the fcience with a fuperfluous mafs of
' moods. As to Mr. De Morgan's ftatement in our correfpondence (p.
' 21) of the principle on which (by his later fyftem) fuch fyllogifms
' proceed, this, to ufe his own expreffion, " I did not comprehend at

' all ;" nor do I now,* having, to speak with the Rabbis, " reserved it
' for the advent of Elias." I saw however, that, be it what it might,
' it had no analogy with mine ; indeed, even from the fuller exposition
' of his doctrines, contained in the body of the Cambridge Memoir and
' its Addition, which I afterwards received, I can find no indication
' of his having generalised either, 1° *the comprehensive principle of all*
' *inference, that the two quantifications of the middle term, should, to-*
' *gether, exceed it as a single whole ; or,* 2°, *under a non-distributed*
' *middle, the* two *exclusive forms of its quantification.* On receipt,
' however, of Mr. De Morgan's Cambridge Memoir, I saw, or thought
' I saw, in the body of the paper, on his *old* view, some manifestation of
' a less erroneous doctrine upon this point, than that afterwards contained
' in his Letters and Addition, upon his *new.* Accordingly, to obviate
' all misconstruction, I wrote immediately the following letter,† of which
' an account has been previously given (p. 26, note).

EDINBURGH, 30th March, 1847.

' Your paper read to the Society I have cursorily perused ; but though
' opposed to many of its doctrines, I admire the ingenuity which charac-

---

* The passages which Sir William Hamilton does not understand, are the following,
and also that relating to the effective terms, in C of the preceding appendix.

" Now suppose propositions in which the quantitative part of the preceding is made
more definite. Say that

$$X_t \quad Y_u \quad | \qquad \text{and} \quad X_t : Y_u \quad |$$

mean

| Every one of t Xs | No one of t Xs |
| is one or other of u Ys | is any one of u Ys |

Let the *effective number* of cases in a proposition be the number which it makes effective in inference. Then the effective number in a positive proposition is the number of cases of the *subject.*

The effective number in a negative proposition is the number of cases of the *middle
term.*

And the criterion of inference being possible, is that the sum of the effective numbers of the two premises (not both negative) is greater than the *whole number of cases*
of the *middle term.*

And the excess is the number of cases involved in the inference, of all which are
mentioned in the conclusion-term (or terms) of the positive premiss (or premises).

For instance, let b be the whole number of Ys in existence : I ask whether we can
infer anything from

$$X_t \quad Y_u \qquad \text{effective number} \qquad t$$
$$Z_w : Y_v \qquad \ldots \ldots \qquad v$$

Answer, if $t + v$ be greater than b, we can infer

$$X_{t+v-b} : Z_w$$

Or, if each of t Xs be one or other of u Ys, and no one of w Zs be any one of v Ys,
then if t and v together are more in number than there are Ys, we may infer that no
one of $t + v - b$ Xs is any one of the w Zs just spoken of."

† This letter (the first paragraph of which is omitted, as not relevant to *this* appendix,) was addressed to me, and was sent open to my friend Dr. Sharpey, to be delivered to me. Dr. Sharpey refused to deliver (and, as it happened, I was as much
prepared to refuse to receive) any thing on the literary subject matter of the controversy
which did not contain a retraction of Sir W. Hamilton's then subsisting charge against
me. Accordingly, I never saw it till it appeared in print.

' terifes it throughout. On one point, I find we coincide, in principle,
' at leaft, againft logicians in general. They have referred the quantify-
' ing predefignations *plurimi*, and the like, to the moft oppofite heads ;
' fome making them univerfal,—fome, particular,—and fome between
' both ; (for you are not correct in faying, (p. 6), that logicians are
' unanimous in regarding them as particular, [though moft do]). This
' confliction attracted my attention ; and a little confideration fhowed
' me, that befides the quantification of the pure quantities, *univerfal* and
' *particular*, (which I call *definite* and *indefinite*,) there are two others of
' thefe, mixed and half developed, which ought to be taken into account
' by the logician, as affording valid inference ; but which, without fcien-
' tific error, cannot be referred either to *univerfal*, (definite,) or to *par-
' ticular*, (indefinite) quantity, far lefs left to vacillate ambiguoufly be-
' tween thefe. I accordingly introduced them into my modification, in
' Englifh doggerel, of " *Afferit A*," &c., which [in the original caft] I
' formerly faid was at your fervice ; and as it affords a brief view of my
' doctrine on this point, I may now quote it.

' A, it affirms *of this, that, all,*＊
  Whilft E denies of *any*;
I, it affirms, whilft O denies,
  Of *fome* (or few or many).

' Thus A affirms, as E denies,
  And definitely either ;
Thus I affirms, as O denies,
  And definitely neither.

' A *half*, left femi-definite,
  Is worthy of its fcore ;
U, then, affirms, as Y denies,
  This, neither lefs nor more.

' Indefinito-definites,
  To UI, YO, laft we come;
And that affirms, and this denies,
  Of *more*, *moft*, (half plus fome).

' " The rule of the logicians, that the middle term fhould be once at
' " leaft diftributed [or indiftributable,] (*i.e.* taken univerfally or fingu-
' " larly, = definitely,) is untrue. For it is fufficient, if, in both the
' " premifes together, its quantification be more than its quantity as a
' " definite whole. (Ultratotal)" - - - - - - - " It is enough for a
' " valid fyllogifm, that the two extreme notions fhould (or fhould not),
' " of neceffity, partially coincide in the third or middle notion ; and
' " this is neceffarily fhown to be the cafe, if the one extreme coincide

＊ Better : ' A, it affirms of *this, thefe, all*.'

' " with the middle, to the extent of a half, (dimidiate quantification) ;
' " and the other, to the extent of aught more than a half, (ultradimi-
' " diate quantification). - - - - -
' " The firſt and higheſt quantification of the middle term (. .) is
' " ſufficient not only in combination with itſelf, but with any of all the
' " three inferior. The ſecond (. ,) ſuffices, in combination with the
' " higheſt, with itſelf, and with the third, but not with the loweſt.
' " The third (.) ſuffices, in combination with either of the higher, but
' " not with itſelf, far leſs with the loweſt. The fourth and loweſt (,)
' " ſuffices only in combination with the higheſt." [1. Definite ;
' " 2. Indefinito-definite ; 3. Semi-definite ; 4. Indefinite.]" '
    Of the effect of this new ſyſtem of quantification in amplifying the
' ſyllogiſtic moods, (which in all the figures remain the ſame,) I ſay no-
' thing. It ſhould be noted, however, that the letters A, E, &c. do not
' mark the quantification [and qualification] of *propoſitions*, (as of old)
' but of *propoſitional terms.* The ſentences within inverted commas are
' taken from notes for the " Eſſay towards," &c.
    ' Before concluding, I ought to apologiſe, in the circumſtances, for
' the details, that have inſenſibly lengthened out, of a part of my doc-
' trine, which I have found, to a certain extent, coincident with what
' appears in your paper. I was anxious, however, that you and others
' ſhould have no grounds for ſurmiſing, that I borrowed any thing from
· my predeceſſors without due acknowledgment.—On ſecond thoughts,
' however, I deem it more proper to make this communication through
' a third party.'
    The diſcuſſion between Sir William Hamilton and myſelf called a
very able third party into the field, who addreſſed the following letter
to the editor of the *Athenæum,* in which journal it was publiſhed, June
19, 1847.
    ' Sir,—As two great logical innovations—the one due to Sir William
' Hamilton, the other due to Mr. De Morgan—uſed in conjunction, have
' led me to the ſimpleſt and moſt general formulæ of ſyllogiſm that ever
' have been given (formulæ which correct a ſerious miſtake into which
' both Sir William Hamilton and Mr. De Morgan have fallen), I think
' it will gratify thoſe intereſted in logical ſcience if you would give them
' publicity through your columns.
    ' n^{I}, n^{II}, n^{III}, &c. are any numbers. When placed before a term, as
' $n^{II}xs$, n^{II} marks the total number of the claſs x; placed before a pro-
' poſition, it marks the number of things of which we mean to ſpeak.
' Thus, n^{I}, of $n^{II}xs$ are of $n^{III}ys$, means that a number of things n^{I} are
' alleged to have both the characteriſtics x and y; and are to the whole
· claſs of xs as n^{I} to n^{II}, and to the whole claſs of ys as n^{I} to n^{III} : ſimi-
' larly with the negative propoſition n^{I} of $n^{II}xs$ are not of $n^{III}ys$, n^{I}
' things being here ſaid to have the characteriſtic x, and to want the
' characteriſtic y. It is clear, from the nature of a propoſition, that in
' affirmatives, n^{I} can never be greater than the leaſt extenſive of the
' terms, and in negatives never greater than the number of the claſs
' whoſe characteriſtic it is ſaid to have. But within theſe limits the pro-

' portion n^I to n^{II} may be wholly undetermined; we then mark it with
' the word *fome*,—we call this, with Sir William Hamilton, indefinite
' quantity. It may be perfectly determined; as of equality when we mark
' it with *all*, *every*, or, following Mr. De Morgan, any other arithmeti-
' cal proportion—as a half. (Sir William Hamilton has erred in calling
' a half, femi-definite; it is thoroughly definite). All this we call defi-
' nite quantity. Laftly, the indefinitude may be reduced within limits
' —indefinito-definite, as *moft*, &c.

' The firft formula contains all fyllogifms with an affirmative conclu-
' fion, without any exception.

'I.　　　　n^I of $n^{II}xs$ are of $n^{III}ys$

n^{IV} of $n^V zs$ are of $n^{III}ys$

$(n^I + n^{IV} — n^{III})$ of $n^{II}xs$ are of $n^V zs$

' As Sir William Hamilton's principle takes away all diftinction of
' fubject and predicate in affirmative propofitions, it will be feen that, by
' varying the proportions of the fymbols, n^I, &c., every poffible affirma-
' tive logical inference, in whatever mood or figure, emerges.

' The fyllogifms with negative queftions or conclufions, are not fo
' fimple. They fall into two divifions, according as, in the negative
' premifs, the things fpoken of have the characteriftic of the extreme, or
' of the middle; and from each of thefe, *two* conclufions, not *one*, are
' drawn, according as the things to be fpoken of in the conclufion have
' the characteriftic of the extreme in the affirmative premifs, or of that
' in the negative premifs.

'II.　n^I of n xs are of $n^{III}ys$

' n^{IV} of $n^V zs$ are not of $n^{III}ys$ concludes;

' doubly 1° $(n^I + n^{IV} — n^V)$ of $n^{II}xs$ are not of $n^V zs$

' 2° $(n^I + n^{IV} — n^{II})$ of $n^V zs$ are not of $n^{II}xs$.

' It is to this formula I referred as correcting a ferious error into which
' Sir William Hamilton and Mr. De Morgan have fallen—of holding, as
' a general principle of all inference, that the two quantifications of the
' middle term fhould exceed it as a whole; for this fyllogifm proceeds
' wholly irrefpective of the total quantity of the middle, which is excluded
' from our fymbolic conclufion.

'III.　n^I of $n^{II}xs$ are of $n^{III}ys$

' n^{IV} of $n^{III}ys$ are not of $n^V zs$ concludes; alfo,

' doubly 1° $(n^{IV} + n^I — n^{III})$ of $n^{II}xs$ are not of $n^V zs$

' 2° $(n^{IV} + n^I + n^V — n^{III} — n^{II})$ of $n^V zs$ are not of $n^{II}xs$.

' Such are the three fymbolical formulæ of every poffible logical infe-
' rence. I have the demonftrations that thefe are in all their extent valid,
' and are the only poffible forms; but it is fufficient to give here the re-
' fults.

' It will furprife no one who confiders that the negative propofition is
' not converted in the fame fenfe as the affirmative, that the negative
' fyllogiftic formulæ are not reducible to one. For the rule of negative

' conversion changes the things spoken of, and is as follows : n^1 of $n^{11}xs$
' are not of $n^{111}ys$; converts $(n^{111} + n^1 - n^{11})$ of $n^{111}ys$ are not of $n^{11}xs$. The
' consequence of a form universally true, $(n^{111} - n^{11})$ of $n^{111}ys$ are not of $n^{11}xs$.
' As to the two conclusions, they are but the converse of each other.

' It will not be difficult to interpret these, by $n^1 = n^{11}$ as *every* or n^1:
' n^{11} indefinite $= $ *some*, &c. The usual Aristotelic forms will be seen to
' be derived from them. Thus the mood Cesare, and the corresponding
' indirect mood (or, if you will, the mood of the fourth figure, call it at
' another time Celantes or Cadere at will, but let it be Celantes for the
' nonce), come forth from the third formula.

$$' n^{IV} = n^{III} \text{ gives no } y \text{ is } z \ldots n^{IV} : n^{V} \text{ indefinite}$$
$$' n = n^{11} \quad \text{every } z \text{ is } y \ldots n^1 : n^{III} \text{ indefinite.}$$

' Hence in Cesare, no x is z from our first,
' and in Celantes, no z is x from our second conclusion, and so of all
' the others.

' I owe it to Sir William Hamilton and Mr. De Morgan to say that
' without their improvements I could not have advanced one step. Mr. De
' Morgan has even attempted a like reduction to general formulæ, and has
' failed, chiefly through a misapprehension of Sir William Hamilton's prin-
' ciple of quantified predicate. He has introduced a superfluous quantity,
' —one logically useless, or worse than useless, as the result has shown.
' This confusion explains his errors. Had it not been for this circum-
' stance, I should not have had the honour of presenting these formulæ
' to logicians.

' Permit me to add what I think also of some value. I am not of those
' who think with Sir William Hamilton that the syllogism always pro-
' ceeds in the two counter wholes of intension and extension—that it
' must always be an involution or evolution in respect of classification.
' This is, no doubt, true in the most important reasonings of science; but
' it is not scientifically accurate to assert this universally.

' Quality, which is the comprehensive element, is of three kinds—not
' two, as heretofore affirmed; for since Kant, the division of affirmatives
' into analytic and synthetic, or (as Sir William Hamilton wishes) expli-
' cative and ampliative, has been established. James Bernouilli has puz-
' zled himself to reduce these two to the same form, but without success;
' for that contains an immediate relation of part to whole, and only a re-
' mote one of part to part, while this contains an immediate relation of
' part to part, and remote of part to whole. These, as distinct kinds of
' quality, are erroneously elided in language. As the words *ampliative*
' and *restrictive* are generally opposed in logic, perhaps we might replace
' the old division of propositions, according to quality, into affirmative and
' negative—by one into *Explicative, Ampliative*, and *Restrictive*.

' Where, then, both premises are ampliative, the syllogism proceeds
' purely by force of extension. There is neither involution nor evolution
' —neither induction nor deduction—but a passage or transition from one
' mark to another, or from class to class. Of this kind are all singular,
' or, as Ramus calls them, proper syllogisms. Let us call this new

' clafs of fyllogifms traductive, to contraft it with the inductive and de-
' ductive.

' The ufe of thefe in philofophy as independent modes of inference will
' eafily appear. When we collect the fcattered fragments of our know-
' ledge into unity of fcience, we ufe *induction and inductive fyllogifm;*
' when we apply the principles of fcience to fpecial events of things, we
' ufe *deduction and deductive fyllogifm;* but when, abandoning one fcheme
' of claffification, we transfer our knowledge *directly* to another, we ufe
' *traduction and traductive fyllogifm.* Thus, in political fcience, what
' has been predicated by hiftorians of men claffed geographically is tranf-
' ferred to men claffed according to conftitutions of government by tra-
' duction. This laft efcapes Sir William Hamilton's rule, and never
' concludes through a comprehenfive containing and contained.

' I fhall not add, at prefent, any attempt to prove *à priori* the exclufive
' validity of fyllogiftic inference.

' I admit that I ought not, without good ground, to diffent from a ma-
' tured opinion of Sir William Hamilton in any part of philofophy, ftill
' more in logic ; but I obey the force of demonftration,—and, as Ludo-
' vicus Vives faid in refpect to Ariftotle, *Verecundè diffentio.*

' Yours, &c.

' JAMES BROUN.

' *Temple, June* 9, 1847.'

My reply to this confifted in forwarding, on the fame 19th of June,
to the editor of the *Athenæum*, a fummary of the refults of chapter VIII,
then written. This fummary appeared on the 26th : I do not infert it,
becaufe the chapter in queftion is a better anfwer; and though the publication faved my rights, the republication is unneceffary. Mr. Broun's
three forms are the firft (without the contranominal), the ninth, and the
eleventh, of page 161. Mr. Broun was wrong in deducing from the
two latter forms that the principle of the middle term was erroneous :
for in thefe very forms the two quantifications exceed the whole : being
the whole (in premife one) plus fome (in the other). As to the fuperfluous quantity, it only becomes fuperfluous when fuch quantifications
are introduced as diftinguifh fpurious from admiffible propofitions : fee
pages 145, 146, in which it is fhown that the forms are correct.

Nothing but clofe comparifon, and that after practice, would detect
the accordance of the two fymbolic modes of expreffion in pages 145
and 161. I am not therefore furprifed that Mr. Broun fhould, having
obtained cafes of that in page 161, pronounce that in page 145 erroneous.

In the anfwer which I made, I promifed to ftate diftinctly how much
of the chapter was written before Mr. Broun's letter appeared. This
I now do. With the exception of pages 145, 146, the matter of which
is moftly from my Cambridge Memoir, the whole of it was then written,
excepting fuch verbal alterations and occafional introduction of fentences,
as take place at the prefs, or at the laft reading of the manufcript. I had

thought that there would be no neceffity to introduce thofe pages, except flightly, and in anfwer to certain objeftions which feemed likely to
occur. The examination which the affertion that they are erroneous
made me give my previous forms, pointed out the defirablenefs of introducing them as they now ftand.

September 17, 1847. I had finifhed the preceding appendix, when I
became aware of the exiftence of the 'Commentationes Philofophicæ
Seleftiores' of Godfrey Ploucquet, of Tubingen, Utrecht, 1781, quarto.
The laft title (p. 561) is 'De Arte Charafteriftica. Subjicitur Methodus
calculandi in logicis, ab auftore inventa. 1763.' I find by a catalogue*
that this *methodus calculandi* had been previoufly publifhed in 1773, at
Tubingen, at the end of a work entitled 'Principia de Subftantiis et
Phænomenis:' alfo that the 'Methodus demonftrandi direfté omnes
fyllogifmorum fpecies' of the fame author (which is probably the thing
I am going to defcribe) was publifhed at Tubingen in 1763. From the
title of a work which, I am informed, exifts, namely, 'Sammlung der
Schriften welche von logifchen Calcul des Prof. Ploucquet betreffen'
Tubingen, 1773, one would fuppofe that this fyftem had obtained great
local currency. I give a fhort account of it : premifing that Ploucquet
appears to have been a well informed mathematician, much given to
pure fpeculation on mental fubjefts.

The *calculus* (a term which Ploucquet ufes in as wide a fenfe as I do
when I call the contents of Chapter V. a part of the calculus of inference) confifts in the invention of a fimple notation, and the mechanical
fubftitution, in one premife, of an identical equivalent to the middle term
therein contained, taken from the other premife (this laft being one in
which the middle term is univerfal). There is neither ufe of contraries,
nor numerical definition : but there is every variety of quantity of the
predicate which can be produced by fimple converfion of the ordinary
forms. A term ufed univerfally is denoted by the capital letter ; particularly, by the fmall letter : affirmation by juxtapofition ; negation, by
interpofing $<$. Thus X)Y is Xy ; X.Y is X$>$Y ; XY is xy ; X:Y is
x$>$Y. The following is a complete fpecimen :

Sint præ-   Pm
   miffæ   s$>$M

Calculo :   s$>$mP   quoddam s non eft P
   Omnis ducatus eft aureus
   Quædam moneta non eft aurea.

   Da
   m$>$A

Calc. m$>$aD.   feu m$>$D, quædam moneta non eft ducatus.

As Ploucquet feems to think that this aftual application of the *calculus*
to concrete inftances, by aid of their initial letters, is a material part of

* The *fecond edition* of Mr. Blakey's 'Effay on Logic' recently publifhed, contains
a catalogue of upwards of a thoufand works on logic, briefly titled.

his fyftem, I have inferted the cafe entire. The rationale of the fyftem confifts in that fubftitution of identicals for each other, which I under-ftand Sir William Hamilton (with perfect truth) to employ in every cafe. Thus we have in the above 'Some of the Ss are not any Ms, are not thofe Ms which make up all the Ps, are not therefore any Ps.' This demand for identical fubftitutes requires both kinds of quantity for every predicate, and Ploucquet ufes them accordingly, as far as wanted to eftablifh the Ariftotelian fyllogifms. Sir W. Hamilton goes further, and invents fyllogifms for all the kinds of quantity. Thus Ploucquet ufes mP or 'fome Ms are all the Ps' and P $>$ m or 'all Ps are not fome of the Ms;' but not MP or p $>$ m.

At the fame time with the knowledge of Ploucquet I obtained that of the work of a follower and extender, M. W. Drobitfch, author of 'Neue Darftellung der Logik . . . Nebft einen logifch-mathematifchen Anhange,' Leipzig, 1836, octavo. As far as the fymbolic part is con-cerned, Mr. Drobitfch begins by a convention which would reconcile any one to the found, not merely of *Barbara* and *Celarent,* but even of *Baroko* and *Frefifon.* He makes S and P the fubject and predicate of the conclufion and M the middle term; and puts the Ariftotelian vowel between them: thus S)P is SAP, and P:S is POS. Hence his pre-mifes may be *map fam* or *mop fam ;* and one of his fyllogifms is *mep-famfep.* In the *algebraical* part, he ufes large and fmall letters for the univerfal and particular, or for the whole and part extent of a term. He alfo introduces the figns $=$ and $<$ to fignify identity and (what I call) fubidentity. This ufe of the mathematical figns involves an ex-tenfion, which is made by all thofe who fignify the identity of X and Y by X=Y. The mathematician thinks of extent as quantity only: the logician includes both quantity and pofition. Thus when the for-mer fays that five feet are lefs than feven feet, he means any five feet, be they part of the feven feet or not: the latter, when he fays that X is a name of lefs extent than Y, means not only that the former *can be* contained in the latter, but *that it is.* To make negative propofitions, Mr. Drobitfch takes a limited univerfe (call it U, as I have done) an extent greater than the utmoft extent of all the names, otherwife inde-finite. And here he falls into fome confufion: X and Y being the names, he fays U muft be of greater extent than X$+$Y: now if we had X)Y, U need only be of greater extent than Y. If from the genus Y be taken all the fpecies X, the remainder is denoted by Y$-$X. Ac-cordingly, the contrary of X is U$-$X.

Mr. Drobitfch then lays down eight forms of predication, of which, however, he only ufes the ordinary ones. And I cannot find out that the limited univerfe, or the contrary, has any ufe except to furnifh means of notation. The eight forms are;—firft, X$=$y, or my X)Y; fecondly, X$=$Y, or X)Y$+$Y)X; thirdly, x$=$y, or XY; fourthly, u$=$Y, or Y)X; fifthly, X $<$U$-$Y, which tells us that X is all contained in what is left of the univerfe after Y is removed, or is X.Y; fixthly, X$=$z $<$Z $<$U$-$Y, a very roundabout way of faying that X is *fub*con-trary of Y, or X.Y$+$xy; feventhly, x$=$U$-$Y or X:Y; eighthly,

$x = X—Y$, which tells us that Y is a fubidentical of X, or $Y)X + X:Y$.

This is in fact a mixture of two fyftems, both in principle and notation. The forms are A_1, A', O_1 (and O'), E_1, I_1, D, D_1 (and D'), and C_1. Alfo C is virtually given : but E', I', C', do not appear. The ordinary rules under which the mathematicians ufe $=$ and $<$, remain true in this logical ufe of them : and thus there is an elegant mode or exhibiting the inference in fyllogifms. For inftance, in *Cameftres* we have $P = m$, $S < U—M$ $\therefore$ $< U—m$ $\therefore$ $< U—P$; or $S < U—P$.

It would have been more confiftent to have made $=$, $<$, and $>$, (introducing this laft) ferve all purpofes. But it has happened very often that a fyftem of notation, already exhibited, has been extended by a better one, and mended only, inftead of being reconftructed. Ploucquet had ufed the large and fmall letters, and $>$ for denial : the latter fymbol a ftrange one, if mathematical analogy were intended. Mr. Drobitfch has ingenioufly contrived that $<$ fhould reprefent denial, and has been led to what might have ufefully amended all he had to begin with. Taking little x to reprefent a part of the extent of X, &c. and U for the extent of the univerfe, the following notation might have been adopted :

Firft when $<$ and $>$ both include their limit, $=$. We fhould have

A_1	$X < Y$ or $Y > X$	A'	$Y < X$ or $X > Y$
O_1	$x < U—Y$ or $U—Y > x$	O'	$y < U—X$ or $U—X > y$
E_1	$X < U—Y$ or $U—Y > X$	E'	$X > U—Y$ or $U—Y < X$
I_1	$x < Y$ or $Y > x$	I'	is inexpreffible.

To exprefs I', we muft invent a fymbol for a part of $U—X$.

Next, when $<$ and $>$ do not include their limits, we have

D_1	$X < Y$ or $Y > X$	C_1	$X < U—Y$ or $U—Y > X$
D	$X = Y$ or $Y = X$	C	$X = U—Y$ or $U—Y = X$
D'	$X > Y$ or $Y < X$	C'	$X > U—Y$ or $U—Y < X$

P is inexpreffible.

I am inclined to think that the reprefentation of quantity and location both under one fymbol is objectionable, if that fymbol be one already appropriated in mathematics to quantity only. I would on no account accuftom myfelf to read $A < B$ as A is lefs than (becaufe a part of) B. Mr. Drobitfch is much more complete than his predeceffors in his enumeration of the various kinds of forites.

October 29, 1847. While this fheet was paffing through the prefs, I became acquainted with " A fyllabus of logic, in which the views of Kant are generally adopted, and the laws of fyllogifm fymbolically expreffed. By Thomas Solly, Efq." Cambridge, 1839, 8vo. The fymbolical expreffion here given is of a peculiar character : the algebraic figns are adopted in a fenfe which preferves the rules of fign, while the fymbols reprefent the terms of the fyllogifm, or elfe the notions of particular and univerfal. Thus, if p ftand for particular, u for univerfal, and m for one of the terms of a fyllogifm, $m = u$ or $m—u = o$ implies that *m* is a univerfal term, and $(m-u)(n-p)=0$ implies the alternative that either *m* is univerfal, or *n* is particular. By means of fuch alternative relations, the conditions of validity of the various figures are exprefled. Mr. Solly contends for fix forms in each figure, by introducing all forms which have weakened conclufions, and proves *à priori*, from his equations, that fix and no more are poffible in each figure. If I had admitted weakened forms, there would have been fixteen more fyllogifms, which might be deduced, either from the eight univerfals, or from the fixteen particulars.

THE END.

C. WHITTINGHAM, CHISWICK.

Printed in the United States
By Bookmasters